Nuclear Engineering Division

Presentations at the 2010 AIChE Annual Meeting

Salt Lake City, Utah, USA
7-12 November 2010

ISBN: 978-1-61782-159-2

Printed from e-media with permission by:

Curran Associates, Inc.
57 Morehouse Lane
Red Hook, NY 12571

CURRAN ASSOCIATES INC.
proceedings.com

Some format issues inherent in the e-media version may also appear in this print version.

Printed by Curran Associates, Inc. (2011)

AIChE
3 Park Avenue
New York, NY 10016-5991

Phone: (203) 702-7660
Fax: (203) 775-5177

www.aiche.org

Additional copies of this publication are available from:

Curran Associates, Inc.
57 Morehouse Lane
Red Hook, NY 12571 USA
Phone: 845-758-0400
Fax: 845-758-2634
Email: curran@proceedings.com
Web: www.proceedings.com

TABLE OF CONTENTS

(26a) Compact Hydrogen Storage with Extended Dormancy In Cryogenic Pressure Vessels

salvador M. Aceves, Engineering, Lawrence Livermore National Laboratory, Livermore, CA

LLNL is developing cryogenic capable pressure vessels with thermal endurance 5-10 times greater than conventional liquid hydrogen (LH2) tanks that can eliminate evaporative losses in routine usage of (L) H2 automobiles. Cryogenic pressure vessels can be fueled with ambient temperature compressed gaseous hydrogen (CGH2), LH2 or cryogenic hydrogen at elevated supercritical pressure (cryo-compressed hydrogen, CcH2). When filled with LH2 or CcH2, these vessels contain 2-3 times more fuel than conventional ambient temperature compressed H2 vessels. LLNL has demonstrated fueling with LH2 onboard two vehicles. The generation 2 vessel, installed onboard an H2-powered Toyota Prius and fueled with LH2 demonstrated the longest unrefueled driving distance and the longest cryogenic H2 hold time without evaporative losses. A third generation vessel has been installed, reducing weight and volume by minimizing insulation thickness while still providing acceptable thermal endurance.

See more of this Session: Hydrogen Storage System Engineering and Applications: Infrastructure, Delivery Systems and Demonstration Systems
See more of this Group/Topical: Topical 8: Hydrogen Production and Storage

(26b) Evaluation of Sorption System for Hydrogen Storage

Richard Chahine, Hydrogen Research Institute, Trois-Rivieres, QC, Canada

Significant R&D efforts are being pursued globally to improve the efficiency and economics of onboard hydrogen storage systems for mobile applications. The sorption of hydrogen on or in a solid substrate, relying either on physisorption or chemisorption or even a combination of both offers a potential alternative means of storing hydrogen. Hydrogen storage via physisorption on porous materials, such as activated carbon, carbon nanostructures and metal organic frameworks (MOFs) is particularly enticing due to its inherent reversibility and cyclability. Moreover, these materials exhibit fast kinetics and operate at relatively low storage pressures. However, hydrogen physisorption at ambient temperatures on currently available porous materials leads to unacceptably low hydrogen storage densities. Acceptable densities are only attainable at cryogenic temperatures.

We will present the results of our evaluation of an adsorption-based hydrogen storage tank for vehicular application. We will discuss the net storage capacity of the system over wide temperature and pressure ranges, the thermal requirements to charge and discharge hydrogen at the desired flow rates, and dormancy.

See more of this Session: Hydrogen Storage System Engineering and Applications: Infrastructure, Delivery Systems and Demonstration Systems
See more of this Group/Topical: Topical 8: Hydrogen Production and Storage

(26c) JHFC (Japan Hydrogen and Fuel Cell) Demonstration Project – Current Status Report

Jinichi Tomuro, Hydrogen Project Office, Engineering Advancement Association of Japan (ENAA), Tokyo, Japan

JHFC (Japan Hydrogen and Fuel Cell Demonstration) Project originally started in 2002 as a Japanese national project to demonstrate the effectiveness of fuel cell vehicles and hydrogen infrastructure for them. Its first phase lasted for 4 years and succeeded by its second phase which was expected to last for 5 years. Now the project entered into the last fiscal year. Under the JHFC project, 14 hydrogen refueling stations are operated and approximately 60 FCV's are on the public road. Up to the end of 2009, the cumulative driving distance reached 1.4 million kilometers and 29 tons of hydrogen had been refueled. In this paper, the author describes the results obtained from the project mainly in the field of hydrogen infrastructures. It includes various lessons learned, technical data and so on. Unfortunately the market introduction of FCV's seems to require a few more years. The Japanese government and the related industrial organizations began to consider the next demonstration project. Although the project scheme in the future is not clearly fixed yet, the author will try to introduce the upcoming project.

See more of this Session: Hydrogen Storage System Engineering and Applications: Infrastructure, Delivery Systems and Demonstration Systems
See more of this Group/Topical: Topical 8: Hydrogen Production and Storage

(26d) Development of Hydrogen Storage Technique for Fuel Cell Vehicle

Kenji Komiya, Fuel Cell System Development Div., Higashifuji Technical Center, Susono, Japan

Hydrogen is drawing attention as a clean and sustainable source of energy, and fuel cell vehicles that use hydrogen as fuel have the potential to become the ultimate automobiles of the future. However, since the energy density per volume of hydrogen fuel is extremely small (1/3000 of gasoline), it is important to develop technology to enable the storage of hydrogen in a more compact form. We will report development of hydrogen storage techniques for fuel cell vehicle including 70MPa CFRP high-pressure tank system and Multi-cylinder tank system.

See more of this Session: Hydrogen Storage System Engineering and Applications: Infrastructure, Delivery Systems and Demonstration Systems
See more of this Group/Topical: Topical 8: Hydrogen Production and Storage

Development of Hybrid Hydrogen Tank for Fuel Cell Vehicles

Shigeru Tsunokake[1], Tatsuya Fuura[1], Kazuya Takahashi[2], Yoshiki Sakaguchi[2], Etsuo Akiba[3], Kohta Asano[3], Hiroyuki Mitsuishi[4], Masanori Monde[5] and Yuichi Mitsutake[5]

(1)Japan Metals & Chemicals Co.,Ltd.(JMC), Tokyo, Japan, (2)SAMTEC CORP., Osaka, Japan, (3)National Institute of Advanced Industrial Science and Technology (AIST), Tsukuba Ibaraki, Japan, (5)Japan Automobile Research Institute (JARI), Ibaraki, Japan, (6)Saga University, Saga, Japan

Introduction

Hydrogen attracts attention as clean energy of the next generation. In order to put a hydrogen energy system to practical use, the technical development of safe and efficient storage/transportation media is on of the most important subjects. On board hydrogen tank which is under development can be classified into following two types; 1) high pressure hydrogen tank, 2) metal hydride storage tank. However, metal hydride storage tank do not meet the requirement for on board storage of hydrogen in weight densities. Many fuel cell vehicles adopt a high pressure hydrogen tank (35MPa, 70MPa), but have a problem to volumetric density now. This research is the development of a "hybrid hydrogen tank" which is a combination of type 3 high pressure hydrogen tank and metal hydride. "Hybrid hydrogen tank" is superior to a high pressure container of 70MPa in volumetric density. In this presentation, the development status of recent "hybrid hydrogen tank"[1)] [2)]will be introduced.

Target and Current state

This development projects were supported by New Energy and Industrial Technology Development Organization (NEDO). NEDO sets the target shown in Table 1 about the development goal in a hybrid tank.

Table 1. Target of Hybrid Hydrogen Tank System (Hydrogen: 5kg)

	2010(Current state)	2015(Target)	2020(Target)
Cost Target	¥5,000,000	¥2,000,000	¥200,000
Weight density Target	2-2.5wt%	3-4wt%	4-5wt%
Volume density Target	4.0kg/100L (Inner Volume)	5.0kg/100L (Inner Volume)	7kg/115L (Inner Volume)

Hybrid Tank

Hybrid tank is composed of type 3 high pressure hydrogen tank and MH cartridge.

The absorption of hydrogen by metal hydride is an exothermic reaction and desorption is an endothermic reaction, therefore it is necessary for the hybrid tank to have the function of the heat exchanger. MH cartridge consists of Metal Hydride and a heat exchanger. The structure of the hybrid tank is shown in figure 1 and the cut model of the hybrid tank is shown in figure 2.

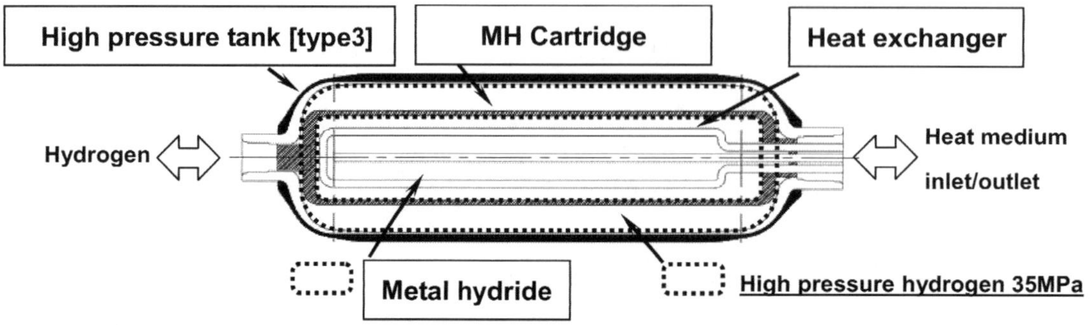

Figure 1. Hybrid tank structure

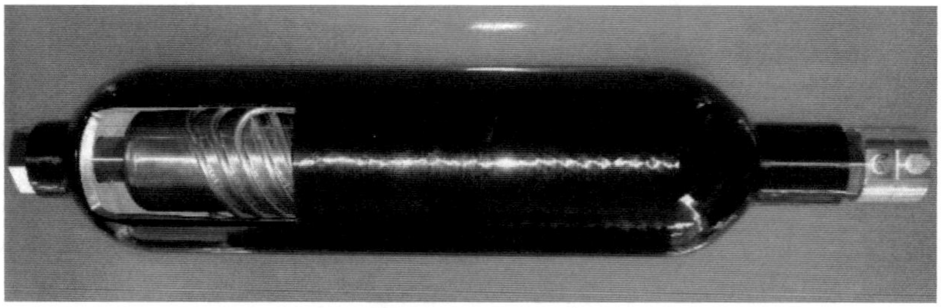

Figure 2. Cut model of hybrid tank

The specification of the hybrid tank is shown in table 2. The composition of metal hydride is V-5at%Ta-15at%Cr. The equilibrium pressure is 1.3MPa at 0deg.C. The value of the heat of formation for V-5at%Ta-15at%[3] is -30.0[kJ/mol H2].

Table 2. The specification of the hybrid tank

Inner volume	30.5 [L]
Total weight	61[kg]
(1) Type-3 tank weight	21[kg]
(2) heat exchanger weight	18.4[kg]
(3)Metal hydride weight	21.6[kg]
Hydrogen volume	1.1[kg]
Weight density	1.8[wt%]
Volume density	36[g/L]

Safety Test

We carried out a test of the safety along JARI-S001 which is standard of the safety of the type-3 container for Fuel cell vehicle. The safety test result is shown in table 3. Burst test, hydraulic cycle test , environmental cycle test and fire test was passed. But drop test was failure. The hybrid tank was heavier than the high pressure container (type-3), the damage was big. The strain for hybrid storage tanks was increased about 14%. For it the strain for the high pressure container (type-3) was increased about 6%. A fall of the performance of cycle test was predicted.

Table 3. Safety test result

Test	Test Result	Pass or Failure
Burst Test	Burst Pressure is 112MPa.	>78.75MPa Pass
Hydraulic Cycle Test	Leak was detected at 25,224 cycles.	>11,250 cycles Pass
Environmental Cycle Test	No Leakage	Pass
Fire Test	No Explosion (PRD activated)	Pass
Drop Test	-	Failure

Hydrogen charge test

We carried out hydrogen charge test of hybrid tank. The test carried out at Hydrogen & FCV Safety Evaluation Facility (Hy-SEF). The cooling medium temperature is -20deg.C and the flow rate is 60 L/min. The hydrogen pressure is 35 MPa. The test result is shown in figure 4.

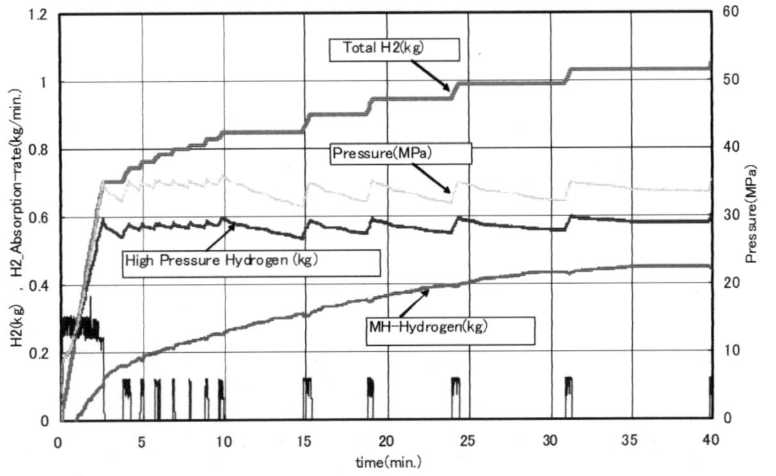

Figure 4. Hydrogen charge test

Acknowledgment

This development projects were supported by New Energy and Industrial Technology Development Organization (NEDO). This research has been conducted as a part of the "Development for Safe Utilization and Infrastructure of Hydrogen Research and Development of Vehicle Components" and "Development of Technologies for Hydrogen Production, Delivery and Storage Systems".

References

1 D.Mori et al. "High-pressure Metal Hydride Tank for Fuel Cell Vehicles",2005 MRS Spring Meeting

2 N.Takeichi et al./International Journal of Hydrogen Energy 28(2003) 1121-1129

3 NEDO "Development for Safe Utilization and Infrastructure of Hydrogen Research and Development of Vehicle Components" Report 2005

(26f) Testing of Hydrogen Storage Tank Based On 8 Kg NaAlH4

Oliver Metz, José M. Bellosta von Colbe, Gustavo A. Lozano, Julian Jepsen, Martin Dornheim and Thomas Klassen, Institute of Materials Research, Materials Technology, GKSS Research Centre Geesthacht, Geesthacht, Germany

Abstract

In the frame of the EU project STORHY (2004-2008) [1] an 8 kg-$NaAlH_4$ [2] based hydrogen storage tank has been designed and built. In parallel a new laboratory for storage tank testing and characterization has been constructed. According to the safety classification, this laboratory is divided in three parts: 1st part is a explosion proof chamber, 2nd part is a hydrogen certified laboratory and the 3rd part is used for the thermal auxiliary equipment. This paper will present basic features of the experimental testing facility, the storage tank investigated in this study as well as experimental results concerning the performance of the storage tank under different pressure/temperature and flow conditions like measured hydrogen charging and release rates as well as capacities measured under different conditions. An outlook on an improved second generation storage tank which is currently under development in the frame of the EU project NESSHY is given.

References:

[1] Ranong CN, Hohne M, Franzen J ,Hapke J, Fieg G , Dornheim M , Eigen N, von Colbe JMB, Metz O, accepted by CHEMICAL ENGINEERING & TECHNOLOGY, 32 (8): 1154-1163 AUG 2009 [2] Eigen, N; Keller, C; Dornheim, M; Klassen, T; Bormann, accepted by R, SCRIPTA MATERIALIA, 56 (10): 847-851 MAY 2007

See more of this Session: Hydrogen Storage System Engineering and Applications: Infrastructure, Delivery Systems and Demonstration Systems
See more of this Group/Topical: Topical 8: Hydrogen Production and Storage

(44a) Kinetic Study of Ion Exchange Between Multivalent Cations and Zeolite-4A in a Molten Salt

Michael R. Shaltry, Nuclear Engineering Program, University of Idaho, Idaho Falls, Idaho Falls, ID, Supathorn Phongikaroon, Department of Chemical Engineering and Nuclear Engineering Program, University of Idaho, Idaho Falls, ID and Michael F. Simpson, Pyroprocessing Technology Department, Idaho National Laboratory, Idaho Falls, ID

One of the important steps in pyrochemical spent fuel processing is removal of spent electrolyte from electrorefiners. This electrolyte is a molten chloride salt consisting of LiCl, KCl, NaCl, actinide chlorides, and fission product chlorides. Current technology being used for processing spent fuel from Experimental Breeder Reactor-II (EBR-II) at Idaho National Laboratory (INL) involves absorption of the salt into zeolite-4A followed by thermal consolidation of the material into a ceramic waste form. An alternative salt treatment process involving ion exchange into zeolite-A has been proposed and studied to allow for the recycle of most of the LiCl-KCl back to the electrorefiner and, thus, minimize waste volume generated from the process. Previous experimental work and model development on ion exchange between zeolite-A and molten chloride salts has been focused on the equilibrium condition.[1-4] However, an understanding of the fundamental kinetics of ion exchange for this system is needed for optimal design and operation of this process.

Ion exchange kinetics experiments have recently been performed and are reported in this paper. Testing was performed using a lab-scale furnace operating at 500°C in an argon atmosphere glovebox with the moisture and oxygen contents below 0.1 ppm to prevent contamination of the zeolite or salt. Cesium (Cs), strontium (Sr), and neodymium (Nd) were selected as the surrogate fission product species to capture monovalent, divalent, and trivalent cation behavior for ion exchange between zeolite-4A and molten eutectic LiCl-KCl salt. Concentrations of Cs ternary salt (CsCl-LiCl-KCl), Sr ternary salt ($SrCl_2$-LiCl-KCl), and Nd ternary salt ($NdCl_3$-LiCl-KCl) were varied from 1.0 to 5.0 wt%. Zeolite-4A pellets were loaded in a rotating stainless steel mesh basket and lowered into a magnesium oxide crucible containing the molten salt mixture. The ion exchange experiments were run from 2 to 12 hrs. Samples of the zeolite phase were taken at prescribed intervals of time. Those samples were then dissolved with various acids and analyzed using an inductively-coupled mass spectrometer (ICP-MS) to measure ion loading in the zeolite. Exchange of each cation was measured using this technique as a function of time.

Preliminary results of this study reveal that Cs loading in the zeolite reaches its equilibrium level after approximately 30 minutes of contact time, similar to the results reported in literature.[3-4] However, Nd exhibits unexpected behavior. It reaches its maximum loading in the zeolite at approximately 100 minutes of contact time and then decreases over subsequent time intervals. It is expected that there is a possibility of neodymium oxy-chloride formation during this process. Further detailed studies and results, including ion exchange of Sr, will be presented and discussed.

References:

1. S. Phongikaroon and M. F. Simpson, "Two Site Equilibrium Model for ion exchange between multivalent cations and zeolite-A in a molten salt," AIChE Journal, Vol. 52, 1736-1743 (2006).

2. M. F. Simpson, and M.L. Dunzik-Gougar, "Two-site equilibrium model for ion exchange between monovalent cations and zeolite-A in a molten salt, Industrial & Engineering Chemistry Research, 42, 4208-4212 (2003).

3. D. Lexa, and I. Johnson, "Occlusion and ion exchange in the molten (lithium chloride-potassium chloride-alkali metal chloride) salt + zeolite 4A system with alkali metal chlorides of sodium, rubidium, and cesium," Metallurgical and Materials Transactions B, Vol. 32B, 429-435 (2001).

4. M. L. Dunzik-Gougar, M. F. Simpson, and B. E. Scheetz, "Two-site equilibrium model for ion exchange between multi-valent cations and zeolite-A in a molten salt," Microporous and Mesoporous Materias, Vol 84, 366-372 (2005).

See more of this Session: Ned Division Student Award Competition
See more of this Group/Topical: Nuclear Engineering Division

(44c) Modeling of Zone Freezing for Pyrochemical Process Waste Minimization

Ammon N. Williams[1], Supathorn Phongikaroon[1] and Michael F. Simpson[2], (1)Department of Chemical Engineering and Nuclear Engineering Program, University of Idaho, Idaho Falls, ID, (2)Pyroprocessing Technology Department, Idaho National Laboratory, Idaho Falls, ID

Treatment of spent nuclear fuel using pyroprocessing technology results in accumulation of fission product chlorides in electrorefiner (ER) salt. This process is currently being carried out at Idaho National Laboratory (INL) to treat spent fuel from Experimental Breeder Reactor-II to prepare it for permanent disposal. Several purification techniques for this salt were proposed and studied by INL and Korea Atomic Energy Research Institute (KAERI) under an International Nuclear Energy Research Initiative (I-NERI) project. One of the most promising methods revealed in that project was zone salt freezing. In zone freezing, the ER salt is slowly cooled by advancing it up from a hot region through an adiabatic region and into a cold region. This causes the melt to cool from the top down. As the melt cools, pure LiCl-KCl will solidify first, causing the fission products to concentrate in the melt. Once the salt has completely cooled, the lower portion of the salt crystal containing the bulk of the fission products is removed, and the upper portion is recycled back into the electrorefiner. To better understand and optimize the zone freezing process, a mathematical model has been developed based on fundamental thermodynamic, heat transport, and mass transport theories. With this model, the temperature and concentration distributions can be determined as a function of time, providing valuable information on the zone freezing process. The temperature distribution can be found by using an implicit finite-difference method and basic laws of thermodynamics. The temperature profile was modeled by dividing the melt region into small elements and solving for the temperature of each element at a given time via the heat equation. To model the phase change phenomena, the thermodynamic property of enthalpy was used in correlation with the finite-difference method. In addition, a moving boundary condition technique was applied for each time step to simulate the advancement for the temperature distribution as a function of time of the melt into both the adiabatic and cold regions. The concentration distribution can be found using the temperature distribution along with mass transfer theory. The impurity distribution in zone freezing should be a function of both the diffusion flux of the impurity and the cooling rate. A correlation will be used between the diffusion equation and the segregation coefficient k. Details of this phase of the model are still being investigated. Preliminary results for the temperature distribution have been calculated and indicate that the temperature for a given element should decrease steadily with time until the phase change region was reached. Through the phase change region, the temperature is predicted to remain constant until the entire volume of the salt element solidifies, at which point the temperature once again should begin to drop. The temperature and concentration distribution calculations will be presented for the LiCl-KCl salt with impurities of strontium (Sr) and cesium (Cs) as a function of time.

See more of this Session: Ned Division Student Award Competition

(44d) Comparison of Anodic Signals From the Mark-IV Electrorefiner Using Statistical Invariance and Direct Computational Modeling

Robert O. Hoover[1], Supathorn Phongikaroon[1], Tae-Sic Yoo[2] and Michael F. Simpson[2], (1)Department of Chemical Engineering and Nuclear Engineering Program, University of Idaho, Idaho Falls, ID, (2) Pyroprocessing Technology Department, Idaho National Laboratory, Idaho Falls, ID

The Mark-IV electrorefiner (ER) is currently being used at the Idaho National Laboratory to treat sodium-bonded spent fuel from the Experimental Breeder Reactor-II (EBR-II). As part of this treatment process, spent uranium-zirconium alloy driver fuel is chopped and loaded into steel fuel dissolution baskets, which become the anode within the ER. Uranium and other active fission products are electrochemically dissolved from the anode into a molten LiCl/KCl eutectic salt electrolyte, while pure uranium is deposited onto a stainless steel mandrel cathode. One major goal of the process is to maximize uranium dissolution while maximizing the retention of noble metals (e.g. stainless steel cladding and zirconium).

Voltage and applied current signals obtained from experimental runs in the Mark-IV ER as reported by Phongikaroon and Yoo (Ref. 1) have been used to calculate anodic resistance as a function of time based on Ohm's Law, $\Omega(t_i) = V(t_i)/I(t_i)$. These values can be used to infer dissolution progress as different species will be electrochemically oxidized and dissolved depending on the anode potential. The resistance values can be analyzed using a histogram method, which shows the distribution of resistance values during a given experimental run. This arrangement shows a very large peak at ~0.003 ohms, representing uranium dissolution with much smaller peaks at higher resistance values representing possible dissolution of other species.

A computational model of the Mark-IV ER anode and cathode reactions involving uranium, zirconium, and plutonium (which collectively constitute greater than 90 wt% of the EBR-II spent fuel) has been developed. This model can be used, along with the experimental data from Ref. 1, to help interpret and confirm the statistical invariance method. The ability of the model to accurately calculate the presented experimental anode voltages has been determined. From the model, the partial currents of the three important species (U^{3+}, Zr^{4+}, and Pu^{3+}) were examined and compared with the resistance values predicted to correspond with the preferential range for the dissolution of uranium and other species.

Preliminary results from a different experimental run as reported in Ref. 2 have been analyzed. For data points where primarily uranium is being electrochemically dissolved, a mean resistance value has been calculated to be, $\langle \Omega_m \rangle = 2.599$ m•ohms, which compares well with the overall mean resistance in the proposed range of preferential uranium dissolution presented in Ref. 1, $\langle \Omega \rangle = 2.856$ m•ohms. Analysis and comparison of the statistical histogram method and the direct model will be presented for four experimental runs.

References

1. S. Phongikaroon and T. Yoo, "Identification of Statistical Invariance for Anodic Signals of Mk-IV Electrorefiner," Proceedings of GLOBAL 2007; Boise, Idaho, September 9-13, 2007.

2. R.O. Hoover, S. Phongikaroon, S.X. Li, M.F. Simpson, and T. Yoo, "Development of Computational Models for the Mark-IV Electrorefiner – Effect of Uranium, Plutonium, and Zirconium Dissolution at the Fuel Basket/Salt Interface," *Nuclear Technology*, in press.

See more of this Session: Ned Division Student Award Competition
See more of this Group/Topical: Nuclear Engineering Division

NUCLEAR ENERGY BEYOND THE CORE: A WIDER PERSPECTIVE

STEVEN R. SHERMAN
Savannah River National Laboratory, Aiken, SC, USA

Introduction

Within the United States, there are 104 commercial nuclear power plants which are located at 65 sites. Of these reactors, 10 reactors have been operating between 10-19 years, 42 reactors have been operating between 20-29 years, and 52 have been operating for between 30-39 years [1].

Nuclear plants in the United States are licensed by the U.S. Nuclear Regulatory Commission for a period of twenty years, and most nuclear plants are operating under extensions of the original 20-year license. Of the total, 59 plants have received 20-year license extentions, 19 plants have filed for 20-year license extensions, and another 20 plants are expected to file for 20-year license extensions by 2017.

In adding up the number of existing, pending, and desired license extensions, it is clear that not all of the existing commercial nuclear plants will still be operating in twenty years, and new commercial nuclear reactors will need to be built if nuclear energy is to at least maintain its current share of the electrical generation market into the future. There is growing enthusiasm in the power generation industry for new nuclear reactors, and there are plans for 19-20 new commercial nuclear plants to be located at existing nuclear sites. Table 1 provides a summary of planned new nuclear reactor projects.

Table 1: **New Planned Commercial Nuclear Reactor Projects**

State	Company	Reactor Type	Early Startup	# of units
MI	DTE Energy	GE-Hitachi ABWR	2018	1
PA	PPL/UniStar	Areva EPR	2017	1
MD	Constellation/UniStar	Areva EPR		1
VA	Dominion	MHI US-APWR	After 2017	1
NC	Progress Energy	Westinghouse AP1000		2
SC	Duke Power	Westinghouse AP1000		2
SC	SCE&G	Westinghouse AP1000	2016/2019	2
GA	Southern Company	Westinghouse AP1000	2016/2019	2
FL	Progress Energy	Westinghouse AP1000		2
AL	TVA NuStart	Westinghouse AP1000?		1 or 2
TX	NRG Energy/STPNOC	GE-Hitachi ABWR	2016/2017	2
TX	Luminant	MHI US-APWR		2

Although the actual number of commercial nuclear reactors may increase in the United States in response to renewal of existing operating licenses and new builds, nuclear's share of the electrical generation market is project to remain relatively flat over the next two decades due to increasing power consumption. According to statistics provided by the IAEA [2], nuclear power's share of the electrical generation market in North America is approximately 18%. This contrasts with other parts of the world, particularly in Asia, Africa, and the Middle East, where nuclear power is expected to increase its market share in electricity production for the foreseeable future. For example, China is currently constructing 24 nuclear power plants, and is projecting a ten times increase in nuclear generating capacity by 2040.

The differences in the growth rate of nuclear power's market share between North America and other parts of the world are striking, and have mainly non-technical origins. While there are many options for producing electricity, some regions of the world are choosing nuclear power with increasing frequency as the means to do it, while in North America, electricity produced from fossil fuels remains entrenched, with more than half of the U.S. electricity still produced from the burning of coal. In spite of the world view that nuclear power is an increasingly desireable method for producing electricity around the world, nuclear power plants in North America appear to be tolerated at best in spite of their safety record, reliability, and their ability to provide enormous amounts of baseload power at acceptable cost, as evidenced by electrical utilities' enthusiasm for nuclear plant license renewals.

It is offered that the differences between the acceptance of nuclear power in North America and other parts of the world are based on deep-down impressions of risks versus benefits of existing technologies, and people in the United States, especially non-engineers, tend to harbor a negative impression of nuclear power, which affects all aspects of the decision process of whether to support or hinder expansion of nuclear power. Emotionally, if a person does not see the benefits of nuclear power, or places extra emphasis on the perceived risks of deploying nuclear power plants, then no strictly technical argument(as is usually offered by nuclear engineering professionals) is going to change a closely held opinion. On the surface, such deepheld beliefs may be expressed in the form of technical objections (e.g., excess cost, safety, security, and waste concerns,), but these technical objections, if not based on solid mathematics, economics, and measured data, are just expressions of underlying mistrust, and no amount of opposing technical argument will affect the underlying beliefs, at least as far as the existing set of nuclear plants are concerned. The underlying set of beliefs must be changed, and this can only be accomplished by offering something new, different, and qualitatively "better" than what exists commercially today, and to clearly communicate how these advancements improve the benefits and reduce the risks.

Advanced Reactor Designs

In the near term, advanced light watar reactors, otherwise known as Generation III/III+, are under construction or are being commercialized throughout the world. These advanced light water reactors are more technologically advanced versions of existing commercial nuclear reactors, which offer better safety and reliability at the same power output. Unlike existing plants, these new designs emphasize passive safety, which means that the plant's safety features will continue to operate during loss of power to the cooling pumps, or even loss of the cooling fluid itself. Nuclear vendors are marketing only a handfull of standardized plant designs (see Table 1), and plants of the same type will all be exactly the same, which should reduce the risk of capital cost overruns and allow for predictable construction schedules after the first few have been built. So, in terms of risks versus benefits, the Generation III/III+ designs offer reduced risk while still providing a similar benefit to utilities.

More advanced designs are in the development stages, which are referred to as Generation IV designs. Generation IV designs emphasize four design principles – economic competitiveness, improved safety and reliability, sustainability, and engineered proliferation resistance and physical protection. In

2002, the U.S. Department of Energy released a technology roadmap for the development of Generation IV technologies [3], and six reactor types were recommended for further development. These six nuclear reactor types include: high-temperature gas-cooled reactor (HTGR), gas-cooled fast reactor (GFR), sodium-cooled fast reactor (SFR), lead-cooled fast reactor (LFR), supercritical water-cooled reactor (SCWR), and molten salt reactor (MSR).

While these more advanced reactor types offer reduced risk (after the first few were constructed and operated), these designs also seek to offer greater benefit. In the case of the HTGR, LFR, and MSR, these reactors offer higher outlet temperatures which can be used for higher efficiency electricity production, or to drive industrial processes that require high-temperature heat input (i.e., endothermic chemical processes). The fast reactor concepts (LFR, SFR, GFR) provide the ability to transmute long-lived fission products into shorter-lived ones, and to even breed fuel for more efficient fuel utilization. In the case of the MSR, the fuel rods are eliminated, and thorium can be used as a fuel with little modification to the basic design, which is more plentiful in the earth's crust than uranium, and also less exploitable as fuel souce for nuclear weapons.

In the United States, the U.S. DOE selected the HTGR as the reactor of choice, and has invested more than $500M in the concept since the Generation IV Technology Roadmap was released. In its current form, the HTGR is referred to as the Next Generation Nuclear Plant, and is targeted towards a dual mission of electricity and high-temperature heat, which would be used to support co-located chemicals and materials production. In its role as a high-temperature heat supplier, it is intended to replace fossil fuels as a source of high temperature heat. Development of the NGNP is authorized in the U.S. Energy Policy Act of 2005 [4], and construction of the NGNP is ordered by law to be completed by the end of 2021.

More recently, the concept of a Generation IV reactor has evolved in the private sector to include reactors of smaller size that can be manufactured in a plant and transported by truck or train to where it is needed. The small modular reactors currently undergoing development will offer longer core life, higher outlet temperatures, and greater proliferation resistance and security than today's Generation III/III+ designs. For most of the small modular reactor designs, the fuel is not easily removed without destroying the reactor vessel, and refueling of a small modular plant will be accomplished by trade-out of the entire modular reactor by the manufacturer.

Table 2 lists advanced nuclear reactor designs that are currently on the U.S. Nuclear Regulatory Commission's Pre-Application Review list, or that are actively being pursued by private companies.

Non-Power Applications

Among the advanced nuclear reactor designs, the high-temperature nuclear reactor designs offer the potential to provide high-temperature heat to industrial processes. Currently, all high-temperature large-scale industrial processes such as chemical manufacturing, the refining of metals, and the manufacture of cement, are largely heated by the burning of fossil fuels. If one or more of these large-sccale industrial processes could be heated using a high-temperature nuclear reactor instead, then the release of copious quantities of CO_2 might be prevented, and fossil fuels might be preserved for use in making higher-value products other than heat. The NGNP, PBMR, Hyperion, and EM2 designs are all conceived as reactor types that can support both electricity production and the production of high-temperature heat for industrial processes.

Table 2: **Privately Funded Advanced Reactor Concepts**

Reactor	Company	Power Output, MWt	Coolant
IRIS	Westinghouse	1000	Water
NuScale	NuScale Power Inc.	150	Water
PBMR	PBMR Ltd.	400	Helium
4S	Toshiba	30	Sodium
Hyperion	Hyperion Inc.	70	Pb-Bi Eutectic
PRISM	GE-Hitachi	840	Sodium
mPower	Babcock and Wilcox	400	Water
EM2	General Atomics	500	Helium
Traveling Wave	Terrapower		Sodium

Another potential application of nuclear reactors that has been a missed opportunity by all current commercial nuclear plants in the United States is the production of radioactive isotopes for medical purposes. In medicine, gamma emitters such as Tc-99m, I-123, Th-201, Ga-67, F-18, and In-111 are used for medical imaging, while beta emitters such as Y-90 and I-131 are used for cancer therapies. These isotopes are short-lived, and cannot be stored for long periods before decaying into more stable and less useful isotopes. At this time, five research nuclear reactors (government-operated) generate 90% of the world's medical isotopes. All of these reactors are more than 40 years old, and there is no redundancy in the system. So, when one reactor is closed for maintenance or experiences a problem and must be shut down, this is almost immediately expressed as a shortage in the medical isotope supply. Certainly the general public recognizes the utility and need for medical isotopes, and generating medical isotopes at commercial nuclear plants would help improve the acceptance of such plants in the public.

It would be beneficial to the commercial nuclear industry in the United States to explore non-power applications as a means to enhance the benefits of nuclear power and to distinguish it as a unique provider of multiple, higher-value products. While electricity can be produced from any number of sources, high-temperature heat can only be obtained from a high-temperature source, which reduces the options to the burning of fossil fuels, solar concentrators (which operate only when the sun is shining), and high-temperature nuclear reactors. For medical isotopes, nuclear processes are the only option for manufacturing them, and on other non-nuclear process can be used to make them.

Waste and Fuel Cycle

A major source of frustration on all sides of the arguments for and against the expansion of nuclear power lies with the handling and storage of used nuclear fuel and nuclear waste. While there are technological solutions to the handling and storage of these materials (e.g., fuel reprocessing, Yucca Mountain), the implementation of nearly all of these solutions have been blocked by opponents of nuclear power in the U.S., and the opposition is more strongly driven by the politics of fear (e.g., fear of personal economic loss, injury, or death) than by arguments on the technological shortcomings of any particular engineering solution.

With such deepheld beliefs by opponents to nuclear power, a fundamental change must be made in how proponents of nuclear power approach this argument. A change in the ethical understanding of nuclear power must be effected in the public, so that people will be less afraid of nuclear energy, and more willing to tolerate it in their communities. One way to do this is to adopt the environmental mantra "Reduce, Reuse, Recycle" that is so familiar to everyone as an underlying philosophy in the nuclear industry, and to promote this ethic publicly. While there is no way to avoid generating radioactive materials, certainly the volume of waste materials can be reduced, and the long-term toxicity of nuclear waste can be reduced, by implementing efficient and safe used nuclear fuel reprocessing plants. It must be recognized that when a nuclear reactor is built, it will someday be shutdown and decommissioned, and nuclear reactors must be designed to facilitate decommissioning in order to avoid expensive clean-up efforts at the end of life.

The seeds of change in the U.S. have been planted by the U.S. DOE, and a plant is being contructed on the Savannah River Site to recycle plutonium from decommissioned Russian nuclear warheads and to convert it into mixed oxide fuel for commercial nuclear reactors. This plant, called the MOX facility, will process 3.5 metric tons PuO_2/year starting in 2016, and is a first step. Research and development on advanced fuel recycling plants is also underway, and seeks to improve upon the costs and efficiency of fuel reprocessing plants that exist in the UK, France, Japan, and Russia, and are based on technologies invented in the U.S. in the 1950's and 1960's.

It is known that high neutron fields can transmute and destroy long-lived radioactive materials, and several of the advanced reactor designs such as the 4S, Hyperion, and EM2 are being targeted to perform the transmutation of long-lived waste materials (e.g., Pu) into radioactive materials with much shorter half-lives (e.g., Cm, Cs, Sr, Y), so that the required storage time of waste materials can be reduced from thousands of years to hundreds of years.

Security

Security involves direct threats and indirect threats. Direct threats involve fears that a nuclear plant can cause direct harm to an individual by exposure to radiation or contamination. Indirect threats generally involve the ability of individuals to steal nuclear materials from the plant which could be used at some point later in time to build destructive nuclear devices. Both types of threats must be defended against rigorously by multiple means, otherwise known as defense-in-depth.

In the chemical industry and within the U.S. DOE, the safety culture emphasizes a four-tiered approach to safety. These tiers are: Tier 1 – Eliminate the item of concern, Tier 2 – Engineering Controls, Tier 3 – Administrative Controls, and Tier 4 – Personal Protection. These types of controls can be adapted and applied to nuclear systems in order to design and build a secure nuclear system.

In nuclear systems, Tier 1 controls are the most effective because a non-existent threat can harm no one, and an example of this would be the U.S. program to remove weapons-grade plutonium from the former Soviet Union and converting this material into MOX fuel. Tier 2 controls in regard to nuclear security involve placing nuclear materials behind engineering barriers such as sealed reactor vessels or containment areas, or in high-radiation fields that prevent a person from coming too close to the materials of interest. Tier 3 controls involve restrictions in access to materials through specific handling procedures and other such controls. The final tier, Tier 4, are the guards posted at the gates.

In order of preference, Tier 1 controls are the most effective, but this is not possible when it comes to the material at the core of a nuclear reactor. So, the next best thing is to reduce the inventory of nuclear material that is desirable to a nuclear proliferator (e.g., more efficient reactors, use of uranium with a lower enrichment, smaller nuclear reactors with smaller inventories), and to build reactors that are less vulnerable to theft through stronger Tier 2 controls (e.g., sealed reactor vessels). The Tier 3 and Tier 4 controls are already in place at nuclear plants, but these controls may be bolstered by stronger Tier 1 and Tier 2 controls.

As these design improvements are developed, they should be communicated to the public to show that the nuclear industry is aware of their concerns, and is working hard to make nuclear systems even safer than they are now.

Technology and Policy

Interaction is essential between technology developers, the public, and policy makers to ensure greater acceptance of nuclear technologies and nuclear power. Such relationships are built upon trust and mutual understanding. Figure 1 illustrates this relationship. Double-headed arrows are used to represent the two-way communication that must take place between all parties in order to be successful. The message here is to know and understand the stakeholders, and to approach them with respect.

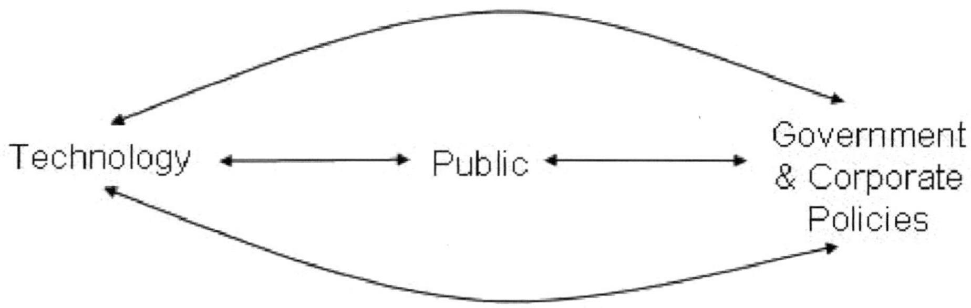

Figure 1: The Nuclear Industry, the Public, and Government are linked.

Summary

New reactors are planned, but there is still considerable public resistance to the expansion of nuclear power in the U.S. This resistance is largely based on an unbalanced emphasis on perceived risks versus perceived benefits. The nuclear industry, as technology developers, can influence these underlying beliefs by working to continue advanced reactor designs that are more efficient, safer, secure, and that offer additional benefits, such as the generation of high-temperature heat or medical isotopes, or that have the ability to transmute long-lived nuclear isotopes into shorter-lived ones. A public emphasis can be placed in the paradigm of "Reduce, Reuse, Recycle" which is already so familiar to the environmental movement, and to incorporate these philosophies in ongoing and future design efforts. The ability to proceed with new nuclear projects will require the input and involvement of all stakeholders, including the public and policy makers, if the efforts are going to be successful in the long run.

References

[1] See www.nrc.gov.

[2] Energy, Electricity and Nuclear Power Estimates for the Period up to 2030. Reference Data Series No. 1, International Atomic Energy Agency (IAEA), 2008.

[3] A Technology Roadmap for Generation IV Nucleaer Energy Systems. GIF-002-00, U.S. DOE Nuclear Energy Research Advisory Committee and the Generation IV International Forum, December 2002.

[4] Energy Policy Act of 2005. Public Law 109-58.

(100a) Adsorbents for Hydrogen Delivery

Pascal Tessier[1], Frederic Barth[2], Sylvain Houle[3], Yudong Chen[1] and Richard Chahine[4], (1)Delaware Research and Technology Center, Air Liquide, Newark, DE, (2)Air Liquide Hydrogen Energy, Jouy-en-Josas, France, (3)Air Liquide Canada, Boucherville, QC, Canada, (4)Hydrogen Research Institute, Trois-Rivieres, QC, Canada

Hydrogen storage systems based on adsorbents are analyzed starting with a review of adsorbents and their properties. The storage capacity is dependent not only on the gravimetric capacity of adsorbents but also on their bulk density, the tank weight, and other parameters related to their operation. Since most adsorbents have a high capacity at low temperature, it is necessary to operate tanks at low temperature and preferable to fill them with cold hydrogen. Filling systems that can accomplish this are discussed. The potential performance of adsorbent-based storage and delivery systems are compared to commercial technologies.

See more of this Session: Hydrogen Storage System Engineering and Applications: Materials and Storage Risk Reduction
See more of this Group/Topical: Topical 8: Hydrogen Production and Storage

(100b) Overview: Development and Safety Test of Low Pressure and Hybrid Metal Hydride Tanks Under Japanese National Projects

Etsuo Akiba[1], Kohta Asano[2], Yoshinori Tsushio[3], Hiroyuki Mitsuishi[4], Kazuya Takahashi[5], Yoshiki Sakaguchi[5], Tatsuya Fuura[6] and Shigeru Tsunokake[7], (1)Etri, National Institute of Advanced Industrial Science and Technology (AIST), Tsukuba Ibaraki, Japan, (2)National Indtitute of Advanced Industrial Science and Technology (AIST), Tsukuba Ibaraki, Japan, (3)Mazda, Japan, (4)Japan Automobile Research Institute (JARI), Japan, (5)SAMTEC, Japan, (6)R&D Department, Japan Metals & Chemicals (JMC), Tokyo, Japan, (7)Japan Metals & Chemicals (JMC), Japan

Hydrogen storage materials contain hydrogen higher volume density than compressed gas and liquid hydrogen. Under Japanese national projects, there have been investigated hydride tanks with maximum pressure below 1 MPa (low pressure metal hydride tank) and those with much higher pressure such as15-35 MPa (high pressure metal hydride tank or hybrid tank). For the tank systems to be applied to real applications, not only performances of the system but also safety issues are critical. In this presentation, both performances and safety test of the tanks developed under the national projects will be introduced.

See more of this Session: Hydrogen Storage System Engineering and Applications: Materials and Storage Risk Reduction
See more of this Group/Topical: Topical 8: Hydrogen Production and Storage

(100c) Mitigation Technologies for Hydrogen Storage Systems Based On Reactive Solids

Daniel E. Dedrick[1], Michael P. Kanouff[1], Joseph G. Cordaro[1] and Yehia (John) F. Khalil[2], (1)Sandia National Laboratories, Livermore, CA, (2)Chemical Engineering Group, Physical Sciences Department, United Technologies Research Center (UTRC), East Hartford, CT

This paper describes mitigation technologies that are intended to enable the deployment of advanced hydrogen storage technologies for early market and automotive fuel cell applications.

Solid State hydrogen storage materials provide an opportunity for a dramatic increase in gravimetric and volumetric energy storage density. Systems and technologies based on the advanced materials have been developed and demonstrated within the laboratory [1,2], and in some cases, integrated with fuel cell systems. The R&D community will continue to develop these technologies for an ever increasing market of fuel cell technologies, including, forklift, light-cart, APU, and automotive systems.

Solid state hydrogen storage materials are designed and developed to readily release, and in some cases, react with diatomic hydrogen. This favorable behavior is often accomplished with morphology design (high surface area), catalytic additives (titanium for example), and high purity metals (such as aluminum, Lanthanum, or alkali metals). These favorable hydrogen reaction characteristics often have a related, yet less-desirable effect: sensitivity and reactivity during exposure to ambient contamination and out-of-design environmental conditions. Accident scenarios resulting in this less-favorable reaction behavior must also be managed by the system developer to enable technology deployment and market acceptance.

Two important accident scenarios are identified through hazards and risk analysis methods. The first involves a breach in plumbing or tank resulting from a collision. The possible consequence of this scenario is analyzed though experimentally based chemical kinetic and transport modeling of metal hydride beds. An advancing reaction front between the metal hydride and ambient air is observed to proceed throughout the bed. This exothermic reaction front can result in loss of structural integrity of the containing vessel and lead to un-favorable overheating events. The second important accident scenario considered is a pool fire or impinging fire resulting from a collision between a hydrocarbon or hydrogen fueled vehicle. The possible consequence of this scenario is analyzed with experimentally-based numerical simulation of a metal hydride system. During a fire scenario, the hydrogen storage material will rapidly decompose and release hydrogen at high pressure. Accident scenarios initiated by a vehicular collision leading a pipe break or catastrophic failure of the hydride vessel and by external pool fire with flame engulfing the storage vessel are developed using probabilistic modeling. The chronology of events occurring subsequent to each accident initiator is detailed in the probabilistic models.

Technology developed to manage these scenarios includes: (1) the use of polymer supports to reduce the extent and rate of reaction with air and water, (2) thermal radiation shielding. The polymer supported materials are demonstrated to provide mitigation of unwanted reaction while not impacting the hydrogen storage performance of the material. To mitigate the consequence of fire engulfment or impingement, thermal radiation shielding is considered to slow the rate of decomposition and delay the potential for loss-of-containment. In this paper we explore the use of these important mitigation technologies for a variety of accident scenarios.

[1]D. Dedrick, "Heat and mass transport in metal hydride based hydrogen storage systems", HT2009-88231 Proceedings of HT2009 ASME Summer Heat Transfer Conference July 19-23, 2009, San Francisco, CA [2]D. Mosher, "High Density Hydrogen Storage System Demonstration Using NaAlH4 Complex Compound Hydrides, 2007 DOE H2 Annual Peer Review, Arlington, VA May 16, 2007

See more of this Session: Hydrogen Storage System Engineering and Applications: Materials and Storage Risk Reduction
See more of this Group/Topical: Topical 8: Hydrogen Production and Storage

(100d) Potential Diffusion-Based Failure Modes of Hydrogen Storage Vessels for On-Board Vehicular Use

Yehia F. Khalil, Chemical Engineering Group, Physical Sciences Department, United Technologies Research Center (UTRC), East Hartford, CT, Norman L. Newhouse, Lincoln Composites, Inc., Lincoln, NE, Kevin L. Simmons, Energy Materials Division, Pacific Northwest National Laboratory, Richland, WA and Daniel E. Dedrick, Sandia National Laboratories, Livermore, CA

Hydrogen-fueled fuel cell vehicles offer an environmentally attractive technology alternative to reduce greenhouse gas emissions driven, in part, by the transportation sector. The use of hydrogen/fuel cell integrated systems could also benefit other mobile applications such as forklifts and yard tractor utility vehicles. This research aims at quantifying the risks associated with gaseous diffusion through the walls of on-board vessels containing hydride materials as hydrogen storage media. The storage vessels of interest to this study include Type-III (made of carbon fiber reinforced composite wrapping with an inner metallic liner) and Type-IV (same as Type-III but with a polymeric liner instead of the metallic liner). In Type-III vessels, the liner could be aluminum or stainless steel and in Type-IV, high density polyethylene (HDPE) is commonly used as the liner material. Two failure modes related to gaseous diffusion are considered in this research. The first failure mode represents diffusion of the contained hydrogen gas through the vessel walls to the environment. Hydrogen could also leak through the vessel's welds, end fittings such as O-rings or other seals. The severity of consequences of this potential failure mode varies depending on the degree of confinement in which the permeated hydrogen will gradually accumulate, hydrogen leak rate, liner thickness, liner temperature, vessel internal pressure and the proximity to an ignition source. The second failure mode represents potential diffusion of oxygen from the surrounding air to the contained hydride inside the vessel. Most of the hydride materials, such as NaAlH4, MgH2, LiBH4, and LiAlH4, are air reactive (pyrophoric) and the severity of consequences depends on the amount of permeated oxygen inside the vessel as well as the pyrophoricity of the hydride. The consequential failures that may result from this failure mode may include over-temperature, vessel over-pressurization, or loss of containment integrity. It is also possible that the diffused oxygen gas may react with the contained hydrogen gas to form water vapor which would exacerbate the severity of hydride reactions. Also in Type-III vessels, there is a chance that the diffused oxygen could react with the metal liner and form oxides. In this work, consequence analysis using experiments and modeling is utilized to evaluate risk associated with these oxygen diffusion scenarios. Detailed event tree models are presented for each gaseous diffusion failure mode as an accident initiating event. The frequency of occurrence of the generated risk sequences will be quantified. The safety insights to be drawn from this study should assist risk analysts, OEM, Codes & Standards development teams, and other stakeholders in making risk-informed decisions with respect to identifying the optimum type of liner material, liner wall thickness and other design components of the hydride storage vessels for on-board vehicle applications.

(100e) Fundamental Reactivity Analysis of Hydrogen Storage Materials

David A. Tamburello[1], Charles W. James Jr.[2], Jose A. Cortes-Concepcion[2] and Donald L. Anton[2], (1) Computational Sciences, Savannah River National Laboratory, Aiken, SC, (2)Savannah River National Laboratory, Aiken, SC

Hydrogen is seen as the future automobile energy source due to its increased efficiency over conventional means and its ready utilization in fuel cell applications. In searching for ever higher gravimetric and volumetric density hydrogen storage systems, it is inevitable that higher energy density materials will be used. To make safe and commercially acceptable condensed phase hydrogen storage systems, it is important to understand quantitatively the risks involved in using and handling these materials to develop appropriate risk mitigation strategies to handle unforeseen accidental events. This will allow for the identification of potential risks and implementation of risk mitigation strategies.

A rigorous set of environmental reactivity tests have been developed based on modified testing procedures codified by the United Nations for the transportation of dangerous goods. Potential hydrogen storage materials—$2LiBH_4 \bullet MgH_2$, NH_3BH_3, AlH_3, and $8LiH \bullet 3Mg(NH_2)_2$—have been tested using these modified procedures to evaluate the relative risks of these materials coming in contact with the environment in hypothetical accident scenarios, including the identification of self-reactive substances, pyrophoric substances, and gas-emitting substances with water contact. An ignition event is most likely to occur if both a flammable concentration of hydrogen and sufficient thermal energy are available to ignite the hydrogen gas mixture. In order to predict hydride behavior for hypothesized accident scenarios, an idealized finite element model has been developed for a breached tank system resulting in dispersed hydride. Empirical thermodynamic calculations based on precise calorimetric experiments have been performed in order to quantify the energy and hydrogen release rates and to quantify the reaction products resulting from water and air exposure. Both thermal and compositional predictions have been made with respect to environmental scenarios identified as potential ignition events.

See more of this Session: Hydrogen Storage System Engineering and Applications: Materials and Storage Risk Reduction
See more of this Group/Topical: Topical 8: Hydrogen Production and Storage

(100f) Environmental Reactivity of Solid-State Hydrogen Storage Systems: Fundamental Testing and Evaluation

Charles W. James Jr., Jose A. Cortes-Concepcion, David A. Tamburello, Kyle S. Brinkman, Joshua R. Gray and Donald L. Anton, Savannah River National Laboratory, Aiken, SC

There is extensive research being conducted in developing new hydrogen storage systems that have higher gravimetric and volumetric capacities. However, less is understood about the hazards and risks associated with using and handling these materials. Therefore, it is critical to develop appropriate risk mitigation strategies to handle unforeseen events such as a breach-of-tank scenario in order to design commercially safe condensed phase hydrogen storage systems. A crucial aspect in risk identification is the development of rigorous environmental reactivity testing standards and procedures. One way to accomplish this is by adopting a modified testing approach from the United Nations[1] testing procedure for the transportation of dangerous goods to evaluate a potential hydrogen storage material reactivity to air and water. The modified U.N. procedures include identification of self-reactive, pyrophoric, and gas-emitting substances with water contact. The results of these tests for air and water contact sensitivity will be compared for such hydrogen storage candidates as $2LiBH_4{\cdot}MgH_2$, NH_3BH_3, $8LiH{\cdot}3Mg(NH_2)_2$, and AlH_3. The water contact tests are divided into two scenarios dependent on the hydride to water mole ratio and heat transport characteristics. Air contact tests were conducted to determine whether a substance will spontaneously react with air in a packed or dispersed form.

Additionally, thermodynamic calculations and substantiating calorimetric experiments were performed in order to quantify the energy released, energy release rates and to quantify the reaction products resulting from water and air exposure of various solid state systems. These calorimetric measurements, performed on the "small scale", were compared with standardized United Nations (UN) based tests for air and water reactivity and used to develop quantitative kinetic expressions for hydrolysis and air oxidation in these systems. Furthermore, insight gained from these U.N. tests as well as isothermal calorimetry will allow the formulation of additional tests geared toward safety engineering, numerical simulation, and integration.

[1] DOT/UN Doc., *Recommendations on the Transport of Dangerous Goods, Manual of Tests and Criteria*, 3rd Revised Ed., ISBN 92-1-139068-0, (1999).

See more of this Session: Hydrogen Storage System Engineering and Applications: Materials and Storage Risk Reduction
See more of this Group/Topical: Topical 8: Hydrogen Production and Storage

(151a) A System Modeling Approach for Light-Duty Vehicles and On-Board Hydrogen Storage Systems

Kristin L. Day, Matthew J. Thornton and Aaron D. Brooker, National Renewable Energy Laboratory, Golden, CO

Widespread adoption of hydrogen-fueled vehicles is heavily dependent on overcoming the challenges associated with on-board hydrogen storage. Identifying a means to store enough hydrogen capable of enabling a driving range greater than 300 miles within vehicle-related constraints is critical. High level vehicle, cost, and viability models integrated with detailed, low pressure, material-based hydrogen storage system models provide a means to assess the performance of a particular storage system technology in a vehicle context. The likelihood of this technology meeting the overarching vehicle system goals and thus becoming competitive with current vehicles can therefore be estimated. This paper discusses the methodology developed to evaluate the performance of various on-board hydrogen storage system technologies. Performance results focused on hydride, adsorbent, and chemical storage technologies will be presented and compared to vehicle-related packaging, cost, and performance requirements. Unique characteristics of each storage system – including gravimetric capacity, volumetric density, transient response time, full flow rate, and cost – will be identified and their effect on the viability of a theoretical vehicle will be assessed.

See more of this Session: Hydrogen Storage System Engineering and Applications: System Modeling
See more of this Group/Topical: Topical 8: Hydrogen Production and Storage

(151b) System Simulation Models for On-Board Hydrogen Storage Systems

Sudarshan Kumar[1], Mandhapati Raju[1] and V. Senthil Kumar[2], (1)Chemical Sciences and Material Systems Laboratory, General Motors Global R&D, Warren, MI, (2)India Science Laboratory, General Motors Global R&D, Bangalore, India

System simulation models for automotive on-board hydrogen storage systems provide a measure of the ability of an engineered system and storage media to meet system performance targets. Thoughtful engineering design for a particular storage media can help the system achieve desired performance goals. This paper presents system simulation models for two different advanced hydrogen storage technologies – a metal hydride system and a cryoadsorption system. Sodium alanate and super-activated carbon are employed as representative storage media for the metal hydride and the cryoadsorbent systems respectively. Lumped parameter models incorporating guidance from detailed transport models are employed in building the system simulation models.

Simulation results to test the storage systems' ability to meet fuel cell demand for different drive cycles and varying operating conditions are presented. Systems are engineered to provide the ability to refuel a vehicle in a short time guided by DOE targets. Gravimetric and volumetric hydrogen densities are computed for the engineered systems and compared to the DOE system goals. Finally, sensitivity of system performance to important design parameters is presented.

See more of this Session: Hydrogen Storage System Engineering and Applications: System Modeling
See more of this Group/Topical: Topical 8: Hydrogen Production and Storage

(151c) System Modeling Methodology and Analyses for Materials-Based Hydrogen Storage

Daniel Mosher, Jose Miguel Pasini and Bart van Hassel, United Technologies Research Center, East Hartford, CT

In the global efforts to develop advanced materials-based hydrogen storage, the various on-board reversible hydrides, adsorbents and chemical storage candidate materials and systems each have their individual strengths and weaknesses. An overarching challenge in associated research and development is to devise material/system architectures which satisfy all requirements for viability in a particular application area, such as light duty vehicular transportation. System modeling at the level which encompasses not only the storage material and vessel/reactor, but also integration with a fuel cell and balance-of-plant components, provides a more complete assessment of viability and guides options for improvement. The current work will cover the methodology developed for conducting such system modeling consistently across multiple organizations and will present performance results from studies focused on reversible hydride systems. Connecting this high level modeling to more detailed finite element design simulations will be one aspect of the framework approach discussed. The complex hydride $NaAlH_4$ is representative of novel materials under development and will be used as the basis for properties, such as temperature dependent kinetics, which influence the integrated system configurations and component sizing. While system charging is included, emphasis is placed on the discharge step in the storage process, interrogated through both simplified operation conditions (ex. constant fuel rate demand) and drive cycle transients. Comparisons of performance relative to requirements, including effective gravimetric capacity, effective volumetric density and energy utilization, will be given for the baseline material and for sensitivity studies on key material properties.

See more of this Session: Hydrogen Storage System Engineering and Applications: System Modeling
See more of this Group/Topical: Topical 8: Hydrogen Production and Storage

(151d) On-Board and off-Board Performance of Hydrogen Storage Options

Rajesh K. Ahluwalia[1], Thanh Q. Hua[2] and J.-K. Peng[2], (1)Nuclear Engineering Division, Argonne National Laboratory, Argonne, IL, (2)Nuclear Engineering, Argonne National Laboratory, Argonne, IL

A number of physical and materials-based hydrogen storage options are being developed to simultaneously meet the vehicular targets for gravimetric and volumetric capacity, cost, efficiency and greenhouse gas emissions, durability and operability, fuel purity, and environmental health and safety. The purpose of this paper is to evaluate the currently available physical, complex metal hydride, sorbent, and chemical hydrogen storage methods for their potential to meet these targets.

Our analyses show that hydrogen stored as a compressed gas (cH_2) at 350 bar in Type III or Type IV tanks cannot approach the near-term (2010) volumetric target of 28 g/L. Increasing the storage pressure to 700 bar improves the volumetric capacity (without reaching the target value) but degrades the gravimetric capacity. Dormancy and hydrogen loss are major concerns if hydrogen is stored as a cryogenic liquid (LH_2) in insulated low-pressure tanks, even if advanced vapor-cooled heat shield concepts are deployed. These problems can be mitigated by storing cryogenic hydrogen in pressure-bearing insulated tanks (the cryo-compressed option, CcH_2). Our analyses confirm that CcH_2 can also meet the near-term and intermediate-term (2015) targets for gravimetric and volumetric capacities. However, the cost of the cryo-compressed system is projected to be 3-4 times the target cost of $4/kWh, and the well-to-tank efficiency for the common pathway of producing hydrogen by central steam methane reforming (SMR) and subsequent hydrogen liquefaction is only about 41%, significantly lower than the efficiency target of 60%.

We have evaluated an on-board regenerable complex metal hydride ($NaAlH_4$, sodium alanate) system and an off-board regenerable metal hydride (AlH_3, alane) system. Our analyses show that, because of the large heat of reaction (ΔH = 37 kJ/mol-H_2) and slow hydrogenation and dehydrogenation kinetics, the doped sodium alanate system needs a combustor for releasing hydrogen, a bulky heat exchanger for refueling in 3-5 min, and a hydrogen buffer tank to handle startup and rapid demand transients. On the other hand, our analyses show that alane is an attractive hydrogen carrier (10 wt% H_2 content, ΔH = 6.7 kJ/mol-H_2), if it can be prepared and used as a slurry with >50% solids loading and a volume-exchange tank concept is developed that can store fresh and spent slurry in the same tank with a movable partition. Regenerating AlH_3 is a major problem, however, since it is metastable and it cannot be directly regenerated by reacting the spent Al with H_2. We have analyzed different indirect methods of regenerating AlH_3, including the organometallic route. We have constructed a flowsheet for this three-step method in which AlH_3 is formed as a stable adduct to an amine, the adduct is destabilized by transamination, and the AlH_3 is released by the catalytic decomposition of the transaminated adduct by decomposing under vacuum. We concluded that fuel cycle efficiencies in 40-55% range are possible with this organometallic route, if low-grade waste heat is available for the decomposition steps and the electricity requirements for the vacuum distillation steps match the estimates.

We have analyzed two sorbent-based hydrogen storage systems, one using AX-21, a high surface-area superactivated carbon, and the other using MOF-177, a metal organic framework material, Zn_4O(1,3,5-benzenetribenzoate) crystals. We found that moderate pressures and cryogenic temperatures are needed for reasonably high storage capacities, and that a temperature swing must be imposed in addition to the pressure swing to desorb hydrogen. Using off-board liquid N_2 for removing the heat of sorption is not very attractive because of the poor thermal conductivity of the sorbent bed and the unfavorable fuel cycle (well-to-tank) efficiency. We also evaluated the adiabatic refueling option, in which the sorbent bed is cooled evaporatively by charging it with liquid hydrogen, and warmed hydrogen is recirculated during discharge to supply the heat of desorption. The results from our analyses indicate higher storage capacities with MOF-177 than with AX-21, but these capacities are lower than those available with the cryo-compressed option.

We have also evaluated several exothermic and endothermic chemical methods of storing hydrogen. Releasing hydrogen by hydrolysis of sodium borohydride ($NaBH_4$, SBH) has been demonstrated in the laboratory and on-board an experimental vehicle. Our analysis of the SBH system pointed to a number of shortcomings, including the limited solubility of the product of reaction ($NaBO_2$) in the aqueous alkaline medium, difficulty in rejecting the large amount of heat that is liberated in the exothermic reaction ($\Delta H = 52.5$ kJ/mol-H_2), and the low well-to-tank efficiency in regenerating $NaBH_4$ from $NaBO_2$. We concluded that, for reasonable fuel cycle efficiencies, formation of B–O bonds should be avoided. We have evaluated the option of using an organic liquid carriers (LCH2), such as N-ethylcarbazole, which can be dehydrogenated thermolytically on-board a vehicle and rehydrogenated efficiently in a central plant by well-established methods and processes. Our analysis indicated low gravimetric (<3.5 wt% H_2) and volumetric (<25 g/L) capacities of this system, due to large ΔH (51 kJ/mol-H_2), pressure-limited equilibrium conversion, and slow dehydrogenation kinetics. As a result, the on-board system requires a mineral oil-based high-temperature coolant, a combustor, and a buffer H_2 storage tank for startup and rapid transients. Using reverse engineering, we have also determined the attributes that a prospective organic liquid carrier must have for such an on-board system to meet the H_2 storage targets.

Recently, there has been much interest in ammonia borane (NH_3BH_3, AB) as a hydrogen carrier because it has a high H_2 content (19.6 wt%), it undergoes exothermic thermolysis to release the H_2 ($\Delta H = 33$ kJ/mol-H_2), and it decomposes at relatively low temperatures. Like AlH_3, AB is thermodynamically unstable and its decomposition rate is independent of the backpressure. We are evaluating the concept of using AB dissolved in an organic liquid solvent (1-butyl-3-methylimadazolium, bmimCl) as an on-board storage medium. We have devised reactor configurations for controlling the peak temperature through heat transfer and recycle of the product stream. Like AlH_3, regenerating AB efficiently from the spent products is particularly challenging; it would be quite impractical if all three equivalents of H_2 are released from AB, allowing boron nitride (a ceramic) is allowed to form. We have constructed and analyzed flowsheets for off-board regeneration of spent AB using process chemistry developed by others. We have identified process steps that are energy intensive and are the largest contributors to inefficiencies if thiol or an alcohol is used to digest the spent AB. We concluded that, on a primary energy basis, regeneration efficiencies ranging from 17 to 35% are possible with these digestion schemes.

Figure 1 summarizes the volumetric and gravimetric capacities of the different on-board hydrogen storage options evaluated to date and compares them with the near-term, intermediate-term and ultimate targets. Figure 1 also compares the well-to-engine (WTE) efficiencies of these options for the pathway in which H_2 is produced in a central SMR plant. The paper will discuss others aspects of the storage options as well, including high-volume manufacturing cost projections, greenhouse gas emissions, durability and operability, and environmental health and safety issues (dormancy and hydrogen loss).

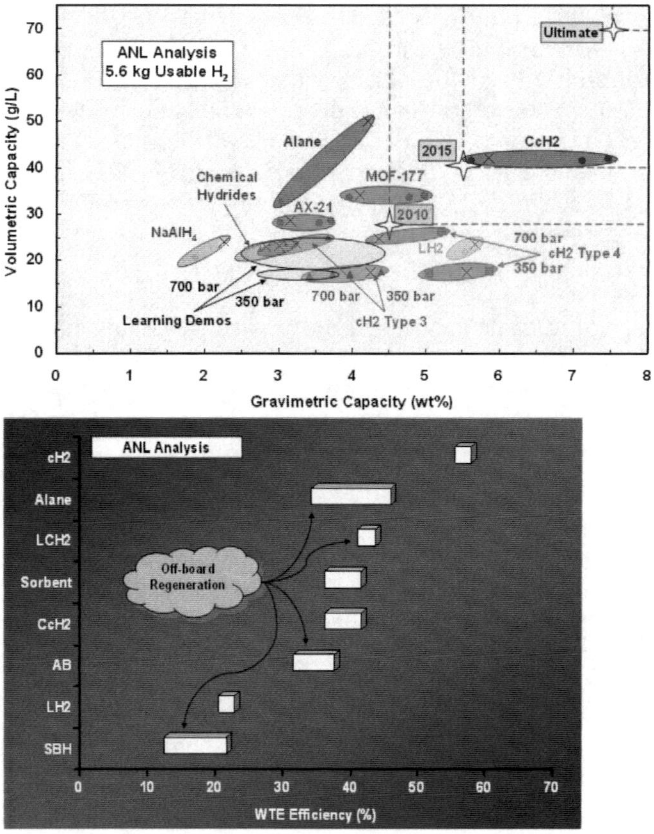

Figure 1 System capacities and fuel cycle efficiencies of different hydrogen storage options

See more of this Session: Hydrogen Storage System Engineering and Applications: System Modeling
See more of this Group/Topical: Topical 8: Hydrogen Production and Storage

(151e) Optimization of Metal Hydride Based Hydrogen Storage Bed Designs

Mandhapati Raju, General Motors Company, Warren, MI and Sudarshan Kumar, Chemical Sciences and Material Systems Laboratory, General Motors Company, Warren, MI

Design of metal hydride storage bed is a key factor that influences the storage capacity, gravimetric hydrogen storage density, and refueling time for automotive on-board hydrogen storage systems. The choice of storage bed design incorporating the heat exchanger, and corresponding design parameters is not obvious. A systematic study is presented to optimize the bed heat exchanger design using CFD modeling. Three different shell and tube heat exchanger designs are chosen. In the first design, metal hydride is present in the shell and heat transfer fluid flows through straight parallel cooling tubes placed inside the bed. The cooling tubes are interconnected by conducting fins. In the second design, the heat transfer fluid flows through helical tubes in the bed. The helical tube design permits use of a specific maximum distance between the metal hydride and coolant for removing heat during refueling. In the third design, the metal hydride is present in the tubes and the fluid flows through the shell. An automated tool is generated using COMSOL-MATLAB integration to arrive at the optimal geometric parameters for each design type. Using sodium alanate as the reference storage material, relative merits and demerits of each design type are analyzed in detail and a comparison of the gravimetric and volumetric hydrogen storage densities for the three designs is presented.

See more of this Session: Hydrogen Storage System Engineering and Applications: System Modeling
See more of this Group/Topical: Topical 8: Hydrogen Production and Storage

An NMR Study of Sodium Borohydride Hydrolysis in Concentrated Aqueous Solution

AIChE Meeting November 2010

Lin Yu[1], Amy M. Beaird[1], Ping Li[2], J. Karl Johnson[2], Rajasree Retnamma[3], Michael A. Matthews[1]

1. Department of Chemical Engineering, University of South Carolina

2. Department of Chemical and Petroleum Engineering, University of Pittsburgh

3. National Laboratory for Energy and Geology, Lisbon, Portugal

1. Introduction

Sodium borohydride is the most widely studied material among chemical hydrides and has been identified as a promising hydrogen storage media for portable applications, because of advantages such as high theoretical energy density (10.8wt%), low cost, safe handling and ease of H_2 release. Traditionally, studies focus on dilute sodium borohydride solutions with the aid of catalysts at room temperature. Various kinetic models for catalyzed hydrolysis have been reported accordingly. However, no kinetic data are available for non-catalyzed, concentrated $NaBH_4$ solutions. This work reports an experimental kinetic study of non-catalyzed sodium borohydride hydrolysis in highly concentrated solutions at elevated temperatures. The concentrations of boron containing species were characterized in-situ by a novel ^{11}B NMR technique. The effects of temperature and pH are reported.

2. Experimental

A Varian Inova 500 spectrometer operating at 160.487Hz for ^{11}B was used to acquire spectra in real time. The NMR operates in the temperature range of 25-90°C, with an accuracy of +/-0.1 °C. ^{11}B NMR spectra were acquired as proton-coupled, with 15s relaxation time and 8 scans. The NMR spectrometer was tuned to increase the signal-to-noise ratio each time before each measurement was carried out.

A concentric cylinder arrangement was used for obtaining real-time measurements of the rate of appearance of $NaB(OH)_4$. Sodium borohydride powder was placed in an 8 mm NMR tube, which was then sealed in a glove box. The 8mm tube was placed inside a 10mm NMR tube containing a standard solution of boric acid dissolved in D_2O. Prior to beginning the NMR scan, a predetermined volume of DI water was injected to the 8mm tube to dissolve the $NaBH_4$. Hydrogen bubbles were immediately seen after DI water was introduced. The concentric tubes were immediately placed in the spectrometer. The elapsed time between solution addition and first NMR spectrum acquisition was in the range of 180 to 240s.

The peak intensities of the reactant $NaBH_4$ and product $NaB(OH)_4$ were collected as a function of time. The concentrations of $NaBH_4$ and $NaB(OH)_4$ were quantified with reference to the standard boric acid solution, utilizing the fact that the concentration is proportional to the ratio of tube cross-sectional areas and the ratio of peak intensities. The data were obtained at temperatures of 25°C, 60°C and 80°C and at $NaBH_4$ concentrations of 10wt%, 15wt% and 20wt%. Additional data were obtained for the 20wt% $NaBH_4$ solution at 80°C, using NaOH at concentrations from initial 1wt% to 15wt% to adjust the pH. Another set of data were obtained for 10wt% $NaBH_4$ at 25°C, using acetic acid and HCl to adjust the pH in the range of 2.85-4.69.

3. Representative Results and Discussion

The effect of initial $NaBH_4$ concentration on the rate of formation of the hydrolysis is shown in Figure1.The data presented are concentration of the product $NaB(OH)_4$ (the formula equivalent to $NaBO_2 \cdot 2H_2O$ in excess water) at 25°C; It is shown that increasing the initial $NaBH_4$ concentration (10 to 20wt%) gives increasing product concentration. Comparison of the product concentration after 24 hours to that of 100% conversion is indicated by horizontal lines. As shown in Figure 1, the most concentrated solution results in lower conversion. In order to quantify the effects of initial $NaBH_4$ concentration on reaction rate at 25 °C, we took the derivative of the product concentration with respect to time. We find that the reaction rate drops quickly within first 5 hours; and it increases with the increasing initial $NaBH_4$ concentration from 10wt% to 20wt%.

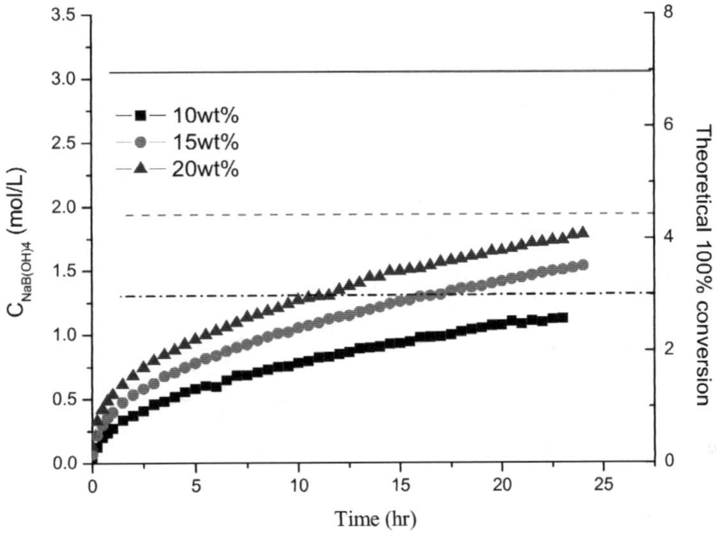

Figure 1. Effect of initial $NaBH_4$ concentration on formation of $NaB(OH)_4$ at 25°C

The effect of temperature within the range 25°C-80°C on the hydrolysis reaction has also been studied with an initial $NaBH_4$ concentration of 15wt%. Data will be presented during the presentation. It is found that at 80°C the reaction rate is the greatest; at around 10 hours, the $NaB(OH)_4$ has already reached theoretical value of 4.89mol/L for 100% conversion. However, at 25°C $NaB(OH)_4$ concentration only is 1.53 mol/L after 24 hours. Therefore, high temperature significantly improves the generation of product in the $NaBH_4$ hydrolysis. The reaction rate was observed to drop quickly within the first five hours and decrease gradually as the reaction continued. The initial reaction rate increases from 0.58 mol/L·hr to 6.73 mol/L·hr as temperature is increased from 25°C to 80°C.

To investigate the effect of increasing pH on the kinetics, alkaline solutions with different initial NaOH concentrations (1wt%, 5wt%, 10wt% and 15wt%) were prepared, with a constant initial $NaBH_4$ concentration of 20wt% at 80°C. NaOH inhibits the hydrolysis reaction. The inhibition effect is the least evident at 1wt% of NaOH, but this effect becomes more pronounced by increasing the concentration of NaOH from 5wt% to 15wt%. For the initial 15wt% NaOH, the hydrolysis rate is constant, very near zero.

The acceleration of the reaction with acid was accomplished by adjusting the initial pH. We used acetic acid and HCl to adjust the initial pH over the range from 2.85 to 4.69 to investigate the hydrolysis kinetics. It is found that with the same initial pH, acetic acid is a better accelerator than HCl at the same conditions.

To fit the experimental kinetic data, we fit the empirical power law model shown in equation (1). We assumed that both the $NaBH_4$ concentration and proton concentration will affect the hydrolysis rate:

$$R=\frac{dCNaB(OH)_4}{dt}=[NaBH_4]^{m}[H^+]^{n} \qquad (1)$$

The kinetic parameters were obtained by a linear regression of the rate versus $NaBH_4$ concentration. The rate equation is based on the assumption of a constant water concentration during the hydrolysis reaction. We are able to correlate the measured pH with the product concentration under the same reaction conditions. For example, Table 1 shows the regression results from the left division algorithm in Matlab. The reaction order with respect to $NaBH_4$ decreases with increasing temperature from 25°C to 60°C; while the order with respect to proton concentration remains unaffected to temperature change in this range. The rate constant also shows an increase with rising temperature.

Table 1 Kinetic Parameters Determined by Linear Regression (neutral pH)

T	m	n	logk
25°C	2.8	1.1	7.5
60°C	1.9	1.1	9.5

We also performed a linear regression for the reaction order with respect to $NaBH_4$ and NaOH at 80°C, respectively. According to the result, the reaction rate model can be expressed by model equation (2):

$$R = k\,[NaBH_4]^{3.3}[NaOH]^{-0.92} \qquad (2)$$

It is seen that the reaction rate increases with the $NaBH_4$ concentration and decreases with NaOH concentration. The negative dependence of reaction rate on the NaOH concentration is consistent with the inhibition effect of NaOH.

4. Conclusion

In this study, we investigated the non-catalyzed hydrolysis kinetics of concentrated $NaBH_4$ and NaOH stabilized solutions for hydrogen generation. The reactions are evaluated using ^{11}B NMR at 25, 60 and 80°C for 24 hrs. It is found that both high initial $NaBH_4$ concentration and high temperature are able to increase the reaction rate of $NaBH_4$ aqueous hydrolysis. Concentrated alkaline solutions will result in severe inhibition of the hydrolysis reaction. In contrast, acids accelerate the hydrolysis rate. We have also developed an empirical kinetic power-law model for the hydrolysis. It indicates that at higher temperature the dependence of hydrolysis reaction rate on $NaBH_4$ concentration decreases; the increasing

NaOH concentration decreases the reaction rate. Future work will take into account the water concentration change in modeling.

Ammonia Borane Dehydrogenation Always Generates Ammonia, How Much and How to Remove It?

Ahmad Al-Kukhun, Hyun Tae Hwang and Arvind Varma, School of Chemical Engineering, Purdue University, West Lafayette, IN

Hydrogen is a widely acknowledged clean fuel for PEM fuel cell (FC) vehicles and, Ammonia Borane (AB) is among the best hydrogen storage materials as it contains 19.6 wt.% and 150 kg/m3 H2 on weight and volume basis, respectively. Current methods to release hydrogen from AB include catalytic hydrolysis, thermolysis and hydrothermolysis [1]. As recently shown experimentally [2], all of these methods generate ammonia gas as a by-product of ammonia borane dehydrogenation. In addition to poisoning the anode catalyst site of PEM FC, ammonia can also cause ion exchange with protons in the polymer electrolyte. It has been reported that as low as 13 ppm NH3 can significantly decrease the FC performance, and that the degradation is irreversible for long-term exposure (15 hr) to 30 ppm NH3 [3]. Thus, to utilize any AB-based method to generate hydrogen for PEM FC applications, a technique for removing ammonia from the product gas stream is required.

In this study, for different operating conditions, we first quantified the ammonia generated from the more promising thermolysis and hydrothermolysis methods. Next, we evaluated different methods for ammonia removal, including membranes, absorption, adsorption, selective oxidation and cracking. The results show that, using a combination of these methods, ammonia can be removed completely. The optimized system configuration for ammonia removal was achieved to minimize the weight penalty of the added system.

[1] Diwan M., Hwang H. T., Al-Kukhun A., Varma A., "Hydrogen Generation from Noncatalytic Hydrothermolysis of Ammonia Borane for Vehicle Applications," AIChE Journal (in press, DOI 10.1002/aic.12240). 2010

[2] Hwang H. T, Al-Kukhun A., Varma A., "Hydrogen for Vehicle Applications from Hydrothermolysis of Ammonia Borane: Hydrogen Yield, Thermal Characteristics, and Ammonia Formation," Ind. Eng. Chem. Research (in press, DOI 10.1021/ie100520r). 2010

[3] Halseid, R., Vie, P. J. S., Tunold, R., "Effect of Ammonia on the Performance of Polymer Electrolyte Membrane Fuel Cells". J Power Sources 2006, 154: 343.

(218b) Critical Loading Level for Low-Temperature, Simultaneous Release of Two Equivalents of Hydrogen From Ammonia Borane Nanocomposites

Junshe Zhang[1], **Yu Zhao**[2], Daniel L. Akins[2] and Jae W. Lee[1], (1)Department of Chemical Engineering, The City College of New York, New York, NY, (2)Department of Chemistry, The City College of New York, New York, NY

Ammonia borane (NH3BH3, AB) has a gravimetric density of 196 g H2/kg AB; however, it slowly releases about 7 wt % hydrogen at temperatures below 85 degree Celsius. Although the dehydrogenation kinetic can be accelerated by compositing AB with mesoporous silica, the hydrogen density decreases significantly. The reduced density of nanocomposites can be offset by initiating the second hydrogen release step. However, this only occurs at low AB loading levels. Here, we report the critical loading level (or the maximum loading level) for the low-temperature, simultaneous release of two equivalents of hydrogen from MCM-41-AB nanocomposites. At a loading level of 0.15 (mass ratio of AB to matrix), 2.1 equivalents of hydrogen releases in 1 hour at 89.5 degree Celsius. Also, we find that nanoconfined AB (AB resides inside mesoporous channels) is amorphous and aggregated AB (AB resides outside mesoporous channels) has a tetragonal structure. Additionally, the first dehydrogenation of aggregated AB peaks at a low temperature compared with pristine AB. The critical AB loading level for nanocomposites involving MCM-41 and AB is found between 0.15 and 0.21. The findings improve understanding of mesoporous matrix-AB nanocomposites and are critical for design of novel AB nanocomposites.

See more of this Session: Fundamentals of Hydrogen Production: Reactions, Separations and Reactive Separations
See more of this Group/Topical: Environmental Division

(218c) Dihydrogen Catalysis Assisted by Molecular Hydrogen Reactions: Storage Linked Processes and Possible N2-Activation

Joseph Bozzelli, Chemistry and Environmental Science, New Jersey Institute of Technology, Newark, NJ and Rubik Asatryan, Chemistry and Env't Science, New Jersey Inst Tech, Newark, NJ

ABSTRACT. Density functional theory is employed in the context of molecular hydrogen assisted processes in order to disclose two types of reactions: (i) symmetric-syn¬chro¬nous double hydrogen-atom abstraction, and (ii) H2-mediated H-transfer. The reaction analysis is specifically applied to the heterogeneous decomposition of diazene, the most important intermediate of N2-activation processes.
KEYWORDS. Diazene, Dihydrogen Catalysis, Nitrogen Fixation, Nitrogenase, Hydrogen Storage.
While developing a comprehensive potential energy surface for hydrazine, N2H4 at CCSD(T)/CBS and CBS-QB3 levels of theory to uncover chemical activation reactions of amidogen radicals (NH2), we have revealed a stereoselective reaction pathway (Eq.1) directly associated to the gas-phase decomposition of cis-diazene, the most important substrate of nitrogenase enzyme and a key intermediate in heterogeneous catalytic processes involving elemental nitrogen and hydrogen, with fairly low gas-phase activation barrier of 22.5 kcal mol-1 (Fig.1): 1,2 H2 + cis-N2H2 $\leftrightarrow$ N2 + H2 + H2 (1) Please Insert Fig.1. Here This reaction (Eq. 1) is the simplest possible model for the symmetric-synchronous double hydrogen-atom abstraction reaction from an H-donor as cis-diazene molecule. or: D2 + cis-N2H2 • 2 DH + N2 D–notation we employed here for clarity, and note that such an isotopic scheme explains the observed formation of HD during N2-reduction in the presence of D2 (vide infra). 3 Intriguingly, the reverse N2-activation reaction (Eq.-1) faces a barrier of only 76 kcal mol-1 which is almost 50 kcal mol-1 lower than the analogous bimo¬le¬cu¬lar nitrogen fixation reaction (ca.125 kcal mol-1).4 The reverse reaction is termolecular and is consequently, significantly un¬favored entropically. The reaction becomes viable at very high pres¬sures5 but would be feasible if the molecular hydrogen is partially bonded to a hetero-surface and the H-H bond is both polarized and weakened. Natural Bond Orbital (NBO) analysis of the reaction (1) at B3LYP/6-311++G(2d,2p) level of theory shows that N-H bonds in isolated cis-N2H2 are polarized (natural charges on H and N atoms are QH= +0.14e and QN= -0.14e, respectively). However, the electron density in the transition state ring (Fig.1) is almost completely delocalized, with partial polarization of newly formed H-H bonds (D-H in Eq.1): QH= +0.08e and QN= -0.07e. Terminal hydrogen atoms (D atoms) posses small negative charges, QD = -0.01e. Natural electron configuration of nitrogen atoms in TS is 2s0.782p1.77. As expected, 2p-orbital electrons are responsible for the partial electron donation from the isolated diazene (2s0.762p1.86) to the approaching hydrogen. Hence, promoting role of the initial activation of attacking H2 can be anticipated. In this report, we have studied analogous metalorganic systems and discuss several possible implications. The fixation of N2, generally considered as its reduction to NH3, is an important topic throughout the natural sciences studied over several decades; yet the detailed molecular mechanism remains an open issue.6-11 The N≡N bond is exceptionally strong (ca. 225 kcal mol-1) and hence the

first step activating this triple bond is extremely unfavorable despite the overall process of ammonia synthesis being exothermic by ca 22 kcal mol-1. Recent theoretical calculations at G2M(MP2)//MP2/6-31G** level suggest4a that the lowest barrier associating H2 and N2 into iso-diazene is 125.4 kcal mol-1, consonant with the MRCI(+Q)/aug-cc-pVQZ//FVCAS/aug-cc-pVQZ value of 5.39 eV.4b Chain reactions are therefore considered as the dominant gas-phase and surface mediated mechanisms for hydrogenation of nitrogen.4 It is known, that very high pressures and temperatures (200 atm and 500oC) with iron-based catalysts are employed to hydrogenate nitrogen in the industrial Haber-Bosch process, which is believed to proceed via initial dissociative adsorption of reagents in a (Fe, or Ru based) catalyst surface. 4b In contrast, the activation of N2 occurs at ambient conditions in biological media, supported by enzymes, such as nitrogenase.6 The con¬ven¬¬tional bio-catalytic mechanism considers binding of N2 to the Fe-[H2] comp¬lex, reduction of N2 to diazene (diimide, N2H2) and hydrazine (N2H4) and release of NH3.5 Hydrazine as an inter¬me¬diate has been detected ex¬pe¬rimentally, while diazene, which is also proven to be a substrate for nitrogenase, 7 is not observed due to its high reactivity. 6b It is further important to note that molecular hydrogen is always produced when nitrogenase reduces N2 to NH3 (a turnover cycle); evolution of hydrogen by nitrogenase is uniformly observed, even under high nitrogen pressures (e.g., 50 atm).8 Most N2 fixing bacteria contain hydrogenase, the another relevant enzyme, to recycle excessive hydrogen.9 The equation for conversion of N2 on nitrogenase is considered as N2 + (6+2x)H+ +(6+2x)e- • 2NH3 + xH2 (2) The value of x varies from 1.0 - 7.5 depending on the type of enzyme used; for main MoFe enzyme it is considered unity. 6,8,10 The nitrogenase also reduces other substrates such as C2H2, HCN and N2O to C2H4, CH4 + NH3 and N2 + H2O, respectively. Note that hydrogen is not evolved during enzymatic reduction of acetylene and interestingly, C2H2 is specifically converted to cis-C2H2D2 in presence of D2O.6 Iron plays a vital role in variety of enzymatic and heterogeneous catalytic processes.6,9,11 We have therefore, studied several simple heterogeneous models based on iron-organic complexes to further evaluate the relevance of assisted by molecular hydrogen mechanisms. Calculations were performed at the B3LYP hybrid DFT level12 using the Dunning-Hay double zeta basis set13 and Los Alamos effective core potential14 for Fe as implemented in Gaussian-03. 15 A Dresden/Stuttgart pseudo¬potential is also used to verify key results, not presented here. Different spin-less (S=0) and high-spin (S=4) neutral and ionic systems were considered for comparison. In this short report the reaction of hydrogen molecule with a straightforward (CH3)2FeH2 model as a source of activated hydrogen atoms will be addressed. A five (trigonal bipyramidal) and a four (tetrahedral) coordinated Fe-complexes are also adopted, to generally model some features of the organometallic catalytic active sites. Structures similar to the gas-phase transition state (Fig.1) are observed for reactions between the organome¬tallic¬ model (CH3)2FeH2 with diazene and CH2NH intermediates (Fig2a,b), where a hydride of a tran¬si¬tion metal serves as a donor of the dihydrogen, a "H2-catalyst": R2FeH2 + cis-N2H2 = R2Fe + H2 + H2 + N2 (3) R2FeH2 + CH2NH = R2Fe + H2 + H2 + HCN (4) In Fe-organic models the hydridic H-atoms are pre-activated, and this substantially reduces the activation energy. The barrier height for the catalytic decomposition reaction of cis-N2H2 (Eq.3) with R≡CH3 is fairly low (6.5 kcal mol-1) compared to the 22.5 kcal mol-1 barrier in the gas-phase. Thus, the ring-mechanism is expected to be functioning in both gas and condensed phase systems. Please Insert Fig.2. Here This mechanism dealing with different nitrogenase substrates and intermediates (such as diazene, HCN, N2O, etc) may serve as a simple molecular model involving requisite evolution of hydrogen (an attribute to the turnover cycle of nitrogenase enzyme) and could explain why the activation process is inhibited by H2.8 We demonstrate that a similar ring- transition state endures in different environments, in a variety of neutral and ionic systems. This dihydrogen catalysis phenomenon can be possibly applied to the decomposition processes of high hydrogen content compounds. It opens also a variety of new pathways re¬le¬vant to a number of processes, including remote transfer of H-atoms supported by molecular hydrogen. 16,18 We report one further, me¬¬cha¬nistically important mechanism, viz., a new H_2-mediated H-transfer reactions, which we believe may occur virtually in any systems. We demonstrated this mechanism on a simple Fe-S-Fe backbone organometallic model with neutral CO ligands (Fig.3), which is a rather common moiety in biological systems.16 Fig.3a demonstrates a regenerative cycle for hydrogen transfer from an S-H group to the coordinated N_2H-group, the possible intermediate of N_2-hydro¬ge¬na¬tion. Here, the reactant H_2 molecule is a transferor of an H atom from a bridged S to the distal N atom. It can be considered a catalyst as it is chemically unchanged at the end of reaction, when only one hydrogen isotope is involved. A similar process holds in the case of H-transfer from a Fe-H group in unsaturated tetrahedral Fe1 center to the axially located N_2H group in the trigonal bipyramidal structure (Fig3b), while formed (and yet adsorbed) N_2H_2–group transfers back to the unsaturated Fe1 center. Note, that hydrogenation (protonation with further electron attachment) of a bridged sulfur atom in nitrogenase models is shown to facilitate activation of N_2 adsorbed on the proximate Fe-centers.10,17

See more of this Session: Fundamentals of Hydrogen Production: Reactions, Separations and Reactive Separations
See more of this Group/Topical: Environmental Division

(218d) Energetically Self-Sufficient Coproduction of Hydrogen and Formic Acid

Jorge Pena Lopez and Vasilios I. Manousiouthakis, Chemical Engineering Department, University of California, Los Angeles, Los Angeles, CA

The use of hydrogen has been proposed as a solution to the air quality deterioration in metropolitan areas and as an effective energy carrier for small and medium vehicles. This option has strengthened with the advancement in the development of fuel cells manufacturing and hydrogen storage. To technologically advance this alternative, new hydrogen production designs are to be developed. Several options exist for hydrogen production; one of the most explored has been the reforming of natural gas given that methane contains potentially two hydrogen gas molecules. The main problem with conventional natural gas reformers is the emission of carbon dioxide, a greenhouse whose emissions have led to an increase in its concentration in the atmosphere. Transforming carbon dioxide into a useful product prevents its emission into the atmosphere; however, an inadequate process design might largely increase the cost of producing hydrogen, making it uncompetitive with gasoline. In this work we present a comparison between three different hydrogen and formic acid coproduction process. The first design provides a baseline for the production of hydrogen with recovery of carbon dioxide downstream of the steam reforming process, where the carbon recovered in this way is utilized to produce formic acid. The second design recovers the carbon dioxide released at the source of energy generation, thus producing a larger amount of formic acid and eliminating carbon emissions to zero. A third design implements the commercially established formic acid production process into a design including hydrogen production. Given the structure and feedstock of this formic acid plant, the hydrogen production can no longer by carried out by a steam reforming process but rather from a combination of dry reforming and water gas shift. These three processes go through heat and power integration techniques that allow the maximum utilization of energy in favor of hydrogen production, maintaining a self-powered plant and maintaining the carbon dioxide emissions at zero. We show that there exists an optimal themodynamical and operational point for the operation of these plants.

See more of this Session: Fundamentals of Hydrogen Production: Reactions, Separations and Reactive Separations
See more of this Group/Topical: Environmental Division

(221a) Research and Development of Hydrogen Storage Materials for Lightweight Tanks

Yoshitsugu Kojima and Takayuki Ichikawa, Institute for Advanced Materials Research, Hiroshima University, Higashi-Hiroshima, Japan

Fuel cell vehicles (FCV) feature 35-70 MPa hydrogen storage tanks and can travel 620 km (pressure: 35 MPa, inner volume:171 L) -780km (pressure: 70 MPa, inner volume:156 L) on full tanks. The tank size of the vehicle is large compared to those of gasoline vehicles. Hydrogen can be stored in many different forms as compressed or liquefied hydrogen in tanks, as hydrogen adsorbed carbon materials, as hydrogen-absorbing alloys, as metal hydrides with light elements such as NaBH4, NaAlH4, MgH2, LiBH4 and metal nitrides, or as organic hydrides (methylcyclohexane, decalin)[1]. In this study, we will review our experimental results on metal hydride with high dissociation pressure (Ti1.1CrMn), light weight Li-Mg-N-H system, NH3-based materials and NaH-NH3BH3 composites. We found that the effective hydrogen capacity of Ti1.1CrMn exhibited the maximum value of 1.8 mass% in the pressure range of 33 MPa and 0.1 MPa at 296 K (dissociation pressure: 11 MPa) [2]. At the low temperature of 233 K, the hydrogen desorption capacity at 0.1 MPa was 1.6 mass%. The standard enthalpy difference of Ti1.1CrMn was -22 kJ/molH2. The absolute value was about 10 kJ/molH2 smaller than those of LaNi5 and Ti-Cr-V. The mixtures of LiH and Mg(NH2)2 with molar ratios of 8 to 3 were milled for 2-20 h under 1 MPa H2 pressure to synthesize the Li-Mg-N-H system by a planetary ball milling [3]. It was found that milling time had a remarkable effect on improving kinetics and then an optimum milling time was 20 h. The H2 desorption capacity close to 6 mass% was obtained after keeping for 8 h at 423 K. For the dehydrided Li-Mg-N-H system (vacuum, 473K, 24h), ball milling treatment was performed at 423 K for 5 h under 15 MPa of H2 pressure, leading to the H2 absorption capacity up to 5.8 mass% . Alkali metal hydrides (LiH, NaH and KH) reacted with NH3 at room temperature by the exothermic reactions [4, 5]. In the hydrogen generating reactions, the activation treatment by ball milling drastically improved the kinetics for the Li, Na, K systems. After these H2 generations, alkali metal amides were able to store H2 at 373-573 K, forming NH3 and the corresponding alkali metal hydrides under 0.5 MPa H2 flow for 4 h. H2 of 8.1 [H2/LiH+NH3] , 4.9 [H2/NaH+NH3] and 3.5 mass% [H2/KH+NH3] can be reversibly stored by this reaction. Sodium hydride-ammonia borane composite systems were prepared by ball milling. The excess hydride (NaH/NH3BH3= 2-3 /mol/mol) suppressed the foaming and ammonia gas release. The dehydrogenation peak temperature decreased with increase in the NaH content. For the composite materials of 3NaH/ NH3BH3, it reaches the dehydrogenation peak at 345 K, which is much lower than the dehydrogenation temperature of Na NH2BH3 (357K). When we filled the tank (inner volume: 100 L) with the Ti1.1CrMn alloy of 50 kg, the gravimetric and the volumetric hydrogen densities at 35 MPa and 298 K were calculated as 6.0 mass% and 3.0 kg/100 L, respectively. The volumetric hydrogen density provides 30 % more capacity than compressed hydrogen at 35 MPa (volumetric hydrogen density: 2.3 kgH2/100L) and the gravimetric and the volumetric hydrogen densities are similar to those of the Li-Mg-N-H system (packing ratio: 50%, packing density : 0.6 g/cm3, gravimetric hydrogen density: 5.8 mass%, volumetric hydrogen density: 3.5 kgH2/100L). We have calculated the gravimetric and the volumetric H2 densities of the LiH-NH3 system. At 70 MPa, the volume of a tank (100L) for gaseous hydrogen allows storage density of 3.9 kgH2. We assumed that the density of LiH was 0.78 g/cm3. When we filled the tank (internal volume 62 L) with liquid NH3 (density: 0.61 g/cm3) of 38 kg and the tank (internal volume 46 L, packing ratio 50 %) with LiH of 18 kg (LiH/NH3=1mol/1mol), the gravimetric and the volumetric H2 densities are calculated as 8.1 mass% and 4.2 kg/100L, respectively. The volumetric H2 density provides 1.8 and 1.1 times of compressed hydrogen at 35 and 70 MPa (3.9 kgH2), respectively. References [1]"Hydrogen Storage". In: J. Garche, C. Dyer, P. Moseley, Z. Ogumi, D. Rand, B. Scrosati, editors. Encyclopedia of Electrochemical Power Sources, Vol 3. Amsterdam: Elsevier; 2009. [2]Y. Kojima, Y. Kawai, S. Towata, T. Matsunaga, T. Shinozawa and M. Kimbara, J. Alloys Compd. 419 256 (2006). [3]Y. Wang, T. Ichikawa, S. Isobe, M. Tsubota, S. Hino, T. Nakagawa and Y. Kojima, Abstracts of MH2008, International Symposium on Metal Hydrogen Systems, June 24-28, Reykjavik, Iceland. [4]Y. Kojima, K. Tange, S. Hino, S. Isobe, M. Tsubota, K. Nakamura, M. Nakatake, H. Miyaoka, H. Yamamoto and T. Ichikawa, J. Mater. Res., 24, 2185 (2009). [5]H. Yamamoto, H. Miyaoka, S. Hino, H. Nakanishi, T. Ichikawa and Y. Kojima, Int. J. Hydrogen Energy, 34, 9760 (2009).

See more of this Session: Hydrogen Storage System Engineering and Applications: Applied Materials Development I
See more of this Group/Topical: Topical 8: Hydrogen Production and Storage

(221b) High Capacity Metal Hydrides – Development towards Scaled up Tanks and Economical Evaluation

José M. Bellosta von Colbe[1], Julian Jepsen[2], Gustavo Lozano[3], Oliver Metz[3], Thomas Klassen[3] and Martin Dornheim[3], (1)Institute of Materials Research, Materials Technology, GKSS Research Centre Geesthacht, Geesthacht, Germany, (2)Institute of Materials Research, Materials Technology, GKSS-Forschungszentrum Geesthacht GmbH, Geesthacht, Germany, (3)GKSS Research Centre Geesthacht, Geesthacht, Germany

Hydrogen storage seems to be a major issue on the way to hydrogen mobility and usage of hydrogen as an energy vector. High capacity metal hydrides and composites, like sodium alanate and the Reactive Hydride Composites (see talks from Lozano and Jepsen) offer promising solutions, especially for applications involving the use of high or medium temperature heat carriers. These materials, some of which are extremely new, need to be developed from the milligram scale in which they were first synthesized to the kg and even Ton scale in order to be made available as applied storage systems. This requires the investigation of properties that are not the first priority of researchers looking for new materials: heat capacity and conductivity, porosity, cyclability, and especially the issues involving aging and deactivation. In this talk, the activities and techniques used to investigate hydrides at GKSS will be described on the example of sodium alanate, but also including other hydrides. Moreover, the final criterion for an energy storage system is its economic viability. Results will be provided showing that hydrides are a cost – effective solution to the issue of hydrogen storage in the medium scale (tens of kg.) Hydrogen storage seems to be a major issue on the way to hydrogen mobility and usage of hydrogen as an energy vector. High capacity metal hydrides and composites, like sodium alanate and the Reactive Hydride Composites (see talks from Lozano and Jepsen) offer promising solutions, especially for applications involving the use of high or medium temperature heat carriers. These materials, some of which are extremely new, need to be developed from the milligram scale in which they were first synthesized to the kg and even Ton scale in order to be made available as applied storage systems. This requires the investigation of properties that are not the first priority of researchers looking for new materials: heat capacity and conductivity, porosity, cyclability, and especially the issues involving aging and deactivation. In this talk, the activities and techniques used to investigate hydrides at GKSS will be described on the example of sodium alanate, but also including other hydrides. Moreover, the final criterion for an energy storage system is its economic viability. Results will be provided showing that hydrides are a cost – effective solution to the issue of hydrogen storage in the medium scale (tens of kg.) [1]

References [1] Jepsen, J. et al. submitted to Int. J. of Hyd. Ener.

See more of this Session: Hydrogen Storage System Engineering and Applications: Applied Materials

(221c) Sorption Kinetics and Catalysis of 1:1 $LiNH_2$:MgH_2

Donald Anton[1], **Christine J. Erdy**[1] and Joshua Gray[2], (1)Energy Security, Savannah River National Laboratory, Aiken, SC, (2)Materials Comp & Weld Tech, Savannah River National Laboratory, Aiken, SC

High-performance on-board hydrogen storage systems are increasingly recognized as critical to implementation of hydrogen fuel cells as clean, efficient automotive power plants [1]. Among the various storage system types: adsorbent materials, chemical hydrides and metal hydrides, the use of complex hydrides is one of the candidates because of their large gravimetric and volumetric storage capacities [2], on-board reversibility, and indefinite ambient temperature storage duration. The destabilized complex hydride system composed of 1:1 $LiNH_2$:MgH_2 is one of the leading candidates of hydrogen storage compositions. In 2007, Lu et al. showed that the reaction between MgH_2 and $LiNH_2$ in a 1:1 molar ratio resulted in a new nitride-binary light metal, LiMgN, for hydrogen storage applications [3]. The dehydrogenation reaction pathway is as follows: $LiNH_2+MgH_2 \rightarrow LiMgN+2H_2$ This 1:1 molar ratio of MgH_2 to $LiNH_2$ mixture has a theoretical hydrogen weight capacity of 8.2 wt%. The rehydrogenation and subsequent dehydrogenation process of LiMgN produces LiH, $Mg(NH_2)_2$, and MgH_2, as seen here: $LiMgN+2H_2 \leftrightarrow LiH+0.5Mg(NH_2)_2+0.5MgH_2$. In a study of more than 300 destabilization reactions using first principle density function theory (DFT), Alapati et al. predicted that the reaction of between MgH_2 and $LiNH_2$ in a 1:1 molar ratio was energetically favorable with an enthalpy of 31.9 kJ/molH_2 [4]. This enthalpy suggests that favorable sorption thermodynamics should be obtainable at temperatures in the range of 160°C to 260°C. Preliminary experiments reported in literature indicate that sorption kinetics are substantially lower than expected in this temperature range despite favorable thermodynamics. Systematic isothermal and isobaric sorption experiments were performed on these materials using a Sievert's apparatus to form a baseline data set by which to compare kinetic results. Various catalytic modifications were performed in attempts to increase the kinetics while lowering the sorption temperatures. In this paper, we investigate the base line sorption kinetics of the 1:1 $LiNH_2$:MgH_2 composition and the effects of catalyst on these rates, the temperature of initial hydrogen release and the amount of ammonia released from the catalyzed and catalyst-free materials. Possible implications for use of this material for hydrogen storage applications are summarized.

1. Pinkerton, F.E., Decomposition kinetics of lithium amide for hydrogen storage materials. Journal of Alloys and Compounds, 2005. 400(1-2): p. 76-82.

2. Matsumoto, M., et al., Hydrogen desorption reactions of Li-N-H hydrogen storage system: Estimation of activation free energy. Journal of Alloys and Compounds, 2007. 439(1-2): p. 358-362.

3. Lu, J., et al., Potential of Binary Lithium Magnesium Nitride for Hydrogen Storage Applications. The Journal of Physical Chemistry C, 2007. 111(32): p. 12129-12134.

4. Alapati, S.V., J.K. Johnson, and D.S. Sholl, Identification of Destabilized Metal Hydrides for Hydrogen Storage Using First Principles Calculations. The Journal of Physical Chemistry B, 2006. 110 (17): p. 8769-8776.

See more of this Session: Hydrogen Storage System Engineering and Applications: Applied Materials Development I
See more of this Group/Topical: Topical 8: Hydrogen Production and Storage

(221d) Engineering Advances in NaAlH4-Based Hydrogen Storage System

Bart A. van Hassel[1], Daniel A. Mosher[2], Jose Miguel Pasini[3], Mikhail B. Gorbounov[4], John E. Holowczak[5], Xia Tang[5] and Ronald J. Brown[5], (1)Physical Sciences, United Technologies Research Center, East Hartford, CT, (2)Applied Mechanics, United Technologies Research Center, East Hartford, CT, (3)System Dynamics & Optimization, United Technologies Research Center, East Hartford, CT, (4) Thermal Management, United Technologies Research Center, East Hartford, CT, (5)Ceramics, United Technologies Research Center, East Hartford, CT

The significant engineering challenges associated with developing lower-pressure, materials-based, hydrogen storage systems for hydrogen fuel cell light-duty vehicles are being addressed by focusing on the role that powder consolidation can play. NaAlH4 with 4 wt. % TiCl3 was selected as the model material. We focused on the changes in the physical (density and thermal conductivity) and mechanical properties (biaxial flexure strength) and on how this impacted the volumetric capacity of the hydrogen storage system. Both the thermal conductivity and the density of the ball milled material improved with applied pressure in an uniaxial press over the range of 14 MPa to 281 MPa. The thermal conductivity reached a value of 1.637 W/m/K, which was a factor seven higher than that of the unconsolidated powder. The volume of the material was reduced with 58% at the highest applied pressure. A method was developed for determining the strength of NaAlH4 pellets before and after hydrogen absorption and desorption cycles. It is based on a biaxial flexure test that was originally designed for determining the strength of green ceramic materials. The tests showed that the pellets were strong with a biaxial flexure strength of 1.4 kpsi which was unaltered over three studied hydrogen absorption/desorption cycles. The increased materials density did not affect the hydrogen absorption and desorption kinetics, which is important in order to benefit from the improved volumetric capacity. The new material properties of the compacted NaAlH4 were used in finite element modeling of a hydrogen storage system that targeted a fast refueling time. The results clearly show an improvement of the volumetric capacity of the system by powder consolidation but the gravimetric capacity remains below target, as expected. A system level study of a light-duty vehicle with such a hydrogen storage system is required in order to determine whether the amount of hydrogen stored in the pore volume of the sodium alanate will still be enough to enable one cold start from room temperature to its operating temperature (120-140°C) or that a buffer volume needs to be installed. While it is recognized that a sodium alanate based hydrogen storage system has its limitations, it has been demonstrated that powder consolidation can address some of those limitations by improving the thermal conductivity and volumetric capacity.

See more of this Session: Hydrogen Storage System Engineering and Applications: Applied Materials Development I
See more of this Group/Topical: Topical 8: Hydrogen Production and Storage

Storing Hydrogen Using (Ti,V)-Based Bcc Materials

Michel Latroche, Blaise Massicot and Jean-Marc Joubert, ICMPE UMR7182 CNRS, Thiais, France

Light vanadium-based alloys crystallizing in the body centred cubic (bcc) structure are promising materials for hydrogen storage. Adding titanium to vanadium improves the kinetics of the hydrogen absorption whereas addition of 3d metals such as iron enhances the reversible absorption capacity [1] and the activation process. The aim of this study is to highlight the influence of alloy composition on the hydrogenation properties (absorption pressure and capacity) of the bcc phase. The isothermal sections at 1000 °C and 1200 °C of the Ti–V–Fe system have been experimentally studied to get a better understanding of the bcc-region boundaries and equilibrium phases. For iron contents below 15 at.%, the isotherm at 1000 °C shows that the bcc phase is the only stable one. Moreover, our results indicate that the ternary extension of the $TiFe_2$ phase with the C14 structure reaches the composition $Ti_{33}V_{33}Fe_{33}$ [2]. The pressure-composition absorption isotherms of several bcc-alloys of the Ti–V–Fe system have been measured at various temperatures. For a constant Ti/V ratio, addition of iron increases the plateau pressure, as well as addition of vanadium for a given Ti/Fe ratio. The dependence of the capacity on the alloy composition has been determined [3] and will be presented as a function of the alloy composition (Fig. 1).

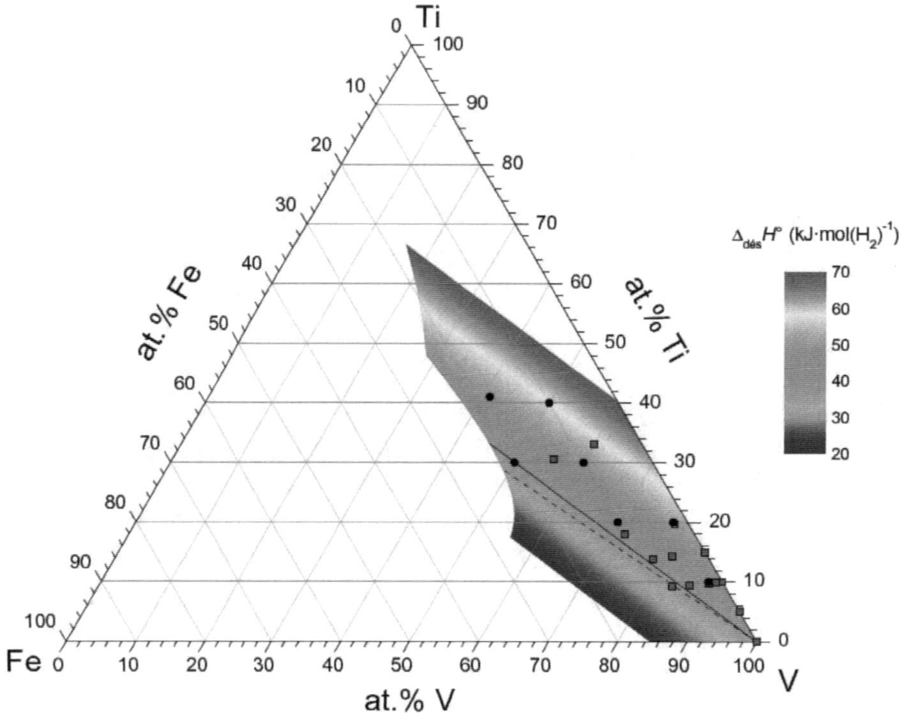

Figure 1 – Map of the enthalpy of dissociation for the dihydride of the Ti–V–Fe bcc system as a function of the composition. The dotted line corresponds to a constant lattice parameter whereas the full line stands for constant enthalpy.

[1] S. Challet, M. Latroche, and F. Heurtaux, J. Alloys Compd., 439, 294, 2007.
[2] B. Massicot, J.-M. Joubert and M. Latroche, Int. J. Mater. Res. (2010), doi: 10.3139/146.110379.
[3] B. Massicot, M. Latroche, J.-M. Joubert, Journal of Alloys and Compounds (2010), doi:10.1016/j.jallcom.2010.09.030.

(336a) Achieving Optimal Hydrogen Storage In MOF-5

Andrea Sudik[1], Jun Yang[2], Justin Purewal[3], Donald J. Siegel[3], Emi Leung[4] and Ulrich Mueller[5], (1) Fuel Cell and Hybrid Electric Vehicle Research, Ford Motor Company, Dearborn, MI, (2)Fuel Cell and Hybrid Electric Vehicle Department, Ford Motor Company, Dearborn, MI, (3)Mechanical Engineering Department, University of Michigan (Ann Arbor), Ann Arbor, MI, (4)Chemicals Research and Engineering, BASF SE, Ludwigshafen, Germany, (5)Chemicals Research and Engineering, Ford Motor Company, Dearborn, MI

Storage of hydrogen in condensed phases has the potential to be more efficient and less expensive than the most widely used storage method, compressed gaseous hydrogen. While substantial progress has recently been made in identifying improved hydrogen storage materials, significant challenges associated with the engineering of the storage system around a candidate storage material persist. In particular, the degree to which materials processing can act as a countermeasure to improve upon the deficiencies of a given storage material is largely unknown. This paper will detail recent efforts aimed at determining the processing-structure-properties (PSP) relationships for an important class of sorbent materials [metal organic frameworks (MOFs)] in order to devise improved packing and processing strategies for their use in a hydrogen storage systems context. In particular, we will detail various mechanical processing routes that have been explored (ranging from powders to pelletization) in an effort to simultaneously maximize packing density, heat and mass transfer, and hydrogen uptake characteristics. We will specifically highlight compaction studies for Basolite Z100H (the commercial form of MOF-5) wherein the relevant hydrogen storage properties have been characterized as a function of processing route for the identification of preliminary PSP relationships. The intent of this investigation is to identify rational pathways by which materials properties can be optimized towards achieving commercially-viable hydrogen storage systems, and to reveal the upper limits of performance achievable with current-generation hydrogen storage materials.

See more of this Session: Hydrogen Storage System Engineering and Applications: Applied Materials Development III

See more of this Group/Topical: Topical 8: Hydrogen Production and Storage

CONFERENCE ABSTRACT

Improving Comparability of Hydrogen Storage Capacities of Nanoporous Materials

V. P. Ting, N. Bimbo, A. Hruzewicz-Kołodziejczyk, L. Fisher, S. Rigby,
A. Burrows and T. J. Mays

One of the primary barriers to the successful and timely implementation of a "Hydrogen Economy" with hydrogen as an energy carrier is the ability to efficiently store hydrogen for later use. The storage of hydrogen via physisorption of H_2 onto porous solid state materials benefits from reversible and rapid adsorption and desorption at moderate temperatures and pressures compared to either chemical storage or storage of hydrogen as a compressed gas or cryogenic fluid. The search for the optimal porous solid state storage medium exhibiting acceptably high volumetric and gravimetric densities has long been the focus of intense research. Due to a lack of standardized data collection and analysis routines, it is difficult, if not impossible, to directly compare the hydrogen storage capacities of such materials, as the real material properties cannot be decoupled from the influence of the different data collection or modeling methods used. We present our investigations into producing a standardized routine for the assessment of the hydrogen storage capacities of different nanoporous materials. This includes the formulation of rigorous and systematic data collection routines for the purpose of obtaining universally reproducible H_2 isotherms under sub-atmospheric, and high pressure conditions, as well as new methods for modelling the resultant isotherms in order to extract information on the interactions of H_2 gas with the porous solid and gain an improved understanding of the physisorption process itself. We will demonstrate these practices, as applied to the analysis of some well-characterized standard materials including commercially-available zeolites and tailored activated carbons materials. With this greater knowledge of the variables that affect the quality of hydrogen sorption data, the resultant routines will permit better control of the myriad contributing experimental factors. Combined with the improved data modelling methods, it will become possible to directly compare the hydrogen storage capabilities of any nanoporous materials, a development which will prove invaluable in the evaluation of hydrogen storage materials investigated in the future.

(336c) Thermal Stability of Nanopores in Palladium Alloys and Their Hydrides

David B. Robinson[1], Markus D. Ong[1], Benjamin W. Jacobs[1], Mary E. Langham[1], Michael S. Kent[2] and Ilke Arslan[3], (1)Energy Nanomaterials, Sandia National Laboratories, Livermore, CA, (2)Sandia National Laboratories, Albuquerque, NM, (3)Dept. of Chemical Engineering and Materials Science, University of California, Davis, Davis, CA

Nanoporous material architectures can offer improved charging and discharging kinetics in metal hydrides due to their high surface area. We have synthesized nanoporous palladium and palladium alloy powders in a scalable fashion by reduction of palladium salts in a concentrated aqueous surfactant. Particle diameters are micrometer-scale, and each particle is perforated by 3 nm pores. In pure palladium, the pores are unstable at 150 C, but when alloyed with rhodium, they are stable up to 400 C on short timescales. These results are expected to be related to pore lifetime at more moderate temperatures. Electron tomography and elemental mapping after in situ or ex situ heating and hydrogen exposure help us understand the mechanism of the improvement. We observe surface segregation of the minor component in alloys. Pore density and stability are also related to the microstructure of the material. Improved stability can be obtained with minimal sacrifice of hydrogen storage properties.

See more of this Session: Hydrogen Storage System Engineering and Applications: Applied Materials Development III
See more of this Group/Topical: Topical 8: Hydrogen Production and Storage

(336d) Facilitating Dehydrogenation of Complex Hydride Hydrogen Storage Materials Using a CNTs-Based Catalytic Matrix

Sean S.-Y. Lin[1], Jun Yang[2], Xin Zhao[1] and Harold H. Kung[1], (1)Chemical and Biological Engineering, Northwestern University, Evanston, IL, (2)Research and Advanced Engineering, Ford Motor Company, Dearborn, MI

Complex hydrides are considered one of the most promising hydrogen storage materials. Hydrogen evolution from complex hydrides is dictated by an inter-related sequence of chemical reaction steps including decomposition of the hydride, diffusion of hydrogen atom/ion and counter ions in the solid lattice, surface diffusion of these atoms/ions, recombination of hydrogen atoms, transformation of lattice phases, and desorption of hydrogen molecules. Using carbon nanomaterials such as carbon fibers, carbon nanotubes, and carbon fullerenes as additive has shown promising result to facilitate dehydrogenation of complex hydrides. Several possible mechanisms have been proposed to explain the effect, including size reduction of hydride particles, confinement effect that prevents phase segregation, and electronic effect that destabilizes the covalent bond associated with hydrogen in the hydrides.

In this study, a light weight, high surface area CNTs-based catalytic matrix was developed and investigated for hydride decomposition using two model complex hydrides, NaAlH4 and Ca(BH4)2. Adding a non-precious metal (Co, Ni, Fe) in the form of nanoparticles decorating the carbon matrix (as confirmed by SEM images) to the complex hydrides facilitated hydride decomposition. The decomposition temperature of NaAlH4 was decreased from 260 oC to 210 oC by adding 10 wt% CNTs only. The temperature was further decreased to 180 oC with a Co-decorated CNTs matrix. In the case of Ca(BH4)2, the decomposition temperature was decreased from 376 to 366 oC. The CNTs-based catalytic matrix was more effective in NaAlH4 decomposition possibly because a better hydride-catalyst interface was created between the melted NaAlH4 and the catalytic matrix by capillary action. Such interface benefits the overall kinetics by facilitating the chemical bond breakage as well as mass transfer of the product nuclei, which can be dispersed homogeneously via the liquid phase. By dispersing such a composite catalyst in a mixture with complex hydride, we can attain a high contact area between the catalyst and the hydride with minimal weight penalty. The contact points could serve both for hydrogen transfer between the catalyst complex and the hydride and as potential nucleation points for phase transformation of the hydride.

See more of this Session: Hydrogen Storage System Engineering and Applications: Applied Materials Development III
See more of this Group/Topical: Topical 8: Hydrogen Production and Storage

Carbon Nanostructures as Catalyst for Improving the Hydrogen Storage Behaviour of Complex Aluminium Hydride

*M. Sterlin Leo Hudson, Himanshu Raghubanshi, D Pukazhselvan and O. N. Srivastava**
Nanoscience and Technology Unit, Department of Physics, Banaras Hindu University, Varanasi, India
* **Corresponding author** : Email: hepons@yahoo.com; Tel: 0542-2368468; Fax: 0542-2369889

Abstract:

The present paper reports the catalytic effect of carbon nanostructures, particularly graphitic nanofibres (GNFs) with different structure morphology, namely helical GNFs (HGNFs) and planar GNFs (PGNFs) as the catalyst for complex aluminium hydride. HGNFs and PGNFs were synthesized by catalytic thermal decomposition of acetylene (C_2H_2). The growth of HGNFs was achieved by employing faceted Ni nanoparticles, whereas spherical Ni nanoparticles produce PGNFs. It has been observed that HGNFs posses superior catalytic activity than PGNFs in improving the desorption kinetics and decreasing the desorption temperature of complex aluminium hydride ($NaAlH_4$, $LiAlH_4$ and $LiMg(AlH_4)_3$). Temperature programmed desorption (TPD) reveals that HGNFs admixed alanates undergo hydrogen desorption at a much lower temperature than PGNFs admixed material. Thus for the heating rate of 2°C/min, the peak desorption temperature corresponds to initial step desorption of $NaAlH_4$ admixed with 8 mol% HGNFs and 8 mol% PGNFs has been lowered to 143.6°C and 152.6°C, respectively. In addition to the enhancement in desorption kinetics, the GNFs admixed $NaAlH_4$ also undergoes rehydrogenation at the moderate condition. In order to get supportive evidence for our experiment, we have carried out ab-initio studies by calculating hydrogen removal energy from alanates with planar and helical model of GNF. It becomes clear that from both experiment and ab-initio calculations, that catalytic effect of GNFs is curvature dependent. Thus HGNF with higher helicity helps to lower the energy needed to remove hydrogen from alanates.

Keywords: Complex hydrides, carbon nanostructures, catalysts, temperature programmed desorption

1. Introduction:

Complex aluminium hydrides, $NaAlH_4$, LiAlH4 and $LiMg(AlH_4)_3$ possess the quite high gravimetric hydrogen capacity of 7.3, 10.6 and 9.7 wt.%H_2, respectively. The fast desorption of hydrogen from these compounds becomes feasible with the use of suitable catalysts. The growing interest on finding an effective catalyst for improving the dehydrogenation/re-hydrogenation behavior of complex metal hydrides after the pioneering work of Bogdanovic et al [1] on the reversibility of $NaAlH_4$ triggers research on various catalysts, including transition and rare earth metal based catalysts. There is an ongoing search for finding an effective catalyst for improving the dehydrogenation/re-hydrogenation behaviour of complex hydrides [2]. Recent studies by Berseth et al [3] reveal that carbon nano-variants are found to possess catalytic effect on complex hydrides. Among the different carbon nanovariants, graphitic nanofibres are very promising due to their large number of localized σ bond electrons around the edges of parallel graphitic sheets, which provides more catalytically active sites [4, 5]. Advantageous features of carbon nanofibre are that it is light weight, not prone to oxidation nor gets substituted in the alanate lattice. In the light of above, we explored the use of graphitic nanofibres of two different morphology namely planar graphitic nanofibre (PGNF) and helical graphitic nanofibre (HGNF) for possible improvement in dehydrogenation characteristics of $NaAlH_4$, $LiMg(AlH_4)_3$ and

$LiAlH_4$. The catalytic activity of GNFs has been proven for binary hydride, MgH_2 [6]. The optimum concentration of GNFs on $NaAlH_4$, $LiAlH_4$ and $LiMg(AlH_4)_3$ for the present study is 8 mol%. The catalytic effect of GNFs on $NaAlH_4$ has been compared with single walled CNTs (SWCNTs). Our investigation reveals that HGNFs possess superior catalytic activity than PGNFs and SWCNTs.

2. Experimental:

As received $NaAlH_4$ (Aldrich, 93%), $LiAlH_4$ (Aldrich, 95%) and $MgCl_2$ (Alfa Aesar, 99%) have been used without further purification. $LiMg(AlH_4)_3$ was synthesized by ball-milling $LiAlH_4$ and $MgCl_2$ in the molar ratio 3:1 together with two steel balls of 12.5 mm each and one ball with 4 mm in a chrome-nickel stainless steel milling vial under argon atmosphere using a locally fabricated attritor ball-miller. The ball-to-powder mass ratio was kept at 10:1. The expected reaction $3LiAlH_4 + MgCl_2 \rightarrow LiMg(AlH_4)_3 + 2LiCl$ has been found to get completed in one hour of ball-milling. All operations on the samples were done under dry argon atmosphere in a glove box to prevent reaction with moisture and oxygen in the air. The graphitic nanofibres are being synthesized in our laboratory through dissociation of hydrocarbon (acetylene/ethylene) over Ni/Fe nanoparticles catalyst at 650°C under flowing He and H_2 gas both with an optimized flow rate of 1500 sccm. In keeping with the known results [7, 8], we have found that morphology of GNF is dependent on the shape of the Ni/Fe nanoparticles. For faceted Ni/Fe nanoparticles helical GNF get formed and for non-faceted Ni/Fe nanoparticles, planar GNF is found to grow. SWCNTs were supplied by MK IMPEX CANADA. Hydrogen desorption behaviour of $NaAlH_4$, $LiAlH_4$ and $LiMg(AlH_4)_3$ samples were measured using computerized pressure composition isotherm (P-C-I) measurement system supplied by Advanced Materials Corporation.

Structural characterization of the samples was carried out through XRD, Philips PW-1710 and X'Pert PRO (PAN*alytical*) X-ray diffractometer equipped with the graphite monochromator employing CuK_α radiation (λ= 1.5402Å). Exposure of the sample to the atmosphere has been avoided by covering the XRD holder by fine layer of Parafilm (Pechiney plastic packing). The microstructural characterization has been done through Transmission electron Microscope (Technai, 200 kV).

3. Results and Discussions:

3.1. Sodium Aluminium Hydride:

$NaAlH_4$ desorbs hydrogen through the following reactions,

$3NaAlH_4 \rightarrow Na_3AlH_6 + 2Al + 3H_2$ (170°C) (1)
$Na_3AlH_6 + 2Al \rightarrow 3NaH + 3Al + 3/2H_2$ (250°C) (2)
$3NaH + 3Al \rightarrow 3Na + 3Al + 3/2H_2$ (>400°C) (3)

Reaction (1) and (2) releases 3.7 and 1.8 wt.% H_2, respectively. Both reaction (1) and (2) liberate 5.5 wt.% H_2, satisfying the US-DOE revised target of 5.5 wt.% H_2 [9]. However, an effective catalyst is required to make reaction (1) and (2) reversible under moderate temperature and pressure. Reaction (3) is generally not considered usable for hydrogen storage, due to its very high temperature. We have employed carbon nanovariants of different types, namely HGNFs, PGNFs and SWCNTs as the catalyst for improving the dehydrogenation and rehydrogenation behaviour of $NaAlH_4$. Representative TEM micrographs of HGNF and PGNF are shown in Figs. 1(a) and 1(b).

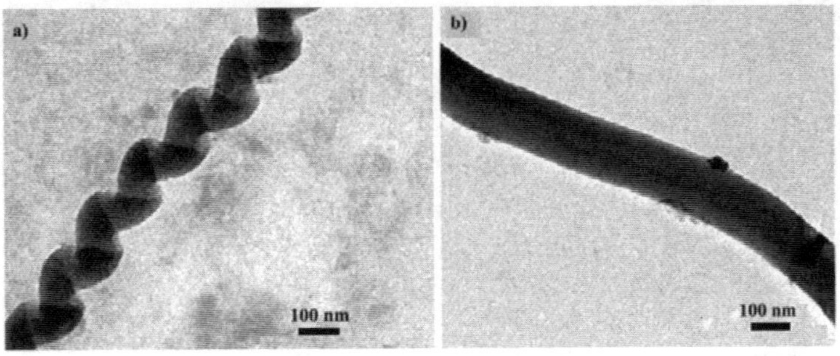

Fig. 1. TEM microgram of GNF a) helical morphology GNF (HGNF) and b) planar morphology GNF (PGNF) (difference between HGNFs and PGNFs are given in table 1)

Table 1: Difference between HGNF and PGNF used as catalyst

Types of GNFs	Catalyst used for synthesis	Surface area (m^2/g)	Average diameter (nm)
Helical GNF (HGNF)	Faceted Ni nanoparticle	125	150
Planar GNF (PGNF)	Spherical Ni nanoparticle	109	200

The temperature programmed desorption (TPD) curve of $NaAlH_4$ admixed with different carbon nanovariants is shown in Fig. 2. As it becomes evident from Fig. 2 that for the heating rate of 2°C/min, the peak desorption temperature of pristine $NaAlH_4$ corresponds to Eq. (1) and Eq. (2) are 163°C and 240°C. However, admixing of 8 mol% carbon nanovariants with $NaAlH_4$ lowers the desorption temperature of $NaAlH_4$. The peak desorption temperature of $NaAlH_4$ admixed with 8 mol% HGNF corresponds to Eq. (1) and Eq. (2) are 143.6°C and 190°C, respectively. Whereas that for $NaAlH_4$ admixed with 8 mol% PGNF corresponds to Eq. (1) and Eq. (2) is 152.6°C and 209°C, respectively. It has been observed that among the different forms of CNTs and GNFs, HGNFs posses superior catalytic effect in lowering the desorption temperature corresponds to both Eq (1) and Eq(2) of $NaAlH_4$. The dehydrogenation kinetics and rehydrogenation kinetics of $NaAlH_4$ admixed with 8 mol% HGNF is found to be superior to that of PGNF and CNT admixed sample.

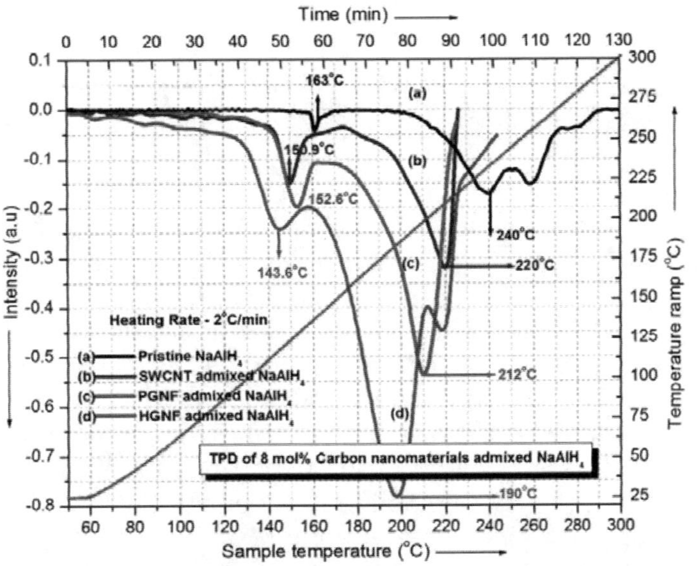

Fig. 2. TPD curve of 8 mol% carbon nanovariants admixed $NaAlH_4$ for the heating rate of 2°C/min (red line represents the heating profile).

3.2. Lithium-magnesium alanate:

$LiMg(AlH_4)_3$ has been synthesized mechano-chemically through the following reaction.

$3Li(AlH_4)_3 + MgCl_2 \rightarrow LiMg(AlH_4)_3 + 2LiCl$ (4)

by using a locally fabricated attritor ball-miller. The above reaction gets completed in ~1 hour of ball-milling. Following the known results [10], the thermal decomposition behavior of $LiMg(AlH_4)_3$ can be described as follows,

$LiMg(AlH_4)_3 \rightarrow LiMgAlH_6 + 2Al + 3H_2$ (~100°C) (5)

$LiMgAlH_6 \rightarrow LiH + MgH_2 + Al + 3/2H_2$ (~150°C) (6)

$2MgH_2 + 3Al \rightarrow Al_3Mg_2 + 2H_2$ (~340°C) (7)

The temperature programmed desorption (TPD) curve was monitored from room temperature to 240°C at the heating rate of 1°C/min. It has been observed from the TPD profile that the peak decomposition temperature of Eq. (5) is 105.5°C and that of Eq. (6) is 149.6°C. For 8 mol% HGNF admixed $LiMg(AlH_4)_3$, Eq. (5) occurs at the time of catalyst admixing. However, for the heating rate of 1°C/min, the peak desorption temperature for 8 mol% HGNF admixed $LiMg(AlH_4)_3$ corresponds to Eq. (6) has been lowered from 149.6°C to 127.9°C. A representative TPD curve is shown in Figs. 3.

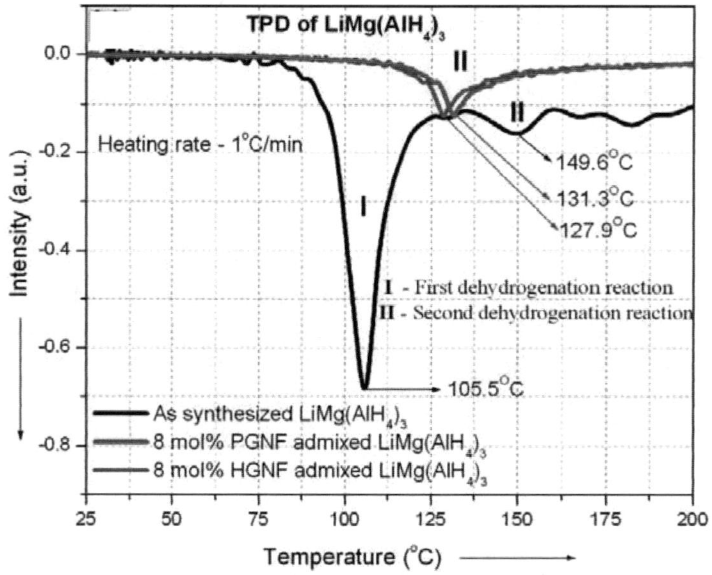

Fig. 3. Temperature Programmed Desorption (TPD) of $LiMg(AlH_4)_3$ for the heating rate of 1°C/min

In order to find the catalytic effect of HGNF in decreasing the activation energy of $LiMg(AlH_4)_3$, we proceed to evaluate this energy from the Arrhenius plot, as described by Janot et al. [11].

The activation energy for the desorption process is estimated by plotting the maximum reaction rate (k) as a function of the temperature (T). The reaction rates are obtained from the slopes of the tangents at the inflection points of desorption kinetic curves. The maximum reaction rates (k) of Eq. (5) at the different temperatures were calculated from desorption kinetic curves as shown in Fig. 4(a). Since the inclusion of LiCl mass in the kinetic measurement of $LiMg(AlH_4)_3$ affects the rate constant value, we have not considered the weight of LiCl for desorption kinetic measurements.

Quantitative estimation of kinetic barrier was then carried out by the determination of the activation energy (Ea) from the well known Arrhenius equation of the form

$$\ln(k) = -Ea/RT + \ln(A) \quad (8)$$

By plotting ln *k* versus *1/T*, *Ea* can be extracted from the linear slope. Fig. 4(b) shows the Arrhenius plot for the first-step dehydrogenation of $LiMg(AlH_4)_3$. A good linearity between ln *k* and 1/*T* is obtained with $R^2 = 0.99$. The activation energy (*Ea*) as calculated corresponds to ~66 kJ/mol.

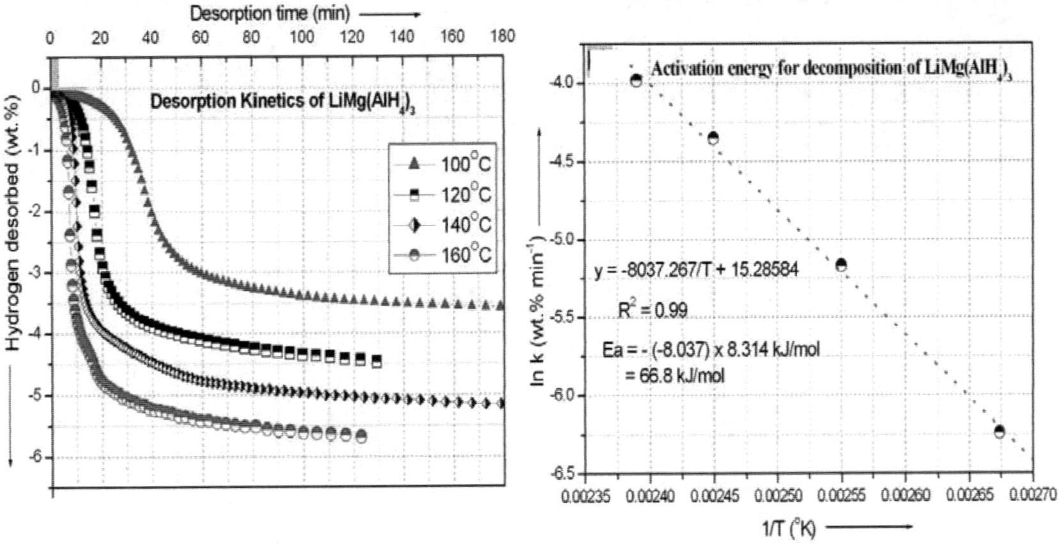

Fig. 4. a). desorption kinetic curve of $LiMg(AlH_4)_3$ at different temperatures b) plot of ln *k* vs 1/T for $LiMg(AlH_4)_3$.

3.3. Lithium alanate:

In general, $LiAlH_4$ has three-stage decomposition, as represented by the following equations:

$3LiAlH_4 \rightarrow Li_3AlH_6 + 2Al + 3H_2$ (~150°C) (9)

$Li_3AlH_6 \rightarrow 3LiH + Al + 3/2H_2$ (~180°C) (10)

$3LiH + 3Al \rightarrow 3LiAl + 3/2H_2$ (>400°C) (11)

Eq. (9) approximately releases 5.3 wt.% hydrogen and eq. (10) and eq. (11) both release approximately 2.6 wt.% H_2 each [12].

The catalyst admixing was carried out through ball-milling for 15 min, using the same ball-miller as mentioned above. The optimum molar concentration of GNFs for $LiAlH_4$ is found to be 8 mol%. Lower concentration of GNF does not produce enhancement in kinetics similar to that from 8 mol% and increasing the molar concentration of GNFs further leads to the decrease in hydrogen storage capacity. Fig. 5 shows the representative TPD profile of $LiAlH_4$ sample. There is a loss of ~2 wt.% H_2 in the 8 mol% GNFs admixed $LiAlH_4$ sample. This was estimated by comparing the TPD profile of the ball-milled and pristine $LiAlH_4$.

A noticeable feature shown in Fig. 5 is that for the heating rate of 1°C/min, the peak desorption temperature for 8 mol% HGNF admixed $LiAlH_4$ sample corresponds to Eq. (6) has been lowered from 158.8°C to 128.5°C and as that for reaction corresponding to Eq. (7) has been lowered from 206.1°C to 165.5°C. It may be mentioned that the effect of an increase in the interface area which will result due to ball-milling, will not be able to explain the significant decrease in desorption temperature. Apparently, it is the catalytic effect of GNF, which leads to decrease of desorption temperature. The exact role of GNF in improving the dehydrogenation behaviour of complex hydride is exactly not known. Based on the known results on the catalytic effect of carbon nanostructures in complex hydrides, it is believed that higher the curvature of catalyst, higher is the catalytic activity. The carbon nanostructures as for example fullerenes have higher curvature than CNTs or graphenes.

Hence fullerene has higher catalytic activity [3]. In contrast to PGNF, HGNF has curved surface. Therefore it will have higher catalytic activity than PGNF.

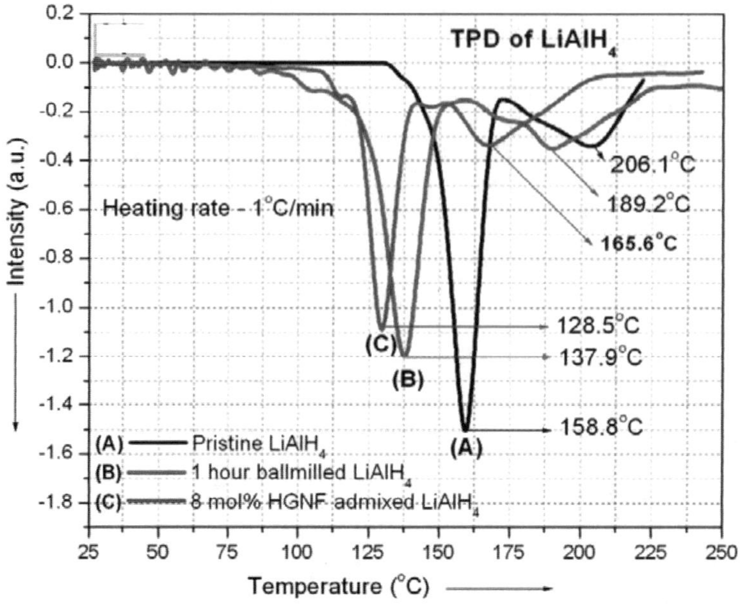

Fig. 5. Temperature programmed desorption of $LiAlH_4$

When admixing 8 mol% HGNF with $LiAlH_4$, desorption kinetics of $LiAlH_4$ get significantly improved. The effect of HGNFs on $LiAlH_4$ is estimated from the activation energy value (*Ea*) calculated for pristine $LiAlH_4$ and 8 mol% HGNF admixed with $LiAlH_4$. In order to find the activation energy of pristine $LiAlH_4$ using Arrhenius relation, the desorption kinetics of pristine $LiAlH_4$ at different temperatures were evaluated. The activation energy calculated for pristine $LiAlH_4$ is 107 kJ/mol. This value is comparable to the recently reported value of 102 kJ/mol by Blanchard et al.[13].

The activation energy (*Ea*) calculated for 8 mol% HGNF admixed $LiAlH_4$ is ~68 kJ/mol. Thus the admixing of 8 mol% HGNF helps to decrease the activation energy for the decomposition of $LiAlH_4$ by 36% (for pristine $LiAlH_4$ calculated *Ea* is 107 kJ/mol, and for 8 mol% HGNF admixed $LiAlH_4$, it is ~68 kJ/mol.

In order to get supportive evidence for our experiment, we have carried out ab-initio molecular simulation studies using density functional theory by calculating hydrogen removal energy from $NaAlH_4$ and $NaAlH_4$ with planar and helical model of GNF. This also confirms that planar model of GNF exhibit the superior catalytic effect due to its increased elecronegativity. The detailed results will be communicated elsewhere.

4. Conclusions:

The catalytic effect of graphitic nanofibres of two morphology, namely HGNF and PGNF on $NaAlH_4$, $LiMg(AlH_4)_3$ and $LiAlH_4$ has been studied. The dehydrogenation behaviour of the samples was studied through TPD and desorption kinetic measurements. We have found that helical morphology of GNF improves the dehydrogenation behaviour of both $LiMg(AlH_4)_3$ and $LiAlH_4$. For the heating rate of 2°C/min, the lowering of peak desorption temperature of $NaAlH_4$ admixed with 8 mol% HGNF as compared to its pristine phase is from 163°C to 143°C for the initial step and from 240°C to 190°C for the second step. Similarly, for $LiAlH_4$ and $LiMg(AlH_4)_3$, HGNF is found to be an effective catalyst for lowering the dehydrogenation temperature. The activation energy calculated for the first step decomposition of HGNF admixed $LiAlH_4$ is ~68 $kJmol^{-1}$, whereas that for pristine $LiAlH_4$, it is ~107 kJ/mol.

Acknowledgement:

The authors would like to thank Prof. T.N. Veziroglu, Prof. C.N.R. Rao, Prof. S.K. Joshi, Prof. S.P. Thyagarajan and Prof. D.P. Singh (VC:BHU) for their encouragement and support. Financial support from the Ministry of New and Renewable Energy, the University Grants Commission, DST: Nano Science Technology Unit (BHU) and the Council of Scientific and Industrial Research are thankfully acknowledged.

References:

1. Bogdanovic B, Schwickardi M. Ti-doped alkali metal aluminium hydrides as potential novel reversible hydrogen storage materials. J. Alloys Compd. 1997;253-254:1-9.
2. Srinivasan S, Escobar D, Goswami Y, Stefanakos E. Effects of catalysts doping on the thermal decomposition behavior of $Zn(BH_4)_2$. Int. J Hydrogen Energy. 2008;33:2268 -72.
3. Berseth PA, Harter AG, Zidan R, Blomqvist A, Araujo CM, Scheicher RH, Ahuja R, Jena P. Carbon Nanomaterials as Catalysts for Hydrogen Uptake and Release in $NaAlH_4$. Nano Lett. 2009;9:1501-05.
4. Serp P, Corrias M, Kalck P. Carbon nanotubes and nanofibers in catalysis. Applied Catalysis A: General. 2003;253:337-58.
5. Kumar LH, Viswanathan B, Murthy SS. Dehydriding behaviour of $LiAlH_4$-the catalytic role of carbon nanofibres. Int. J Hydrogen Energy. 2008;33:366-73.
6. Lillo-Rodenas MA, Guo ZX, Aguey-Zinsou KF, Cazorla-Amoros D, Linares-Solano A. Effects of different carbon materials on MgH_2 decomposition. Carbon 2008;46:126-37.
7. Park C, Baker RTK. Catalytic Behavior of Graphite Nanofiber Supported Nickel Particles. 2. The Influence of the Nanofiber Structure. J. Phys. Chem. B 1998;102:5168-77.
8. Blank VD, Alshevskiy YL, Zaitsev AI, Kazennov NV, Perezhogin IA, Kulnitskiy BA. Structure and phase composition of a catalyst for carbon nanofiber formation. Scripta Mat. 2006;55:1035-38.
9. Yang J, Sudik A, Wolverton C and Siegel D.J. High capacity hydrogen storage materials: attributes for automotive applications and techniques for materials discovery. Chem Soc Rev. 2010;39:656-675.
10. Tang X, Opalka SM, Laube BL, Wu FJ, Strickler JR, Anton DL. Hydrogen storage properties of Na–Li–Mg–Al–H complex hydrides. J. Alloys Compd. 2007;446-447:228-31.
11. Janot R, Eymery JB, Tarascon JM. Investigation of the processes for reversible hydrogen storage in the Li–Mg–N–H system. Journal of Power Sources. 2007;164:496-502.
12. Vittetoe AW, Niemann MU, Srinivasan SS, McGrath K, Kumar A, Goswami DY, Stefanakos EK, Thomas S. Destabilization of $LiAlH_4$ by nanocrystalline MgH_2. Int. J Hydrogen Energy. 2009;34:2333-39.
13. Blanchard D, Brinks HW, Hauback BC, Norby P, Muller J. Isothermal decomposition of $LiAlD_4$ with and without additives. J. Alloys Compd. 2005;404-406:743-47.
   Chen J, Kuriyama N, Xu Q, Takeshita HT, Sakai T. Reversible Hydrogen Storage via Titanium-Catalyzed $LiAlH_4$ and Li_3AlH_6. J. Phys. Chem. B. 2001;105:11214

Hydrogen Storage and Effects of Cycling on Li-Based Hydrides

Wen-Ming Chien*, Joshua H. Lamb and Dhanesh Chandra

Chemical and Materials Engineering / MS388, University of Nevada, Reno, Reno, NV 89557

Abstract

The hydrogen storage capacity was evaluated on lithium imide/amide hydriding/de-hydriding pressure cycling by using hydrogen with low level impurity gases (100 ppm O_2, H_2O, CH_4 or N_2) and industrial hydrogen (H_2O~32ppm, O_2~10ppm and others). Pressure-composition isotherms were obtained at 255°C after cycling between $Li_2NH \leftrightarrow LiNH_2$. Recent research found that Li-based hydrides are candidates for reversible hydrogen storage [2-5], and lithium nitride (Li_3N) has a theoretical capacity of up to ~11 wt% hydrogen when fully converted to $LiNH_2$. Chen et al. [6-7] have shown that Li_3N will absorb up to 11.5 wt% hydrogen with a reversible capacity of 7 wt%. Other research has shown the improvement of the reversible hydrogen storage properties by mixing different ratios of $LiNH_2$ / LiH [8] and $LiNH_2$ / Li_3N [9]. Huq et al. [10-11] have performed structural studies and in-situ deuteration/dedeuteration structural studies of Li_3N using powder neutron diffraction. Chien et al. [12] reported the phase transformation behaviors of Li_2ND, LiD and $LiND_2$ phases during the hydriding/dehydriding of Li_3N.

In this study, the losses of hydrogen storage capacities after pressure cycling were determined. The pressure-composition isotherm results shown that after 560 pressure cycles, 100 ppm O_2 in hydrogen were most detrimental (0.4 wt%H_2 capacity remaining out of ~5.6 wt.%) as compared to 100 ppm H_2O in hydrogen gas (2 wt%H_2). It was found that 1.65 wt.% hydrogen remained after 500 cycles by using industrial hydrogen. Pressure isotherms after ~100 pressure cycles showed 3.6 and 3.3 wt%H_2 capacity remained by using 100 ppm CH_4 and industrial hydrogen, respectively. X-ray diffraction results of the products after pressure cycling showed mainly Li_2NH and LiH phases, and the impurity Li_2O phase. The phase analysis results showed the Li_2NH phase reduced from 77% to 13%, and LiH phase increases from 18% to 57% after 1100 cycles. A summary plot pressure-composition isotherms (at 255°C) before and after Li_2NH-$LiNH_2$ cycling (no catalyst) with industrial hydrogen and 100 ppm levels of O_2, H_2O, CH_4 in UHP H_2 shows in Figure 1. The results showed that the nitrogen addition both allowed more of the theoretical hydrogen capacity to be storage as well as improved the stability of the hydrides after up to 1359 cycles. X-ray diffraction data shows that more Li_2NH phase (less LiH phase) is present in the desorbed hydride after cycling with the nitrogen addition than is present without. The results of the effect of the impurity gases in hydrogen showed that the N_2 addition will improve the hydrogen storage capacity, and the O_2 has the most detrimental effect.

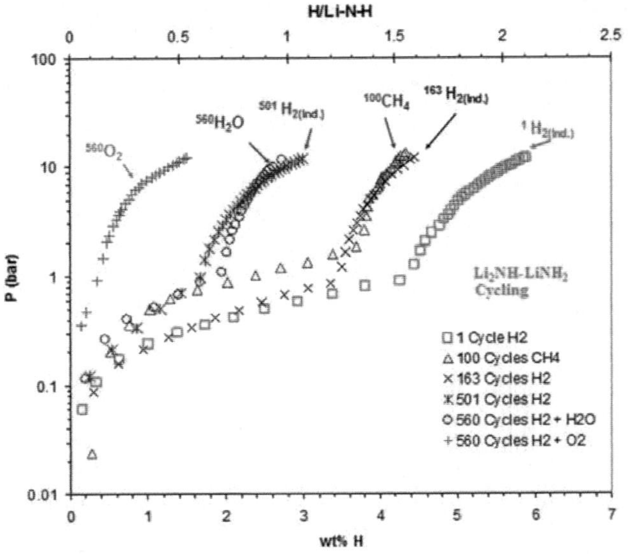

Figure 1. Summary plot of pressure-composition isotherms (at 255°C) before and after Li_2NH-$LiNH_2$ cycling (no catalyst) with industrial hydrogen and 100 ppm levels of O_2 , H_2O, CH_4 in UHP H_2.

References:

[2] Y. Nakamori and S. Orimo, *Materials Science and Engineering*, B108, 2004, p 48-50.

[3] Y. Nakamori and S. Orimo, *J. Alloys Compd.*, 370, 2004, p 271-275.

[4] T. Ichikawa, S. Isobe, N. Hanada and H. Fujii, *J. Alloys Compd.*, 365, 2004, p 271-276.

[5] J.W. Wiench, V.P. Balema, V.K. Pecharsky and M. Pruski, *J. Solid State Chemistry*, 117, 2004, p 648-653.

[6] P. Chen, Z. Xiong, J. Luo, J. Lin and K.-L. Tan, *Nature*, 420, 2002, p 302-304.

[7] P. Chen, Z. Xiong, J. Luo, J. Lin and K.-L. Tan, *J. Phys. Chem.*, B 107, 2003, p 10967-10970.

[8] T. Ichikawa, N. Hanada, S. Isobe, H. Leng and H. Fujii, *J. Phys. Chem.*, B108, 2004, p 7887-7892

[9] Y. Hang and E. Ruckenstein, *Ind. Eng. Chem. Res.*, 44, 2005, p 1510-1513.

[10] A. Huq, J.W. Richardson Jr., E. Maxey, D. Chandra, and W. Chien, *J. Alloys Compd.*, 436, 2007, p 256-260.

[11] A. Huq, J.W. Richardson Jr., E. Maxey, D. Chandra, and W. Chien, *J. Phys. Chem. C*, 111 (28), 2007, p 10712-10717.

[12] W. Chien, J. Lamb, D. Chandra, A. Huq, J. Richardson Jr., and E. Maxey, *J. Alloys Compd.*, 446-447, 2007, p 363-367.

(368b) Hydrogen Absorption-Desorption Cycle Durability and CO Tolerance of Hydrogen Absorbing Alloys Developed in WE-NET Project

Hideaki Tanaka[1], Kenji Aihara[2], Tomoyuki Miyata[2], Naoki Taoka[2], Kozo Mori[2], Teruo Kaneko[2] and **Nobuhiro Kuriyama**[1], (1)Research Institute for Ubiquitous Energy Devices, National Institute of Advanced Industrial Science and Technology, Ikeda, Japan, (2)New Material Center, Osaka Science and Technology Center, Osaka, Japan

Hydrogen absorbing alloys, "traditional alloys", have been studied for the applications to nickel-metal hydride batteries and hydrogen storage tanks for portable power generators and fuel cell vehicles, because of their high volumetric hydrogen density over 90 kg-H2/m3 and good kinetics for the reaction with hydrogen. IMURA, Japan Metals & Chemicals, Japan Steel Works, Sumitomo Steel, and OSTEC had promoted development of high-capacity alloys for room-temperature use in the second phase (FY1999-FY2002) of WE-NET Project, the Japanese national project for hydrogen systems, and Ca-Mg-Ni-based Laves-phase alloys, for example Ca0.25Mg0.66Y0.09Ni1.86, and BCC-structured alloys, Ti1.01Cr1.49V0.30, Ti1.10Cr1.39Mo0.30, V74.5Ti10Cr12.5Mn3, and V75.5Ti10Cr11Mn2Ni1.5. We will report hydrogen absorption-desorption cycle durability and influence of CO impurity of the novel alloys in order to give information for development of hydrogen storage systems. Although Ca0.25Mg0.66Y0.09Ni1.86 indicated considerable decrease, ca 40%, in hydrogen capacity within initial 20 cycles, decrease in prolonged cycles from 20th cycle to 400th was only 10% of the first cycle. This alloy indicated considerable influence of CO in hydrogen even at 1 ppm. The BCC alloys showed stable capacity over 1000 cycles in pure hydrogen and hydrogen containing 1 ppm of CO, although the capacities of Ti1.01Cr1.49V0.30 and Ti1.10Cr1.39Mo0.30 deceased ca. 15% within initial 100 cycles. Those alloy indicated appreciable influence by 10 ppm of CO in hydrogen.

See more of this Session: Hydrogen Storage System Engineering and Applications: Poster Session
See more of this Group/Topical: Topical 8: Hydrogen Production and Storage

(368c) Hydrogen Storage System Using $LiBH_4$

Hiroyuki T. Takeshita, Toyokazu Eguchi, Masanori Hayakawa and Makoto Ikeda, Department of Chemistry and Materials Engineering, Kansai university, Suita, Osaka, Japan

$LiBH_4$ is a promising candidate for hydrogen storage due to its high gravimetric hydrogen capacity, but improvement must be achieved for thermodynamic stability, slow reaction kinetics and reactivity against impurity gas species. We propose that thermodynamic and kinetic problems can be improved by using the mixture composed of $LiBH_4$, MgH_2 and Al and that reactivity against impurity gas can be improved by utilization of excess $LiBH_4$ which reacts with impurity gas species such as water vapor and works as hydrogen source as a result. In this paper, firstly the mixture of $4LiBH_4$-MgH_2-xAl (x=1 or less) is reported on its constituent phases, kinetic hydrogen desorption properties and cyclic de-/re-hydrogenation properties, and then $LiBH_4$ is reported on its reaction against impurity gas species, for example H_2O and O_2, and its hydrogen emission in heating after poisoning.

See more of this Session: Hydrogen Storage System Engineering and Applications: Poster Session
See more of this Group/Topical: Topical 8: Hydrogen Production and Storage

Hydrogen Purification and Storage System Using Metal Hydride

Shinichi Miura[1], Akitoshi Fujisawa[1] and Masayoshi Ishida[2], (1)Kobe Steel,Ltd., Kobe, Japan, (2)University of Tsukuba, Tsukuba, Japan

This work is to develop a new hydrogen purification and storage system for daily start and stop operations. The new system enables us to minimize emissions of carbon dioxide by using compact and highly efficient fuel cells.

As a result of a joint investigation conducted by Kobe Steel and the University of Tsukuba, a new process is proposed. First, this method thoroughly eliminates carbon monoxide using an adsorption material that strongly selects carbon monoxide, and second it introduces a mixed gas to a metal hydride that purifies and stores hydrogen. Thus, the COA-MIB (CO Adsorption Metal Hydride Intermediate- Buffer) method was established.

The selective adsorbent of carbon monoxide used in this process is an adsorbent based on porous alumina supporting Cu (I). Chemical adsorption of carbon monoxide occurs on the supported Cu (I). If carbon monoxide exists in reformed gas, carbon monoxide is adsorbed on the surface of metal hydride and it prevents hydrogen from being absorbed. It is necessary to eliminate carbon monoxide to achieve a hydrogen purification process using a metal hydride. After removing carbon monoxide, the reformed gas is introduced into a metal hydride bed to purify and store hydrogen.

Metal hydride has a high storage volume efficiency. If no poisonous substances are included, metal hydride can be used as a gas purification and storage material that selectively stores hydrogen when a mixed gas is supplied to it. As far as a fixed-type of apparatus is concerned, the existing AB5 type of metal hydride can be used.

Metal hydride absorbs and releases hydrogen in an equilibrium process according to the designed alloy temperature and pressure. When it is maintained at an equilibrium temperature it will not affect the hydrogen maintaining pressure, and it keeps the absorbed amount of hydrogen. On the other hand, when the pressure in a system falls because hydrogen is extracted, the metal hydride releases hydrogen in order to keep the equilibrium pressure. By making the best of this characteristic, a power generation system with a passive adjustment of supply and demand can be realized if a fuel cell is connected to the metal hydride that stores pure and pressurized hydrogen. In other words, a system can be made where the consumed hydrogen is compensated for by the metal hydride.

Some 100 NL/h Laboratory scale apparatus was designed and operated in daily start and stop operations for 100 cycles for a total of 150 hours with quite good efficiency. The new process has achieved an 83% hydrogen recovery ratio in one-month DSS operations.

(403a) Optimization of Hydrogen Storage Tanks: The Case of Sodium Alanate

Gustavo A. Lozano[1], Chakkrit Na Ranong[2], José M. Bellosta von Colbe[1], Rüdiger Bormann[1], Jobst Hapke[2], Georg Fieg[2], Thomas Klassen[1] and Martin Dornheim[1], (1)Institute of Materials Research, Materials Technology, GKSS Research Centre Geesthacht, Geesthacht, Germany, (2)Institute of Process and Plant Engineering, Hamburg University of Technology, Hamburg

Hydrogen is a very promising energy carrier for a comprehensive clean-energy concept in mobile applications. Regarding its use as fuel for the zero-emission vehicle, one main challenge is its storage. Hydrogen storage systems should fulfil the requirements of automotive applications, i.e. high gravimetric and volumetric storage densities, fast charging and discharging rates at moderate operating conditions, and high safety levels. Solid storage of hydrogen as metal hydrides offers a safe alternative to hydrogen storage in compressed or liquid form and has higher volumetric storage density. The light metal hydride sodium alanate, NaAlH4, compared to classical room temperature hydrides, offers a suitable compromise with relatively large gravimetric storage densities at rather moderate operating pressures and temperatures. Design of practical hydrogen storage systems aims at minimizing volume and weight while fulfilling defined criteria such as driving ranges and storage densities of the complete on-board tank system.

In the present investigation, hydrogen storage tanks based on sodium alanate are experimentally investigated, mathematically modelled, simulated and finally optimized. The determination of an optimal tubular storage tank based on sodium alanate material is presented. The optimisation is carried out on the basis of the predictions of a developed simulation tool for the hydrogen absorption, supported on the experimental results. Optimisation is performed for tubular tanks filled with loose powder and with compacted material. Compared to loose powder compacted material has the advantage of higher volumetric storage density and higher thermal conductivity, but the disadvantage of lower permeability to hydrogen flow and higher volumetric heat release during hydrogenation. Beyond compaction of the hydride material, the addition of expanded graphite is also evaluated in the optimisations. Although increasing the inert material of the system and thus reducing the total hydrogen storage capacity, the addition of expanded graphite enhances the effective thermal conductivity of the hydride bed.

See more of this Session: Hydrogen Storage System Engineering and Applications: Heat and Mass Transfer Modeling
See more of this Group/Topical: Topical 8: Hydrogen Production and Storage

(403b) Modeling of Adsorbent Based Hydrogen Storage Systems

Bruce Hardy[1], Claudio Corgnale[2], Richard Chahine[3], Marc-André Richard[4], David A. Tamburello[1], Stephen Garrison[1] and Donald Anton[5], (1)Computational Sciences, Savannah River National Laboratory, Aiken, SC, (2)Hydrogen Technology Center, Savannah River National Laboratory, Aiken, SC, (3)Hydrogen Research Institute, Trois-Rivieres, QC, Canada, (4)Hydro-Quebec, (5)Energy Security, Savannah River National Laboratory, Aiken, SC

Adsorbent based hydrogen storage systems have high potential for meeting the DOE Technical System Targets for light vehicles. However, these systems must charge and be maintained at low temperatures to retain a useable quantity of hydrogen. Heat and mass transfer considerations for these low temperature systems, together with the thermodynamics of the adsorbent media, results in complex behavior that must be addressed for application to on-board vehicular storage. In general, the dynamic behavior of media based hydrogen storage systems is sufficiently complex that numerical models are required for evaluation and design. Numerical models suitable for engineering analyses of media based storage systems must couple the mass, momentum and energy conservation equations to the chemical kinetics equations, or isotherms, for the media as appropriate. Typically, for storage media having very rapid kinetics, state dependent models for isotherms along with mass transfer resistances are used to determine the amount and rate of hydrogen uptake or discharge.

This paper describes the application of numerical models to the evaluation of hydrogen storage systems that utilize MaxSorb MSC-30™ and MOF-5™ adsorbents. The models for these systems incorporate thermodynamic parameters obtained by fitting data to the Dubinin-Astakhov adsorption model. Non-ideal gas properties for hydrogen are required for the range of pressures and temperatures over which these systems operate. The models can be applied in up to 3 dimensions for arbitrary vessel geometries and boundary conditions. Validation is performed against experimental data for MaxSorb MSC-30™.

Designs for temperature control using heat transfer surfaces of various geometries and flow-through cooling by hydrogen gas are investigated. Effects of cooling methods and media compaction on charging rates, capacity, and system performance are assessed. High pressure systems, suitable for hybrid storage, in which a significant amount of hydrogen resides in inter-particle void spaces and pores, as well as in the adsorption volume, are considered.

See more of this Session: Hydrogen Storage System Engineering and Applications: Heat and Mass Transfer Modeling
See more of this Group/Topical: Topical 8: Hydrogen Production and Storage

(403c) Model Based Design of An Automotive-Scale, Metal Hydride Hydrogen Storage System

Terry A. Johnson[1], Daniel E. Dedrick[1], Michael P. Kanouff[1], Gregory H. Evans[1] and Scott W. Jorgensen[2], (1)Thermal Fluids Science and Engineering, Sandia National Labs, Livermore, CA, (2) General Motors Research and Development, Warren, MI

Sandia and General Motors have successfully designed, fabricated, and experimentally operated a vehicle-scale hydrogen storage system using the complex metal hydride sodium alanate. Over the 6 year project, the team tackled the primary barriers associated with storage and delivery of hydrogen including mass, volume, efficiency and cost. The result was the hydrogen storage demonstration system design.

The key technologies developed for this hydrogen storage system include optimal heat exchange designs, thermal properties enhancement, a unique catalytic hydrogen burner and energy efficient control schemes. The prototype system designed, built, and operated to demonstrate these technologies consists of four identical hydrogen storage modules with a total hydrogen capacity of 3 kg. Each module consists of twelve stainless steel tubes that contain the enhanced sodium alanate. The tubes are arranged in a staggered, 4 x 3 array and enclosed by a steel shell to form a shell and tube heat exchanger. Temperature control during hydrogen absorption and desorption is accomplished by circulating a heat transfer fluid through each module shell. For desorption, heat is provided by the catalytic oxidation of hydrogen within a high efficiency, compact heat exchanger. The heater was designed to transfer up to 30 kW of heat from the catalytic reaction to the circulating heat transfer fluid.

The demonstration system module design and the system control strategies were enabled by experiment-based, computational simulations that included heat and mass transfer coupled with chemical kinetics. Module heat exchange systems were optimized using multi-dimensional models of coupled fluid dynamics and heat transfer. Chemical kinetics models were coupled with both heat and mass transfer calculations to design the sodium alanate vessels. Fluid flow distribution was a key aspect of the design for the hydrogen storage modules and computational simulations were used to balance heat transfer with fluid pressure requirements.

An overview of the hydrogen storage system will be given, and examples of these models and simulation results will be described and related to component design. In addition, comparisons of demonstration system experimental results to model predictions will be reported.

See more of this Session: Hydrogen Storage System Engineering and Applications: Heat and Mass Transfer Modeling
See more of this Group/Topical: Topical 8: Hydrogen Production and Storage

(403d) Scaled up LiBH4 - MgH2 Composite Storage System as New Promising Hydrogen Storage

Julian Jepsen, José Bellosta von Colbe, Ulrike Bösenberg, Gustavo Lozano, Thomas Klassen and Martin Dornheim, Institute of Materials Research, Materials Technology, GKSS-Forschungszentrum Geesthacht GmbH, Geesthacht, Germany

The discovery of the reaction mechanism of the Reactive Hydride Composite (RHC) some years ago, where the reaction of two metal hydrides lowers the overall reaction enthalpy [1, 2] gave new impetus to the entire hydrogen storage technology. RHCs combine high hydrogen storage capacities and relatively fast sorption rates, but at more moderate temperature and pressure conditions compared to the conventional high pressure (up to 700 bar) and liquid (-253 °C) storage systems. In particular for the promising combination of LiBH4 and MgH2 many investigations were done for a deep understanding of the reaction pathways (e.g. [3]). However, a scaled up tank system containing the LiBH4-MgH2 composite had not been investigated until now. In this paper we present the first results of such a system concerning the storage capacity, sorption rates and cycling behaviour. The investigated maximum temperature and hydrogen pressure conditions are at 400 °C and 50 bar respectively. The inner tank volume is 600 cm3. The theoretical investigation is performed by calculation of the mass and energy balances. Besides the use of a fluid heat transfer medium we also investigate the possibility of a electrical heat management. Hence we achieve further considerable weight and volume savings. However, the main advantage is the possibility of a free form tank system due to a flexible heat management and the maximum pressure of 50 bar. A free form tank system is advantageous for the automotive industry and most other mobile and stationary applications and can be considered as breakthrough in the hydrogen storage technology for high temperature hydrides.

References [1] Barkhordarian, G., Klassen, T., Dornheim, M., Bormann, R., 2007. Unexpected kinetic effect of MgB2 in reactive hydride composites containing complex borohydrides. Journal of Alloys and Compounds 440, L18-L21. [2] Vajo, J.J., Skeith, S.L., Mertens, F., 2005. Reversible Storage of Hydrogen in Destabilized LiBH4. J. Phys. Chem. B 109, 3719-3722. [3] Bösenberg, U., Doppiu, S., Mosegaard, L., Barkhordarian, G., Eigen, N., Borgschulte, A., Jensen, T.R., Cerenius, Y., Gutfleisch, O., Klassen, T., Dornheim, M., Bormann, R., 2007. Hydrogen sorption properties of MgH2-LiBH4 composites. Acta Materialia 55, 3951-3958.

See more of this Session: Hydrogen Storage System Engineering and Applications: Heat and Mass Transfer Modeling
See more of this Group/Topical: Topical 8: Hydrogen Production and Storage

(403e) Evaluation of Acceptability Envelope for Materials-Based H2 Storage Systems

Claudio Corgnale[1], Bruce Hardy[2], David A. Tamburello[2], Stephen L. Garrison[3] and Donald Anton[4], (1) Hydrogen Technology Center, Savannah River National Laboratory, Aiken, SC, (2)Computational Sciences, Savannah River National Laboratory, Aiken, SC, (3)Savannah River National Laboratory, Aiken, SC, (4)Energy Security, Savannah River National Laboratory, Aiken, SC

One of the most important issues to be addressed to make hydrogen a viable energy carrier is its storage. Among other possibilities (liquid storage, gaseous storage) solid state H2 storage systems, by means of adequate materials, seem to be one of the most suitable solutions being able to achieve high gravimetric and volumetric capacities. The paper shows results of activities carried out at Savannah River National Laboratory to evaluate acceptability envelope for such systems. The aim of the acceptability envelope is to determine the properties of the coupled media and system needed to meet storage technical specifications. To do that, a steady state model has been set up, capable to simulate the behavior of different materials under different operating conditions. Various scenarios have been selected, starting from the DOE technical target specifications, and different H2 storage materials have been analyzed and compared. In addition, several sensitivity analyses have been carried out showing the influence of selected thermo-physical material quantities on characteristic sizes of the storage system.

See more of this Session: Hydrogen Storage System Engineering and Applications: Heat and Mass Transfer Modeling
See more of this Group/Topical: Topical 8: Hydrogen Production and Storage

(403f) Automatic Optimization of Metal Hydride Storage Tanks and Novel Designs

Stephen L. Garrison[1], Mikhail B. Gorbounov[2], David A. Tamburello[3], Bruce Hardy[3], Claudio Corgnale[4], Daniel A. Mosher[5] and Donald Anton[6], (1)Savannah River National Laboratory, Aiken, SC, (2)Thermal Management, United Technologies Research Center, East Hartford, CT, (3)Computational Sciences, Savannah River National Laboratory, Aiken, SC, (4)Hydrogen Technology Center, Savannah River National Laboratory, Aiken, SC, (5)Applied Mechanics, United Technologies Research Center, East Hartford, CT, (6)Energy Security, Savannah River National Laboratory, Aiken, SC

A key technical hurdle to an energy economy focused on clean-burning hydrogen over non-renewable petroleum is on-board storage of hydrogen for automotive vehicular applications. Metal hydrides provide one possible solution to the need for high gravimetric and volumetric capacities. However, metal hydrides release significant amounts of heat during hydrogen adsorption, on the order of 0.1 GJ. The subsequent rise in storage temperature results in decreases in maximum hydrogen storage capacity and loading rate at the short fill times mandated by the Department of Energy (DOE) targets. Integrated tank heat exchangers, e.g., cooling tubes and fins, are needed to maximize heat transfer out of system to maximize storage rate and capacity. Given the high cost of experiments (especially at higher pressures), the non-commodity nature of some metal hydrides, and the possibility of air- or water- reactivity, computational modeling provides a significant contribution to a comprehensive methodology for the design, evaluation, and modification of hydrogen storage systems. While many literature studies focus on maximum capacity assuming an (impossible) infinite fill time, full, detailed modeling of the coupled heat and mass transfer and chemical reaction kinetics is needed. The focus of this presentation will be our work developing detailed numerical models that couple reaction kinetics with heat and mass transfer via the general purpose finite element solver COMSOL Multiphysics® for general metal hydride beds and the automatic optimization of the designs by linking the COMSOL models with routines in MATLAB®. Optimization of novel heat exchanger designs will also be presented.

See more of this Session: Hydrogen Storage System Engineering and Applications: Heat and Mass Transfer Modeling
See more of this Group/Topical: Topical 8: Hydrogen Production and Storage

(427a) Liquid Phase Reforming of Wood Flour to Hydrogen

Timothy D. Davis, Sean C. Emerson, Tianli Zhu, Rhonda R. Willigan, Ying She and Thomas Henry Vanderspurt, United Technologies Research Center, East Hartford, CT

Background

Growing concern over the availability, cost, and environmental impact of traditional petroleum energy sources, has emphasized the need to develop alternative energy technologies that reduce our dependence on non-renewable resources. Lignocellulosic biomass is one such low cost source that is renewable and available in significant quantities. Globally, over 224 billion dry tones of biomass are produced through photosynthesis each year, roughly 10 times the world's energy consumption[1]. Currently, significant levels of research are in the field of converting this sustainable feedstock into many forms of energy, including syngas and several forms of liquid fuels.

Hydrogen is another possible product from biomass. H_2 is useful for numerous applications ranging from fuel cells, chemical processing, or direct consumption as a fuel. The challenge that remains is to provide a low cost, efficient route for its conversion to hydrogen. For such a technology to be economically feasible, a proposed system must be simple, using raw biomass feedstock, have a high selectivity for H_2 production, and not produce toxic byproducts. Significant work has been reported in the literature for liquid phase reforming (LPR) of preprocessed biomass derived feed stocks and related oxygenates like ethylene glycol. Solutions of glycerol, glucose or sorbitol have been reformed to H_2, CO_2, and low molecular weight alkanes over Pt, Ni-Sn catalysts[2,3] and other[4] catalysts. This work concentrates on the production of H2 directly from raw biomass, without the need for intermediate processing steps.

In 2009 we reported on the high efficiency Liquid Phase Reforming of yellow poplar (***Liriodendron tulipifera)*** saw dust over Pt-Re/ceria zirconia catalyst pellets. The striking observation was that the effluent was water white and substantially free of organic matter except for a trace amounts of formic and possibly acetic acid[5]. Thus not only were the hemi-cellulose and cellulose reformed but the lignin as well. This catalyst was selective towards hydrogen co-producing relatively little methane and higher alkanes. It was also robust, but in practice this catalyst would be expensive. Furthermore the high surface area–large pore cubic ceria-zirconia support was not readily available as an item of commerce. Thus one of the next steps before proceeding to a continuous process converting wood to hydrogen was the identification and kinetic characterization of a less expensive, more readily available catalyst.

Materials and Methods

The biomass materials and surrogates used in this work include ethylene glycol and 200 mesh wood flour obtained from P. J. Murphy. The catalyst starting materials was Raney Ni obtained from W. R. Grace. Kinetic experiments were performed in a custom build 0.5" OD inch flow reactor constructed of Inconel 625. The system was heated by means of an infrared furnace with an Isco pump, Model 500D, supplying the feed to the reactor. A catalytic bed bed was placed in the reactor, supported by ceria stabilized zirconia beads. Pressure was controlled by means of a Tescom back pressure regulator equipped with an ER3000 electronic pressure controller. Gas analysis was performed by means of dual Varian 3800 GCs, for hydrocarbon and permanent gas analysis.

Results and Discussion

Kinetic studies on the liquid phase reforming of a biomass surrogate, ethylene glycol, have been carried out in alkaline solutions over several base metal catalysts. Several bases and concentrations have been studied to investigate the effect of basicity and concentration on the system. Although not required for the LPR, the alkalis are for the hydrolysis of the biomass. Of the metals, the study was quickly down selected to Raney Nickel, as it was concluded that the other catalysts tested did not have sufficient C-C bond breaking ability at low temperatures to warrant further study. The complete conversion of wood flour to gas and water white carbonate solution was demonstrated with Raney Ni was confirmed in batch reactor studies.

Flow reactor studies of the LPR of 5 wt.% ethylene glycol over an 8g bed of Raney nickel were performed. Selectivity and yield results are given in Figures 1 and 2. As expected, Raney nickel was capable of completely reforming ethylene glycol to gaseous species at 310 °C in water. In addition, a high level of methanation was evident, an inherent characteristic of Ni reforming catalysts. Increasing amounts of K_2CO_3 slightly decreased the methanation activity as well as reduced the overall yields. However, increasing concentrations of KOH in the system did not detrimentally affect the yield but dramatically decreased the methanation activity, resulting in significantly higher selectivity's. This reduction of methanation activity may be due to homogeneously catalyzed water gas shift reactions, driving any formed carbon monoxide to carbon dioxide before methane can be produced. CO_2 methanation requires higher temperatures to methanate in any significant quantity.

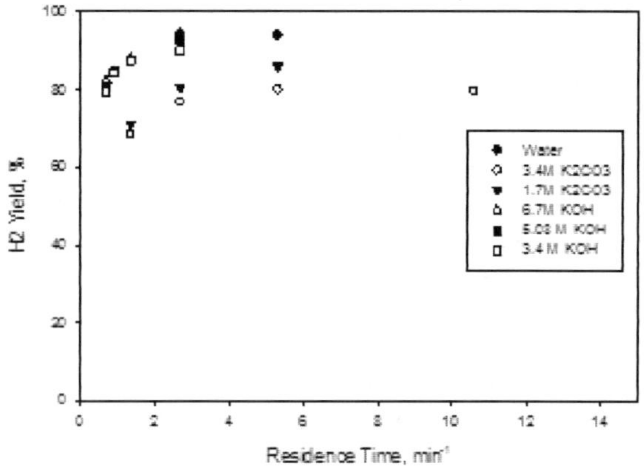

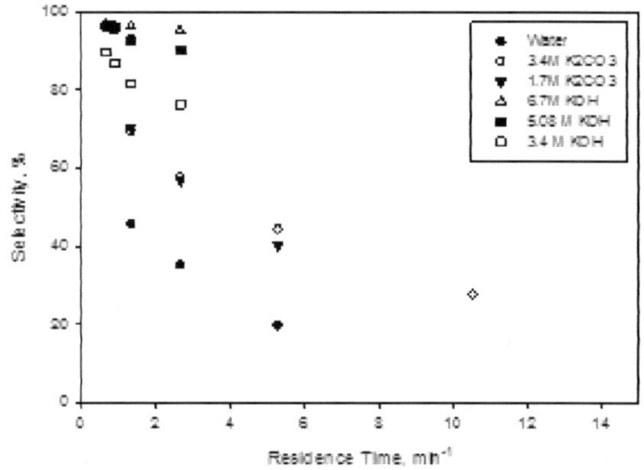

Figures 1 and 2- LPR of 5 wt.% Ethylene Glycol over Raney Nickel, 310 °C, 120 atm.

However, using high levels of hydroxide to improve the selectivity of nickel based catalysts creates a system level energetic issue. The KOH is converted to K_2CO_3 in situ, and would require regeneration. As an alternative, a study was performed of doping the Raney nickel catalyst to promoters to drive down the methanation activity. As shown in Figures 3, a small addition of a promoter was sufficient to drive down the methanation activity without significantly decreasing overall reforming activity. This result enables the use of K_2CO_3 as the base for the wood hydrolysis, which has already been demonstrated as sufficient for the process[5].

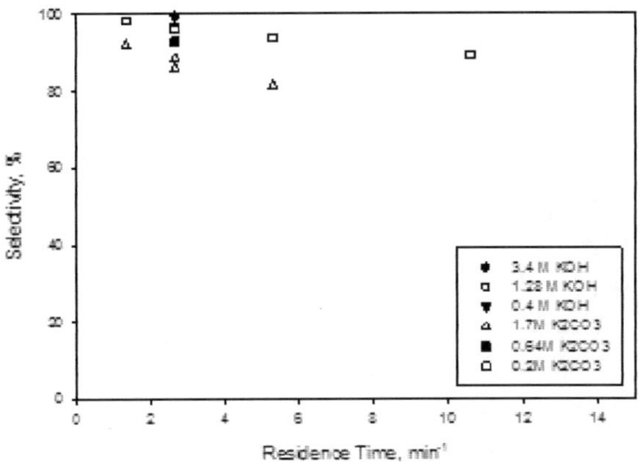

Figures 3- LPR of 2.5 wt.% Ethylene Glycol over Promoted Raney Nickel, 310 °C, 120 atm.

Work is ongoing to understand the LPR mechanism. A combination of experimentation and atomistic modeling will help explain the mechanism. This may lead to catalyst modification enabling faster kinetics and tailoring the selectivity to meet the plant system efficiency targets. Furthermore, the surrogate results will be compared to those obtained from directly pumping and reacting raw biomass into the reactor.

In order to bring the technology one step closer to realization, a prototype power plant will be constructed to demonstrate the feasibility of the system for processing of raw biomass. The system is targeted to produce hydrogen sufficient to power a 1 kWe Fuel Cell by liquid phase reforming of wood flour. As of the time of this abstract, a high pressure slurry pump has been acquired and validated for pumping the feedstock at the desired conditions. Work is ongoing to determine the kinetics and optimal conditions for the wood hydrolysis. The system will use a promoted base metal catalyst for reforming and a palladium membrane for the separation of pure hydrogen from the reformate.

Significance

This work moves the concept of hydrogen production from raw biomass as describe in DOE contract DE-FG36-05GO15042 "A Novel Slurry-Based Biomass Reforming Process" a significant step closer to realization.

See more of this Session: Renewable Hydrogen Production I
See more of this Group/Topical: Topical 8: Hydrogen Production and Storage

NICKEL CATALYZED SUBCRITICAL AQUEOUS PHASE LIQUEFACTION OF BIOMASS FOR HYDROGEN PRODUCTION

Richa Tungal[1], *Rajesh V. Shende*[1], *Lew Christopher*[2]

[1]Department of Chemical and Biological Engineering,
[2]Center for Bioprocessing Research & Development,
South Dakota School of Mines & Technology,
Rapid City, SD 57701 USA

Abstract

This paper reports subcritical aqueous phase liquefaction of office paper waste in presence of homogeneous nickel catalysts for H_2 production. Slurry of 10-20 wt% biomass was prepared in distilled water containing 5 wt% Ni-catalyst and loaded in high-temperature high-pressure PARR reactor and heated to 250°C. The reaction progress was monitored for 2 h at 250°C and during course of reaction; gas samples were withdrawn at regular time intervals and analyzed using Chrompak capillary column on the GC equipped with thermal conductivity detector. The analysis of gas samples indicated the presence of H_2, CO_2, CO, and CH_4. Different Ni-salts that were used as catalyst during liquefaction of the biomass were found to have significant effect on H_2 generation. Specifically, after 1.5 h of reaction, $NiSO_4$, $NiCl_2$ and $Ni(NO_3)_2$ produced H_2 in the vol% of 18%, 40% and 43%, respectively. The effectiveness of a Ni-catalyst on H_2 generation from subcritical hydrothermal processing of cellulose was also investigated as it is the major component present in commonly used office paper.

Introduction

Almost every office around the world uses some kind of paper for multiple reasons. The unwanted used papers are then directly disposed off or shredded and disposed. U.S. produces about 4 million tons of office paper annually, whereas worldwide more than 300 million tons of different variety of papers are manufactured. It has been estimated that for the production of one ton of paper, processing of about 2 - 4 tons of trees are required. Thus significant biomass processing is required for paper production. As the usage of paper and paper products has been increasing day by day, the production of paper waste is also increasing, creating paper waste disposable issues [1]. The production of municipal solid waste in U.S alone is about 3200 pounds annually and out of that about 1500 pounds is a paper waste [2]. Disposing off this large amount of paper waste is posing a great problem due to high costs involved. Currently, most of the paper waste is being discarded through landfill and incineration and only a small portion is being recycled. Worldwide about 30 - 50% of waste paper is disposed by landfilling, majority paper waste is incinerated and only about 10% is recycled. Many industrialized countries recycle paper waste but the recycling efforts vary greatly with different countries, for instance Germany and India recycle about 80% and 20% respectively. As the recycled paper has high content of pulp fiber, it is difficult to get new paper products from recycling [3]. Moreover, due to heavy recycling cost, the market of the recycled paper is limited.

From waste management point of view, it appears that additional energy and efforts are required for the disposal of paper waste. Instead, if energy or fuel is derived from the paper waste, it will open new avenues for energy production as well as waste reduction. There exist thermochemical and biochemical or biological technologies for converting paper waste into gaseous or liquid fuels, which include H_2, lower hydrocarbons and alcohols. Among these technologies, one of the thermochemical processes such as high pressure hydrothermal treatment of waste biomass is attracting great deal of attention [4]. High pressure hydrothermal processes can be either subcritical, where water is operated below critical temperature (T_c) and pressure (P_c) of 374.14°C and 22.064 MPa or supercritical, if water is used above the critical temperature and pressure conditions. Subcritical hydrothermal process has advantages over other thermochemical processes that it can directly process and liquefy wet waste biomass at low temperatures and saves energy from drying operation. This process can be operated under gasification or pyrolysis conditions for liquefaction of a biomass and production of valuable fuels. Recently, rapid conversion of switchgrass in subcritical water under pyrolysis conditions was performed at temperatures of 250- 350°C and pressures upto 20 MPa and this study reported significant presence of H_2 in the gaseous products [5]. In another study, paper waste was processed under subcritical water conditions in presence of charcoal supported ruthenium catalyst at 250-300°C and reported production of methane, H_2 and other gases. In this study, however, very small amount of H_2 generation has been indicated [6]. Many other investigations reported production of different fuels under various hydrothermal conditions.

Among different fuels currently available, H_2 is a green fuel with very high energy density. H_2 can be produced from lignocellulosic biomass using thermal, biological and chemical processes. As U.S is currently producing about 400,000 tons/year of lignocellulosic biomass, there is significant potential exists for H_2 production [7]. For the thermochemical processes, however a great deal of research and development efforts are needed in order for this technology to be cost-competitive over other conventionally used H_2 production technologies. At present R&D activities are focused on the development of the catalytic processes with the goal of achieving highest possible H_2 levels from a biomass processing at moderate temperature and pressure conditions. Therefore, we believe it will be very beneficial if significant level of H_2 is obtained at subcritical hydrothermal processing conditions in presence of suitable catalysts. Few investigators have already performed hydrothermal gasification of glucose in presence of raney nickel and several other organometallic catalysts at 310-350°C and reported higher H_2 levels as compared with non-catalytic conditions [8]. In another study, cellulose was gasified to H_2 rich gas using reduced nickel catalyst at hydrothermal conditions of 350°C and 18 MPa [9]. Antal et al. used glucose as potential feedstock for H_2 production via steam reforming in presence of different metal oxide supported metal (Ni/SiO_2) catalyst and reported H_2 rich syn gas [10]. Most of these studies were focused on H_2 generation from model compounds. However, very little effort is noticed in the literature on the production of H_2 from paper waste.

In this paper subcritical hydrothermal processing of office paper waste was performed using three different homogeneous Ni-catalysts at 250°C. The reaction was monitored for 2 h and it was in progress, the gas samples were withdrawn and analyzed using GC. The results obtained on H_2 volume generation are presented and the effectiveness of catalysts is discussed.

Experimental Methods and Analysis

Materials

$NiSO_4.6H_2O$, $NiCl_2.6H_2O$ and $Ni(NO_3)_2.6H_2O$ were purchased from Alfa Aesar and and used without purification. α-Cellulose was purchased from Sigma Aldrich, WI. N_2 from a cylinder with the minimum stated purity of 99% was purchased from Linweld Inc., SD. The shredded office paper waste or cellulose was made into slurry in distilled water and used for subcritical hydrothermal processing in N_2 environment.

Experimental Set-up and Procedure

High-temperature high-pressure PARR SS316 reactor of 300 ml capacity equipped with heating and cooling assembly, gas inlet and outlet valves, sampling valves, pressure gauge, safety rupture disc, thermocouple, magnetic stirrer and temperature indicator controller was used to perform subcritical hydrothermal processing of a biomass. This reactor is rated for maximum working temperature and pressure of 350°C and 3000 psi, respectively. An additional SS gas sampling vessel was mounted on the reactor. The entire reactor set-up is shown below in Figure 1.

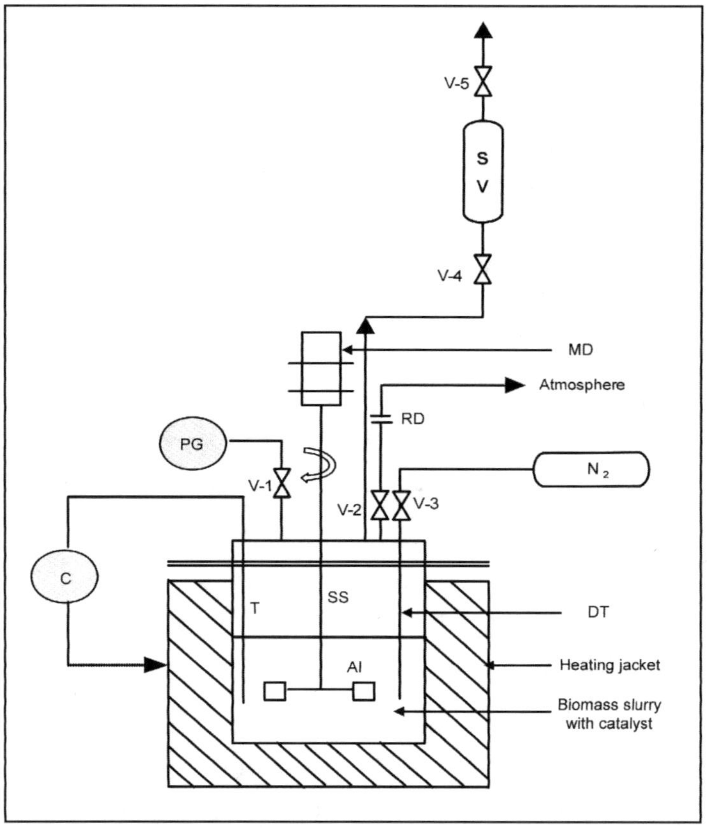

Fig. 1: Reactor set-up for subcritical hydropyrolysis of a biomass for H_2 generation. (RD-rupture disc, T-thermocouple, AI-adjustable impeller, SS-stirring shaft, PG-pressure gauge, MD-magnetic drive, V-valve, DT-dip tube, C-controller, SV-gas sampling vessel).

Accurately weighed 15.0 g of shredded office paper waste or cellulose was made into slurry in 150 ml distilled water containing 5wt% (taken on a biomass weight basis) Ni-catalyst and loaded in the high-temperature high-pressure PARR reactor. The contents in the reactor were flushed continuously with N_2 for 10-15 min to replace air and pressurized to 40 psi. The reactor was heated to 250°C under continuous magnetic stirring of a biomass slurry at 1300 rpm. It took approx. 45 min to attain the set point of 250°C where the total reactor pressure was 700 psi. Once the reactor temperature was attained, a gas sample was withdrawn. This time was taken as 'zero' time for the reaction. The reaction was continued for 120 minutes and during the course of the reaction, gas samples were withdrawn at regular time intervals and analyzed on the GC. At the end of 120 min, the reactor contents were cooled down to room temperature and the liquid obtained was filtered to determine the amount of residue.

GC Analysis and Residue Measurement

Gas samples were withdrawn with a gas tight syringe during the subcritical hydrothermal processing of office paper waste or cellulose and analyzed using Agilent 7890A GC equipped with Chrompak capillary column and thermal conductivity detector. A sample injection volume of 100 µl was used for the analysis. The GC analysis was performed under the isothermal conditions where column and detector temperatures were maintained at 150°C and 205°C, respectively.

At the end of the reaction, the reactor contents were cooled down to room temperature and filtered using Whatman filter paper. The residue was recovered after filtration and dried in the oven at 60°C and weighed accurately.

Results and Discussion

Figure 2 shows the product gas profile obtained during subcritical hydropyrolysis of office paper waste at 250°C in presence of $NiCl_2$ catalyst. The product gas profile indicates primarily H_2, CO, CH_4 and CO_2 gases. In addition to these gases, H_2O vapor should also be present in the gas samples, however, as the gas sampling vessel was held at ambient conditions, the water vapors might have been already condensed. Moreover the Chrompak capillary column that was used for the analysis of product gases was not capable of detecting water vapors. Therefore, the concentrations were determined without accounting for H_2O vapor in terms of vol% of total headspace volume, which was 150 ml. The product gas profile presented in Figure 2 indicates significant concentration levels of H_2 and CO_2 gases followed by lower level of CO, however, negligible volume of CH_4. The figure also indicates that as the reaction time increases, the concentration of H_2 and CO_2 increases and the maximum H_2 generation is approx. 40 vol% at 90 min of reaction time. It is to be noted that in absence of $NiCl_2$ catalyst, no H_2 generation was observed from the office paper waste under the subcritical hydrothermal processing at 250°C. Therefore from the experiments performed we believe that the H_2 generated was solely due to the catalytic action of $NiCl_2$ and that the reactor wall effects were negligible. Our finding is consistent with the study where investigators [11] had reported significant H_2 production in presence of $NiCl_2$ and other metal salts catalyst.

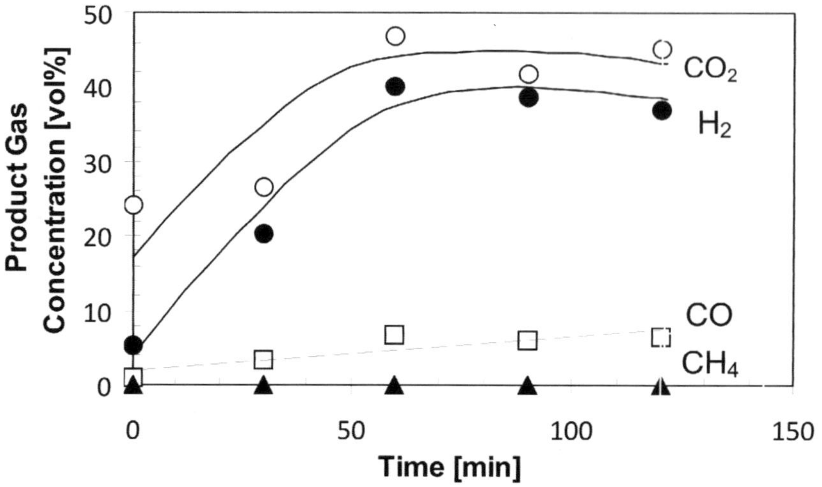

Fig. 2: Product gas profile showing H_2, CH_4, CO and CO_2 gases during subcritical hydropyrolysis of office waste paper at 250°C in presence of 5 wt % $NiCl_2$.

The product gas profile obtained during subcritical hydrothermal processing of office paper waste at 250°C in presence of $Ni(NO_3)_2$ catalyst is presented in Figure 3. Analogous to $NiCl_2$ effect on the product gas profile, the $Ni(NO_3)_2$ catalyst indicated the presence of H_2, CO_2, CO and CH_4 gases. These concentration profiles further reveal H_2 and CO_2 generation in higher volumes as compared with CO; however, CH_4 volume was in negligible amount. The profiles also indicate almost no change in CO concentration. Further observation shows that the H_2 concentration increases to the maximum value of about 43vol% that corresponds to the reaction time of 90 min and further decreases slightly. The CO_2 concentration levels observed with $Ni(NO_3)_2$ catalyst were higher by about 20% than those observed with $NiCl_2$ (Figure 2). This suggests that $Ni(NO_3)_2$ should lead to lower amount of biochar formation as compared with $NiCl_2$ catalyst. The char formation, however, was not studied in detail as it was not the goal of this investigation. Several routes of H_2 formation from a biomass that can be expected include the formation of CO followed by water-gas-shift, direct one step reaction of biomass with high-temperature compressed water, bio-oil formation followed by reaction with water. The bio-oil and its reaction with water generally occurs under pyrolytic conditions. At the end of the reaction, when the reactor contents were cooled down to room temperature, very little amount of the bio-oil was observed in the aqueous solution. Therefore, it is not certain if H_2 was routed via bio-oil formation alone. In the present investigation, two routes appear to be reasonable describing the H_2 generation results. In route one, NO_3^- ions may be accelerating the decomposition of the biomass and Ni^{2+} ions presumably cleaving the C-C bonds and yielding CO and H_2. This may further undergo water-gas–shift reaction and yielding CO_2 and H_2. In route two, after accelerated decomposition of the biomass with NO_3^- ions, Ni^{2+} ions in may be forming water soluble complex with the carbonaceous species derived under hydropyrolysis conditions and yielding H_2 via formation of intermediate compounds.

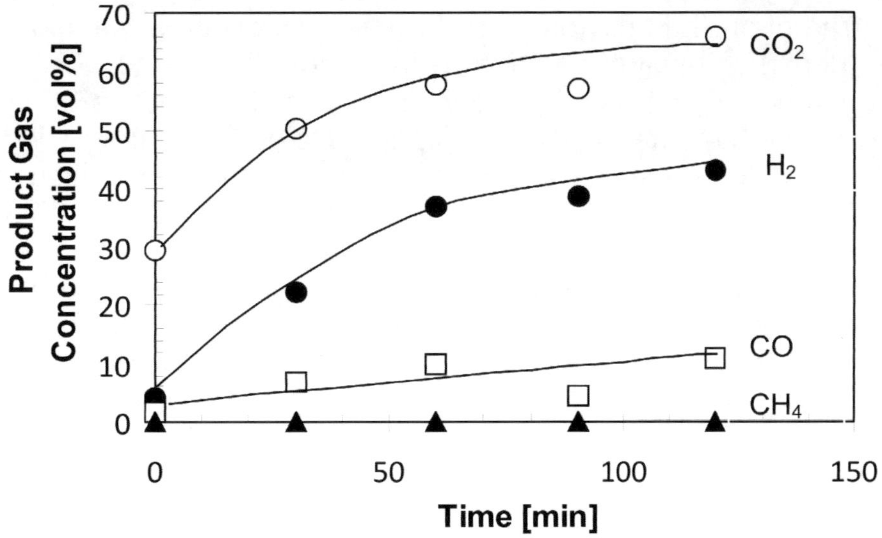

Fig. 3: Product gas profile showing H_2, CH_4, CO and CO_2 gases during subcritical hydropyrolysis of office waste paper at 250°C in presence of 5 wt % $Ni(NO_3)_2$.

The effect of $NiSO_4$ catalyst on the product gas profile during hydrothermal processing of office paper waste at 250°C is shown in Figure 4. Very interesting observations were made. The H_2 formation found to increase almost linearly with the reaction time and reaches to 18vol% in 90 min. This H_2 volume generated is the lowest among all the three Ni-based catalysts investigated here. In 120 min, a total of 24vol% H_2 was observed. The CO_2 formation was almost unchanged from the beginning of the reaction and the volume of CO remains constant after 1 hr of reaction. These trends are significantly different that those observed for the other Ni-catalysts, which suggest different reaction chemistries with $NiSO_4$ catalyst but the biomass liquefaction issue. The office waste paper when treated at different temperatures, a brown colored product slurry containing minimum amount of residue was obtained at 250°C (Figure 5).

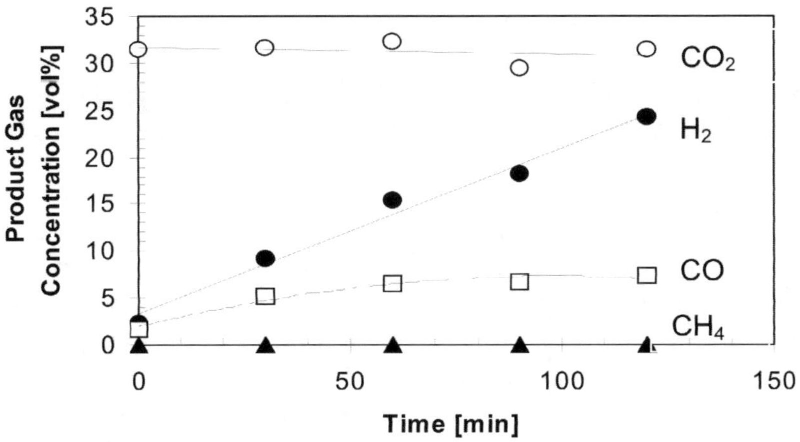

Fig. 4: Product gas profile showing H_2, CH_4, CO and CO_2 gases during subcritical hydropyrolysis of office waste paper at 250°C in presence of 5 wt % $NiSO_4$.

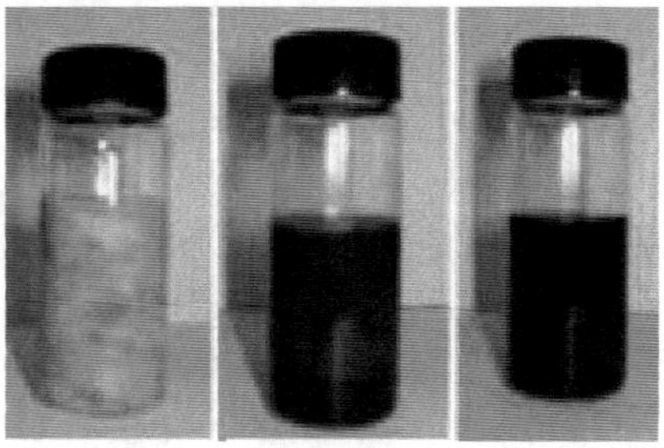

150°C **200°C** **250°C**

Fig 5: Paper waste slurry processed at different temperatures for 2 hrs.

Composition wise the commonly used office paper contains a major portion of cellulose, about 50 - 95wt%. Therefore, it is obvious that for the H_2 profiles presented in Figure. 2 and Figure. 4, a significant portion must of H_2 have been routed from the cellulose present in office paper waste. To verify this aspect, α-cellulose was subjected to the hydrothermal processing at 250°C in presence of $NiSO_4$ catalyst. The H_2 profile obtained over the period of 2 h reaction time is presented in Figure 6, which indicates about 26 vol% of H_2 generation in 90 min, which is higher by about 8% than that of the H_2 volume observed in the case of office paper waste. It is interesting to note that at the end of 2 h, the volume of H_2 generated for both office paper waste and cellulose were similar indicating that in the case of office paper waste, the H_2 must have been routed from cellulose alone. Small deviation in H_2 volume generated from cellulose and office paper waste may be due to the possible interference or slightly different chemistries associated with the presence of impurities (lignin, colors etc) in the office paper waste.

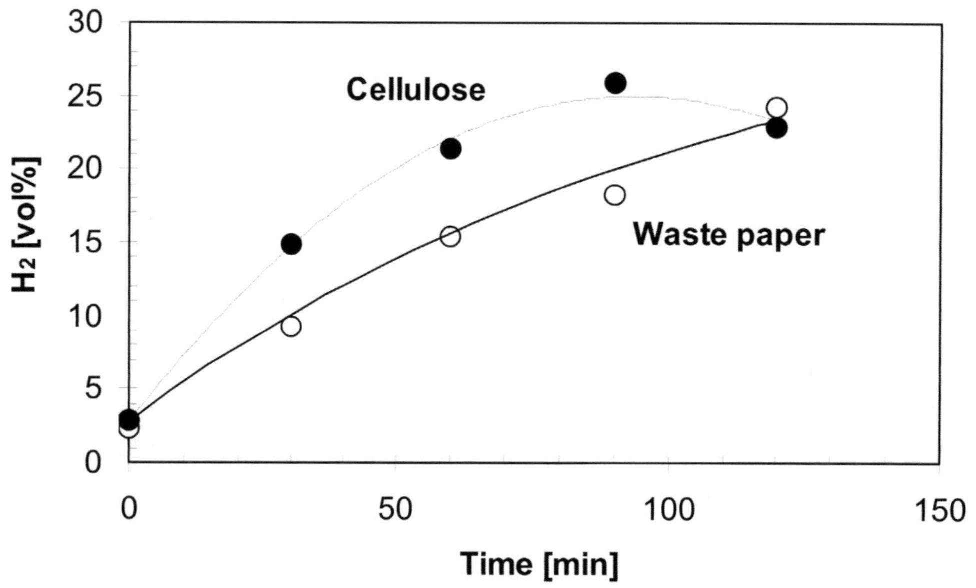

Fig. 6: Comparison of H_2 production from cellulose and waste paper using $NiSO_4$ catalyst at 250°C.

Thus it is possible to achieve significant level of H_2 from the biomass by using homogeneous Ni-catalyst. Such approach does not produce any methane, which is advantageous. From the product gas, the H_2 can be separated using a membrane process and used in the fuel cell for energy generation. The aqueous phase containing Ni-catalyst can be reused again for subsequent biomass processing. Alternatively Ni^{2+} ions can be separated from the processed aqueous stream using ion-exchange resin and the processed water can bedisposed off after recovering the bio-oil and residue. The main advantage of the process investigated here is the minimal amount of char or residue formation.

Conclusion

The subcritical hydrothermal processing of office paper waste and cellulose was successfully performed in high-temperature and high-pressure PARR reactor at 250°C in the presence of homogeneous Ni-catalysts that include $NiCl_2$, $Ni(NO_3)_2$ and $NiSO_4$. Their product gas profiles indicate H_2 rich gas containing CO_2 and CO and CH_4 gases. The concentration of CH_4 was found to be negligible. Almost complete liquefaction of office paper waste was achieved at 250°C with the presence of bio-oil and residue in a very small amount. The maximum H_2 of 43 vol% was produced from office paper waste using $Ni(NO_3)_2$ catalyst whereas $NiSO_4$ generated only 18 vol% after 90 min of reaction at 250°C. In addition to office paper waste, subcritical hydrothermal processing of cellulose was performed at 250°C in the presence of $NiSO_4$. As commonly used office paper contains a major portion of cellulose (approx 50-95%), subcritical hydrothermal processing of cellulose was performed at 250°C in presence of $NiSO_4$ catalyst to determine the extent of H_2 generation. It was observed at the end of 2 hrs that the volume of H_2 generated from both office paper waste and cellulose were very similar indicating that in the case of office paper waste, the H_2 must have been routed from the cellulose.

Acknowledgement

The authors wish to acknowledge the financial supports by the Chemical & Biological Engineering Ph.D. support grant and Center for Bioprocessing Research and Development.

References

1. Li, L., Zhang, H. and Zhuang, X. (2005), "Pyrolysis of waste paper: characterization and composition of pyrolysis oil," *Energy Sources,* 27, pp. 867 – 873.

2. Torri, N., Okai, A., Shibuki, K., Aida, T. M., Watanabe, M., Ishihara, M., Tanaka, H., Sato, Y., Smith, R. L. and Jr. (2010), "Production of D-glucose from pseudo paper sludge with hydrothermal treatment," *Biomass and Bioenergy*, 34, pp. 844-850.

3. Ohtsu, Y., Yamada, R., Urasaki, H., Misawa, T., Popescu, S. and Fujita, H. (2010) "Development of novel hybrid microwave- heater reactor for paper -based waste treatment," *J Mater Cycles Waste Mang*,12, pp. 25-29.

4. Ishida, Y., Kumabe, K., Hata, K., Tanifuji, K., Hasegawa, T., Kitagawa, K., Isu, N., Funahashi, Y. And Asai, T. (2009), "Selective H_2 generation from real biomass through hydrothermal reaction at relatively low temperatures," *Biomass and Bioenergy*, 33, pp. 8-13.

5. Cheng, L., Ye, P. X., He, R. and Liu, S. (2009), "Investigation of rapid conversion of switchgrass in subcritical water," *Fuel Processing Technology,* 90, pp. 301-311.

6. Yamaguchi, A., Hiyoshi, N., Sato, O., Bando, K. K., and Shirai ,M. (2010), "Gaseous fuel production from nonrecyclable paper wastes by using supported metal catalysts in high-temperature liquid water," *ChemSusChem,* 3, pp. 737 – 741.

7. Turner, J., Scerdrup, G., Mann, M. K., Maness, P. C., Kroposki, B., Ghirardi, M., Evans, R. J. and Blake, D. (2008), " Renewable hydrogen production," International Journal of Energy Research, 32, pp. 379 – 407.

8. Azadi, P., Khodadadi, A A., Mortazavi, Y. and Farnood, R. (2009) "Hydrothermal gasification of glucose using raney nickel and homogeneous organometallic catalysts," *Fuel Processing Technology,* 90, pp. 145-151.

9. Minowa, T. and Inoue , S. (1999), "Hydrogen production from biomass by catalytic gasification in hot compressed water," *Renewable Energy*, 16, pp. 1114-1117.

10. Antal, MJ, Jr., Manarungson, S, Mok, WS-L. (1994) "Hydrogen production by steam reforming glucose in supercritical water," *Adv. Thermochemical. Biomass Conversion.,* pp. 367-77.

11. Williams, P. T. and Horne, P. A. (1994) "The role of metal salts in the pyrolysis of biomass," Renewable Energy, 4, pp. 1 – 13.

(427c) Hydrogen Production From Liquefied Switchgrass in Supercritical Water Over Ru, Ni, and Co Catalysts

Adam J. Byrd, Sandeep Kumar and Ram B. Gupta, Chemical Engineering, Auburn University, Auburn, AL

Switchgrass is a renewable high yield energy crop with the potential to offset our fossil fuel usage and reduce carbon emissions. In this work, switchgrass was liquefied in subcritical water (T= 300°C, P=100 bar) in a batch reactor to give an aqueous solution of oxygenated hydrocarbons containing ~1.5 wt% C designated as biocrude. Biocrude was then subject to reforming in supercritical water to produce hydrogen. A suite of nine catalysts designated as Metal/Support (M=Ru, Ni, Co; S=TiO2, ZrO2, MgAl2O4) was prepared and evaluated in the supercritical water gasification of liquefied biomass (T=600°C, P=250 bar, WHSV=9h-1). Activity of the catalysts was evaluated in terms of hydrogen yield, gasification efficiency, and tendency towards char formation, as measured by the degree of closure of a carbon balance around the process. Spent catalysts were characterized by the BET method and XRD to evaluate the stability of the support materials in the SCW environment. The highest gasification efficiency was obtained over Ru/ZrO2, while the highest hydrogen yield was seen with Ni/ZrO2.

See more of this Session: Renewable Hydrogen Production I
See more of this Group/Topical: Topical 8: Hydrogen Production and Storage

Modeling of Biohydrogen Production Using BioWin

Hisham Hafez[a*], Elsayed Elbeshbishy[a], M. Hesham El. Naggar[a], George Nakhla[a,b]

[a]Department of Civil Engineering, The University of Western Ontario, London, Ontario, N6A 5B9, Canada

[b]Department of Chemical and Biochemical Engineering, The University of Western Ontario, London, Ontario, N6A 5B9, Canada

Abstract:

Steady state operational data from the patent-pending integrated biohydrogen reactor clarifier system (IBRCS) during anaerobic treatment of glucose-based synthetic wastewater at HRT of 8 hrs and SRT ranging from 26 to 50 hrs and organic loading rates of 6.5 to 206 gCOD/L-d were used to calibrate and verify a process model of the system developed using BioWin. The model accurately predicted biomass concentrations in both the bioreactor and the clarifier supernatant with average percentage errors (APE) of 4.6% and 10%, respectively. Hydrogen production rates and hydrogen yields predicted by the model were in close agreement with the observed experimental results as reflected by an APE of less than 4%, while the hydrogen content was well correlated with an APE 10%. The successful modeling culminated in the accurate prediction of soluble metabolites i.e. volatile fatty acids in the reactor with an APE of 14%. The calibrated model confirmed the advantages of decoupling of the solids retention time (SRT) from the hydraulic retention time (HRT) in biohydrogen production, with the average hydrogen yield decreasing from 3.0 mol H_2 / mol glucose to 0.8 mol H_2 / mol glucose upon elimination of the clarifier.

Keywords: biological hydrogen production, steady state modeling, dynamic modeling, IBRCS

Introduction

Biological hydrogen production from renewable sources (biomass, water, and organic wastes) [1] has the potential to meet the growing demand for energy. It offers a feasible means for sustainable supply of H_2 with low pollution and high efficiency, thereby considered a promising eco-friendly energy source [2]. Comparing the production rates of H_2 by various biohydrogen systems and the

* Corresponding author Tel.: +1 519-630-8683; fax: +1 519-661-3942
E-mail address: hhafez@uwo.ca

associated operational complexity, suggest that dark-fermentation systems offer an excellent potential for practical applications. Although fermentative biohydrogen technologies are still in their infancy, the sustained efforts for developing more efficient technologies will eventually lead to economically feasible technologies. An important step to achieve these feasible solutions includes the development of a process model in order to perform process sensitivity analysis to facilitate the scale-up the process.

Hafez et al. [3] developed a novel integrated biohydrogen reactor clarifier system (IBRCS) [4] (see Figure 1). The system is comprised of a continuously stirred reactor (CSTR) for biological hydrogen production, followed by an uncovered gravity settler for decoupling of solids retention time (SRT) from hydraulic retention time (HRT). The system was able to maintain a maximum hydrogen yield and production rate of 3.1 mol H_2/mol glucose and 191 L/d, respectively. The performance of the IBRCS underscores its potential as an efficient technology that is economically feasible for biological hydrogen production from organic wastes. However, a process model is necessary for better understanding the process and optimization of the system parameters.

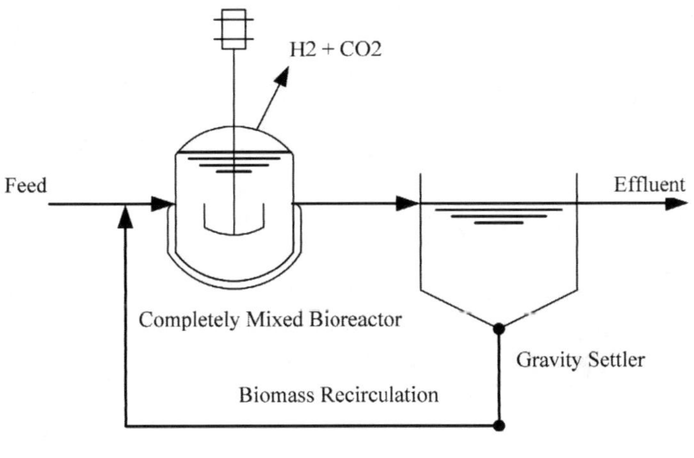

IBRCS [4]

Figure 1. Experimental Setup for the integrated biohydrogen reactor clarifier system

Most studies on biohydrogen production modeling were performed considering batch systems using the Gompertz and Monod equations [5, 6]. The Gompertz equation is an empirical formula where three model parameters (i.e. lag time, H_2 production potential, and H_2 production rate) are adjusted to fit the Gompertz equation to experimental hydrogen evolution data. Even though this curve fitting approach yields high correlation coefficients between the observed and fitted hydrogen evolution data, the three model parameters determined by curve-fitting are restricted to specific experimental conditions and cannot be used in a predictive mode [5]. Due to its empirical nature, the utility of the Gompertz equation is severely limited as it cannot account for relevant process variables such as substrate concentrations, temperature, pH, substrate-types, etc. In some studies, the Gompertz equation has been modified to accommodate typical kinetics of substrate degradation, biomass growth, and hydrogen production [7, 8, 9]. Some biohydrogen studies have utilized the conventional kinetic expressions such as Monod's equation [6, 7, 8, 10, 11]. However, rigorous and multiple simulations, followed by a series of validations may be required to establish the generality of such equations and the associated parameters. Furthermore, comprehensiveness of such models is accomplished only when they can be readily integrated with other complex bioprocesses i.e. hydrolysis, acidogenesis, and H_2 production from complex and particulate organic substrates [5].

A comprehensive literature review demonstrated that no models are readily available that can accurately predict continuous biohydrogen production. On the other hand, most of the available anaerobic digestion models were derived for anaerobic digestion of municipal wastewater biosolids, which employ CSTR and are not capable of modeling a system that decouples the SRT from the HRT (a key feature of the IBRCS). One such model is incorporated in the software BioWin (EnviroSim Associates LTD., Flamborough, Ontario, Canada), which is widely used for modeling wastewater treatment plants. However, it was never used for modeling biohydrogen production systems. Thus, the objectives of this paper are: exploring the use of BioWin model for simulation of biohydrogen production in IBRCS using data from an experimental study [3], evaluating the beneficial effects of decoupling SRT from HRT by comparing the performance of the IBRCS with the conventional CSTR

Materials and Methods

This section provides a brief description of the experimental setup, procedures and operational used in collecting the experimental data for the IBRCS that is later used in the numerical models. Additional information can be found in Hafez et al. [3].

Systems set up and operations

Two lab-scale IBRCSs were considered in the study, each comprising of a continuously stirred reactor (CSTR) for biological hydrogen production (5 L working volume), followed by an uncovered gravity settler (volume 8 L) i.e. open to the atmosphere for the decoupling of solids retention time (SRT) from the hydraulic retention time (HRT). Both systems were operated at 37° C for 220 days (Figure 1), at six different organic loading rates (OLR) ranging from 6.5 to 206 gCOD/L-d. Details of the operational conditions for the six runs are listed in Table 1. It is noteworthy that the systems were run at steady state conditions for at least 20 turnovers of the mean SRT, with the shortest run lasting for 45 days and the longest run for 75 days, excluding the first week of start up. In order to enrich hydrogen producing bacteria, the seed sludges were heat treated at 70° C for 30 minutes prior to startup. Following the completion of each run and the attainment of steady-state conditions, the systems were cleaned and inoculated with pre-treated sludges. OLR-1 and 2 were run simultaneously, followed by OLR-3 and 4, and lastly OLR-5 and 6. The systems were monitored for total chemical oxygen demand (TCOD), soluble COD, volatile fatty acids (VFA), ethanol, lactate, glucose, volatile suspended solids (VSS), total suspended solids (TSS) and biogas composition including hydrogen, methane and nitrogen. The quantity of produced biogas was recorded daily using a wet tip gas meter (Rebel Wet-tip Gas Meter Company, Nashville, TN, USA). Details of analytical methods are reported elsewhere [3].

Table 1. Operational conditions

	Glucose (g/L)	HRT (h)	SRT (h)	OLR (gCOD/L-d)	Final pH
OLR -1	2	8	50 ± 5	6.5	5.5
OLR -2	8	8	45 ± 4	25.7	5.5
OLR -3	16	8	45 ± 6	51.4	5.5
OLR -4	32	8	42 ± 6	103	5.5
OLR -5	48	8	27 ± 3	154	5.5
OLR -6	64	8	26 ± 2	206	5.5

Note. Values represent average ± standard deviation

Inocula and media compositions

Anaerobically-digested sludge from the St.Marys wastewater treatment plant (St.Marys, Ontario, Canada) was used as the seed. The two systems operated in parallel at the same time under two different OLRs for a total of six OLRs (three consecutive runs). The systems were seeded with 5 liters of sludge and started up in a continuous mode with the feed containing glucose at different concentrations as presented in Table 1. The same startup technique was repeated for the three runs. It must be emphasized that there was no sludge wastage from the clarifier throughout the operation, and the values of SRTs presented in Table 1 represent the average ± standard deviation (SD) during steady state operation. It is noteworthy that the reactors operation was consistent over time and accordingly, the average SRT with SD of less than 10% of the mean SRT is representative of the overall SRT during the run.

As expected, the clarifier effluent VSS concentrations were substantially lower than the reactor VSS concentrations and remained unchanged during steady-state operation. The feed contained sufficient inorganics (mg/L): $NaHCO_3$, 2000-16000; $CaCl_2$, 140; $MgCl_2.6H_2O$, 160; NH_4HCO_3, 600; $MgSO_4.7H_2O$, 160; urea, 500-2000; Na_2CO_3, 124-300; $KHCO_3$, 156; K_2HPO_4, 15-20; trace mineral solution, 500; H_3PO_4, 250-1500. Despite the lack of real-time pH control, buffering capacity was sufficient to maintain a final pH of 5.5 in all runs.

Model Formulation

The anaerobic degradation processes in the BioWin model are based on the "four population" model concept (heterotrophs, acetogens, acetoclastic methanogenesis and hydrogenotrophic methanogenesis). To simulate the inhibition of methanogens due to heat pre-treatment of the seed sludge, the methanogens growth was switched off in all modeling runs. A conceptual schematic of the biohydrogen production model is shown in Figure 2. The following points pertaining to the fate of organics in the system describe the key stages:

- Influent biomass undergoes anaerobic decay in the digester. The process rates and stoichiometry are the same as for anaerobic decay in the activated sludge process. The products of decay include unbiodegradable organic, nitrogen, and phosphorus components.

- Hydrolysis of particulate matter is mediated by the (non-polyP) heterotrophs. Particulate matter may be present in the influent, or may consist of products from biomass decay. The products of hydrolysis are phosphate (PO_4-P), soluble organic nitrogen, and readily biodegradable COD.
- Non-polyP heterotrophs ferment the complex readily biodegradable COD to acetic acid, propionic acid hydrogen, and carbon dioxide. There are two model processes for this reaction step. One is for low dissolved hydrogen concentrations while the other is for high dissolved hydrogen concentrations. The stoichiometry of each of these processes is to be calibrated to achieve the appropriate product mix. The calibration can be done by trial and error to achieve the best match between modeled and measured data.
- Propionic acid is converted to acetic acid by acetogenic bacteria. This process also produces hydrogen and is switched off at high levels of propionate. The propionate inhibition constant is 10000 mgCOD/L.

The influent characteristics of the synthetic wastewater containing glucose used in the experimental study was simulated in the model using the influent specifier associated with BioWin model.

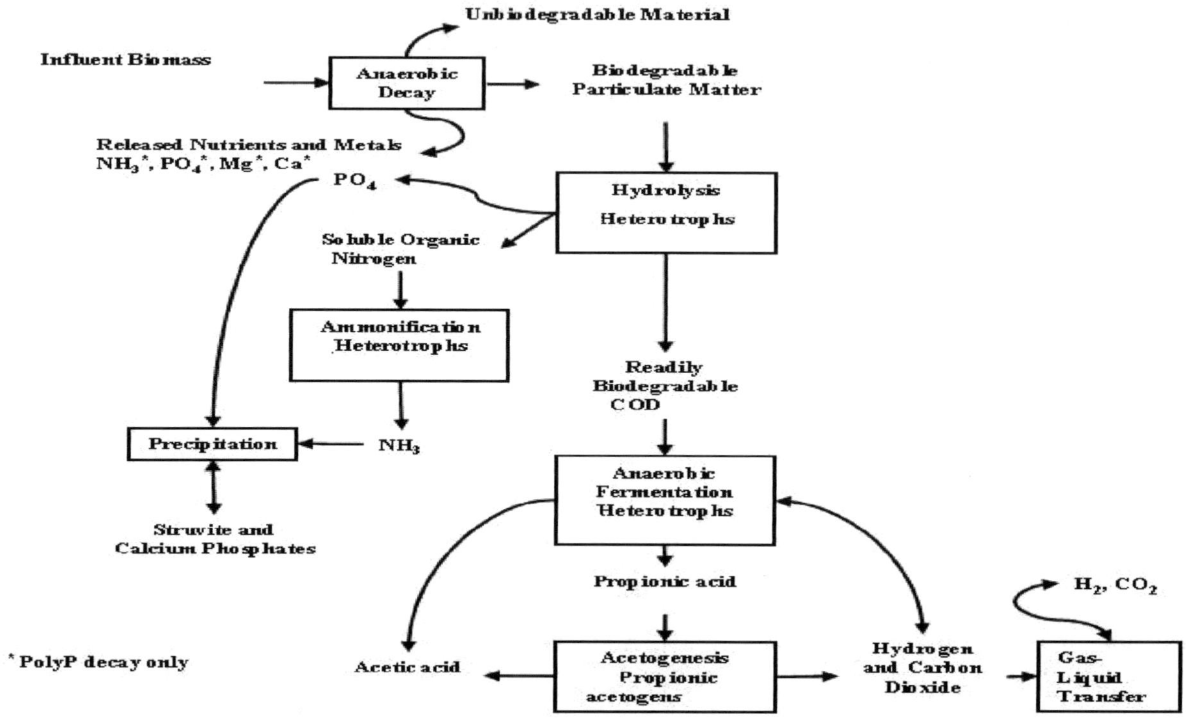

Figure 2. Conceptual schematic for the anaerobic degradation model in BioWin (Adapted from BioWin manual).

RESULTS AND DISCUSSION

Steady State Modeling

Table 2 summarizes the main experimental results under six different organic loading rates, while complete experimental results are reported elsewhere [3]. The model was first calibrated using two sets of experimental data, namely OLR-1 and OLR-2. The calibration involved 180 runs where in tweaking the stoichiometry of the process by trial and error as mentioned earlier was completed to achieve the best fit of the experimental data. It was then validated using the remaining four sets of experimental data (i.e. OLR-3 to OLR-6). Figures 3, 4 and 5 show the correlation between various predicted and measured parameters for the 6 steady state OLRs. The main parameters considered in the comparison were the biomass concentration in the hydrogen reactor measured as volatile suspended solids ($VSS_{reactor}$), biomass concentration in the clarifier supernatant (VSS_{eff}), solids retention time (SRT), main soluble metabolites (i.e. acetate, butyrate vs. propionate), volatile fatty acids (VFAs), soluble chemical oxygen demand in the effluent ($SCOD_{eff}$), hydrogen production rate, hydrogen production yield, and hydrogen content in the biogas.

Table 4. Summary of experimental results

Measured parameters	OLR -1	OLR -2	OLR -3	OLR -4	OLR -5	OLR -6
VSS reactor (mg/L)	1489 ± 116	4190 ± 308	8915 ± 972	15703 ± 926	18472 ± 1404	17038 ± 883
VSS out (mg/L)	247 ± 46	744 ± 50	1578 ± 141	3073 ± 397	5565 ± 581	5240 ± 372
SCOD out (mg/L)	1492 ± 79	6023 ± 194	11922 ± 1230	21267 ± 1627	40960 ± 1624	59091 ± 1358
VFAs (mgCOD/L)	1491 ± 87	5924 ± 257	10344 ± 1114	17976 ± 1444	19232 ± 1156	20582 ± 1686
Hydrogen Gas (%)	71 ± 0.9	73 ± 2.7	65 ± 3.3	67 ± 2.7	43 ± 2.8	39 ± 1.5
Hydrogen Gas (L/d)	12 ± 1.3	48.1 ± 4.7	97 ± 5	179 ± 12	67 ± 5	60 ± 3.4
Yield (mol/mol)	2.8 ± 0.3	2.8 ± 0.3	2.9 ± 0.1	2.8 ± 0.3	1.2 ± 0.1	1.1 ± 0.1

Note. Values represent average ± standard deviation

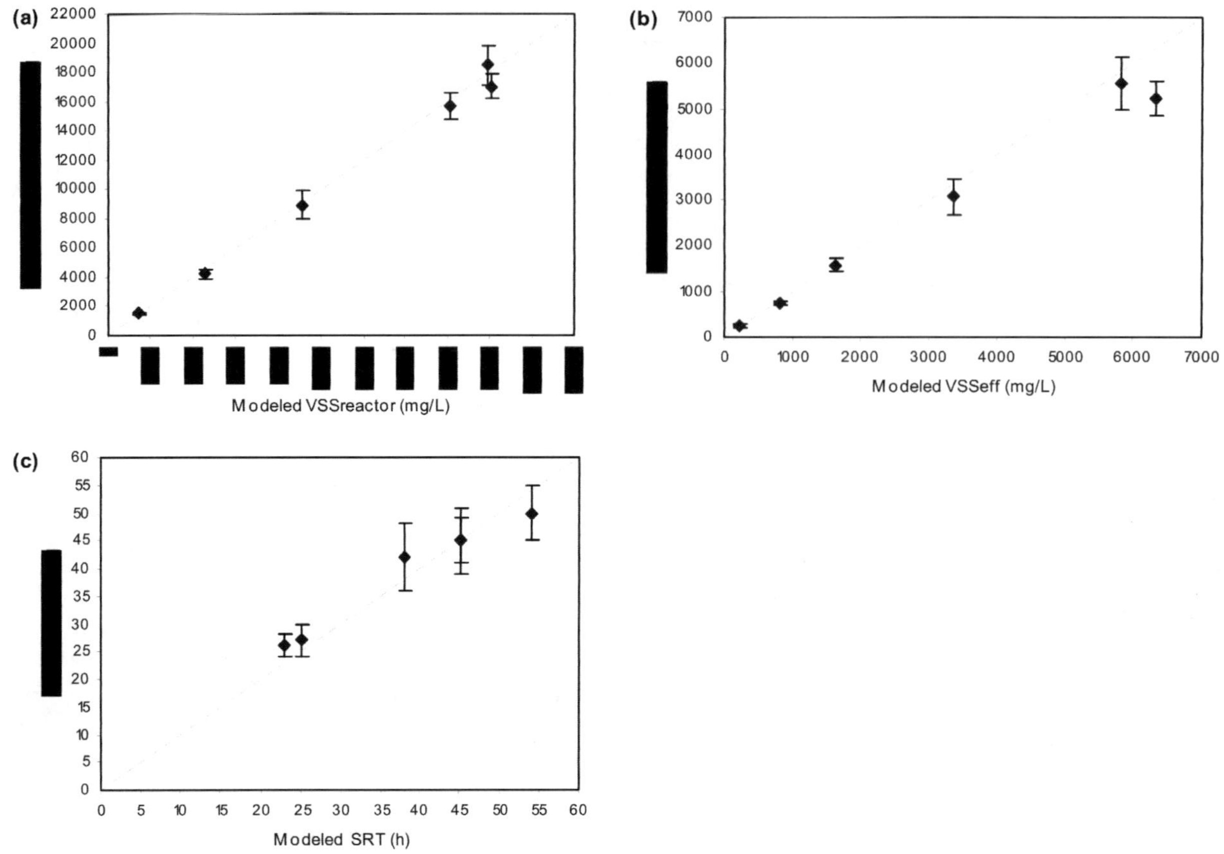

Figure 3. Correlation between modeled and measured parameters for: a) Reactor VSS, b) Effluent VSS, c) SRT (Note: error bars represent standard deviations for experimental results)

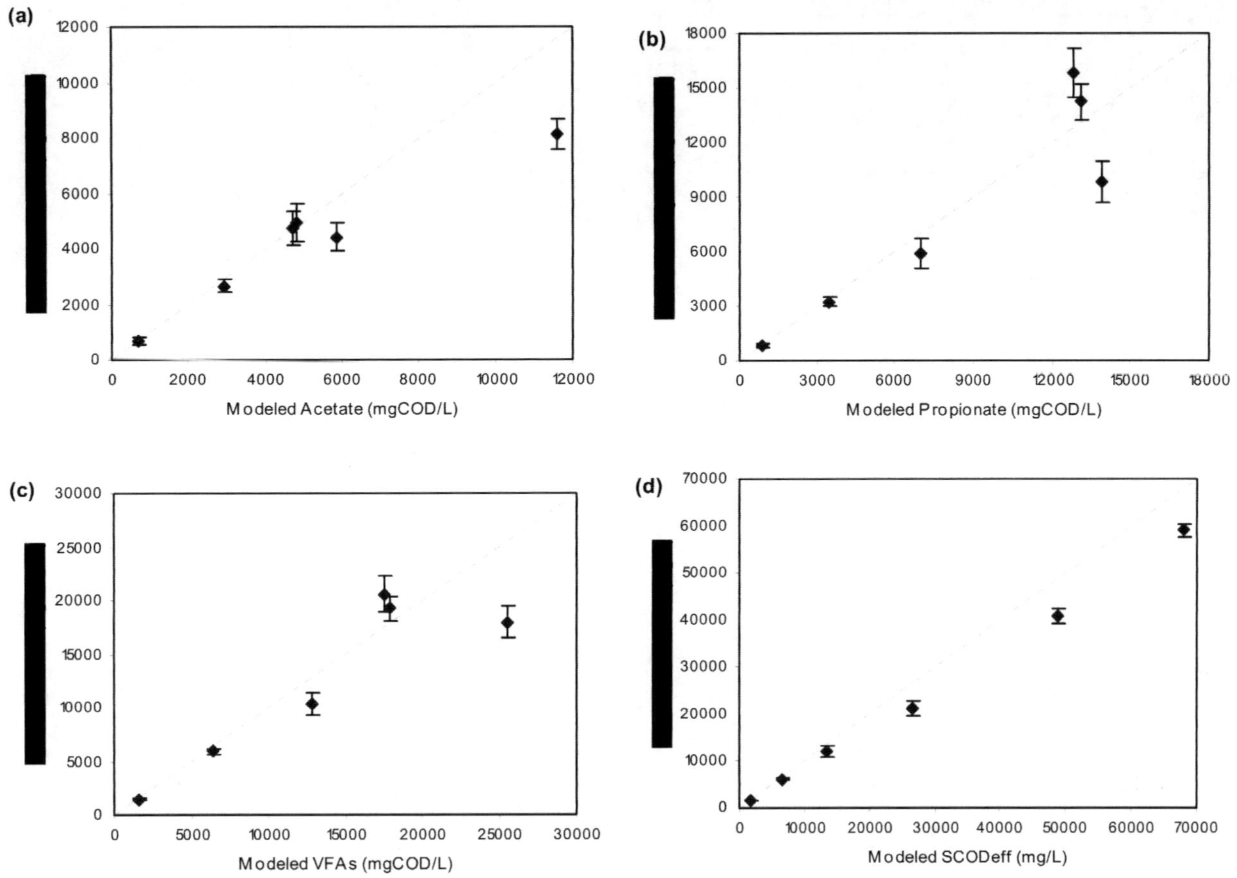

Figure 4. Correlation between modeled and measured parameters for: a) Acetate, b) Propionate vs. Butyrate, c) VFAs, d) Effluent SCOD (Note: error bars represent standard deviations for experimental results)

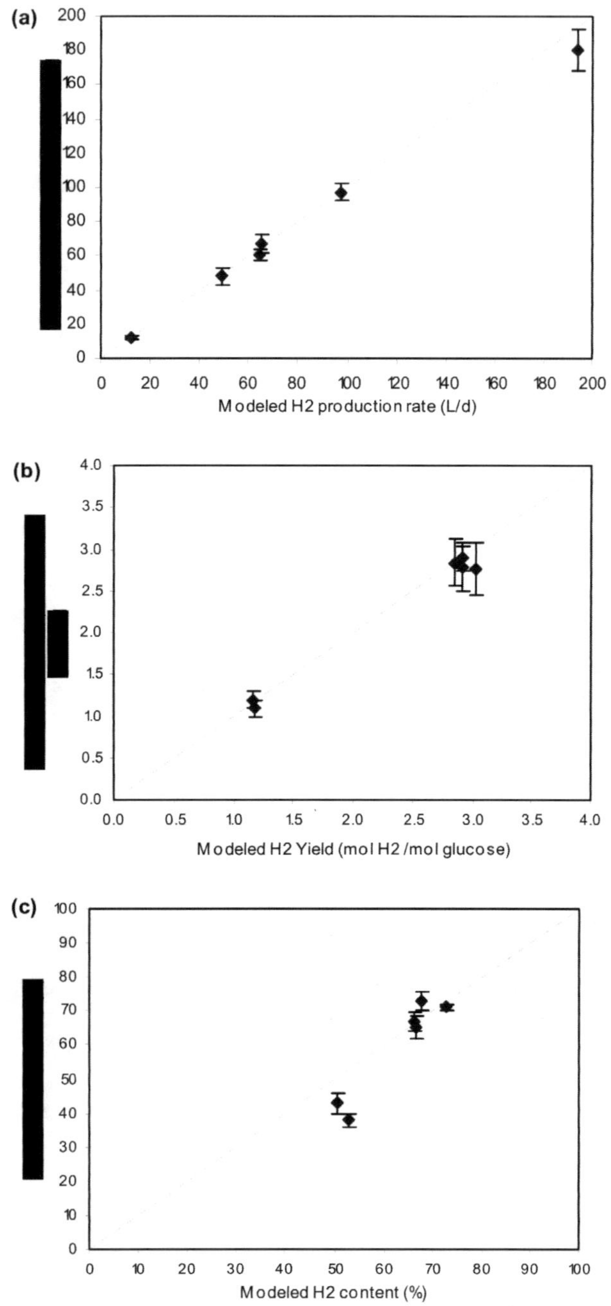

Figure 5. Correlation between modeled and measured parameters for: a) H_2 production rate, b) H_2 yield, c) H_2 content (Note: error bars represent standard deviations for experimental results)

Biomass concentration in hydrogen reactors is a key parameter for the stability and hydrogen production through the process as it directly affects the food-to-microorganisms (F/M) ratio [3]. In the literature, biohydrogen system failures were frequently attributed to marked decrease in biomass content in the hydrogen reactor due to severe cell washout [12, 13]. Figures (3a and 3b) show the correlation between predicted and measured biomass concentrations in both the hydrogen reactor and clarifier effluent for the six experimental runs. As apparent from Figure 3a, the model accurately predicted reactor biomass concentrations throughout the observed range of 1400 to 18000 mgVSS/L at OLRs in the range of 6.5 to 206 gCOD/L-d, with an average percentage error (APE) of 4.6%. The APE is defined as the summation of the absolute difference between the experimental and predicted values divided by the experimental values, averaged over the number of data points. Similarly, the model predictions closely matched the observed biomass concentrations in the clarifier supernatant increase in the range of 200 mgVSS/L to around 6300 mgVSS/L over the same aforementioned range of OLR with an APE of 10% (see Figure 3b). The successful model prediction of the biomass concentrations in both the hydrogen reactor and clarifier effluent was a result of the accurate modeling of the clarifier. As depicted in Figure 3c, the model predicted the solids retention time well for the six experimental runs with an APE of 6.5%.

During the conversion of glucose to hydrogen a mixture of soluble metabolites (i.e. volatile fatty acids) are produced. The two main pathways for glucose conversion are the acetate and butyrate pathways can be given by:

$$C_6H_{12}O_6 + 2H_2O \rightarrow 2CH_3COOH + 2CO_2 + 4H_2 \quad (1)$$

$$C_6H_{12}O_6 \rightarrow CH_3CH_2CH_2COOH + 2CO_2 + 2H_2 \quad (2)$$

Considering Eqs. 1 and 2, the stoichiometric hydrogen yield would be 4 and 2 mol H_2 / mol glucose, respectively. Although BioWin uses an alternative pathway to the butyrate pathway (i.e. glucose → propionate → acetate + hydrogen) (see Figure 2), the predicted and measured acetate and butyrate concentrations agreed well for all experimental runs as can be noted from Figures (4a and 4b), with APEs of 12% and 14.8%, respectively. These results were confirmed by the correlations shown in Figures (4c and 4d) for the total VFAs and the final SCOD with APE of 14% and 13%, respectively.

As apparent from Figure 5, the model accurately predicted both hydrogen production rate and hydrogen yield with a relatively low APE of 4%, while an error of less than 10% was observed when predicting the hydrogen content in the biogas. It is noteworthy that the reactor final pH from both experimental runs and the model were 5.5. The model confirmed the linear increase in hydrogen production rate with the increase of the OLR up to 103 gCOD/L-d. On the other hand, the model predicted an almost constant hydrogen yield of 2.9-3.0 mol H_2/ mol glucose during the same range of OLRs. The steady state model solutions for OLR-5 and OLR-6 have revealed a decrease in hydrogen production rate and hydrogen yield to approximately 65 L/d and 1.2 mol H_2/ mol glucose, respectively.

The high hydrogen yields achieved in the experimental study and corroborated by the model predictions were confirmed by the high biomass concentrations maintained in the hydrogen reactor as well as the operational F/M ratios that fell in the recommended range of 4.4 to 6.4 gCOD/gVSS-d reported in the literature. The authors' earlier work [3] using the experimental data reported here as well as additional data obtained from two CSTRs operated at OLRs of 25.7 and 42.8 gCOD/L-d and HRTs of 8 h and 12 h [13], demonstrated an inverse relationship between the biomass and hydrogen yields (i.e. low hydrogen yields were associated with high biomass yields). On the other hand, the stoichiometric yield of 4 and 2 mol H_2 / mol glucose from Eqs. (1 and 2) are calculated neglecting any biomass yield. Thus, considering the biomass yield would eventually decrease due to energy (COD) utilization for biomass synthesis. Considering COD mass balances, Eqs. 1 and 2 can be modified to calculate the practical ranges of hydrogen yields from glucose conversion to a mixture of acetate and butyrate (neglecting all other factors that might hinder the efficiency), i.e.

$$H_2 \text{ Yield (mol } H_2 \text{ / mol glucose)} = 4\ (1 - Y_{biomass}) \quad (3)$$

$$H_2 \text{ Yield (mol } H_2 \text{ / mol glucose)} = 2\ (1 - Y_{biomass}) \quad (4)$$

where $Y_{biomass}$ is the biomass observed yield in gCOD/gCOD.

Assuming a typical biomass yield of 0.15 $gCOD/gCOD_{converted}$ reported for hydrogen producers [25] and substituting in Eqs. 3 and 4, the practical hydrogen yield should be in the range of 1.7 – 3.4 mol H_2 / mol glucose, while using the widely reported biomass yield of 0.4 $gCOD/gCOD_{converted}$ in CSTRs [3, 13], the practical hydrogen yield ranges from 1.2 to 2.4 mol H_2 / mol glucose. Equations 3 and 4 clearly justify that in fact an inverse relationship between the biomass and hydrogen yields does exist. Furthermore, based on the values of biomass and hydrogen yields from both experimental and predicted results, the IBRCS achieved a maximum efficiency of 88% (a hydrogen yield of 3.0 mol H_2 / mol glucose compared to a maximum practical H_2 yield of 3.4 mol H_2 / mol glucose).

Evaluation of Decoupling of SRT from HRT

After successful model calibration and verification, the clarifier was removed from the IBRCS to assess the performance of CSTR in comparison to the IBRCS at 4 different glucose concentrations (2 g/L, 8 g/L, 16 g/L and 32 g/L) corresponding to the optimal OLRs 1-4 (6.5 to 103 gCOD/L-d). The performance of the CSTR was strongly affected due to biomass washout. Figure 6 compares the IBRCS and CSTR in terms of reactor VSS, hydrogen yield, hydrogen production rate, and total VFAs. The steady state bioreactor VSS concentration in the CSTR dropped drastically to 120, 500 , 1240 and 1970 mgVSS/L for the OLRs, as compared to 1400, 4550, 9000 and 16100 mgVSS/L in the IBRCS. The extremely high F/M ratios of greater than 40 in the CSTR resulted in a low hydrogen yields ranging from 0.6-1.2 mol H_2 / mol glucose. The maximum hydrogen production rate in the CSTR was approximately 53 L/d compared to a maximum of 194 L/d in the IBRCS and they were both achieved at an OLR of 103 gCOD/L-d. The total volatile fatty acids produced in the IBRCS at the 4 different OLRs were at least 3 folds higher than those produced in the CSTR. Thus, the BioWin model emphatically established the superior performance of the novel patent-pending IBRCS relative to the CSTR.

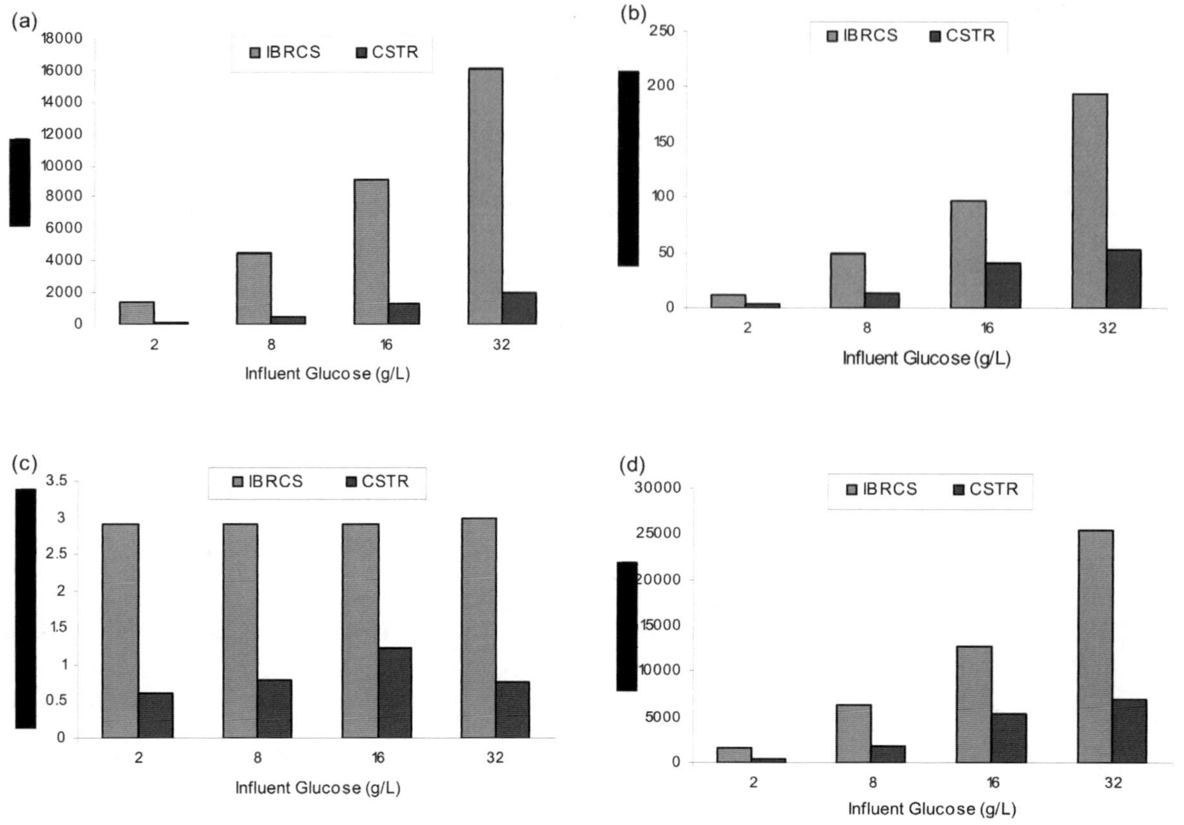

Figure 6. Comparison between the IBRCS and CSTR for: a) Reactor VSS, b) H_2 production rate, c) H_2 yield, d) VFAs

Impact of HRT reduction

To study the impact of HRT reduction on the performance of the IBRCS, three steady state simulation runs at HRT of 4 h, 2 h and 1 h utilizing 32 g/L of glucose were performed in comparison with the steady state data from OLR-4 at 8 h HRT. As depicted in Figure 7, hydrogen production rate decreased by 3%, 13% and 28% with the decrease in HRT from 8 h to 4 h, 2 h and 1 h, respectively. Even though hydrogen production rates seem to be slightly affected by the reduction in HRT, the hydrogen yields were drastically impacted. The hydrogen yield decreased drastically from 3 mol H_2 / mol glucose at an HRT of 8 h to 0.3 mol H_2 / mol glucose at HRT of 1 h. It is noteworthy that the F/M ratio increased from 6.4 gCOD/gVSS-d at HRT of 8 h to 12.8, 26 and 55 gCOD/gVSS-d at HRT of 4 h, 2 h and 1 h, respectively. The increase in F/M ratio was a result of a decrease in the solids retention time due to the increase of solids loading rate to the clarifier which was not sized up accordingly. Thus, it is postulated that sustained higher reactor biomass concentrations with proper clarification would have inevitably improved both hydrogen production rate and hydrogen yield.

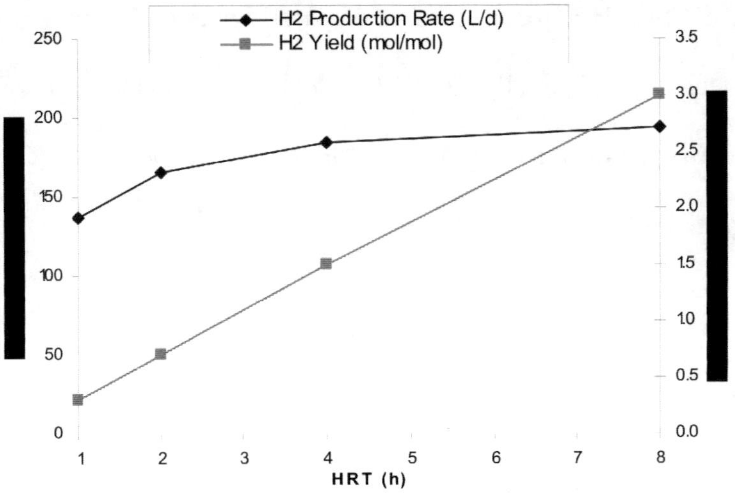

Figure 7. Relationship between hydrogen production rate and hydrogen yield vs. HRT

Summary and Conclusions

Steady state of biohydrogen production in the IBRCS using BioWin was evaluated in this study. The following conclusions can be drawn:

- The model accurately predicted biomass concentrations in both the bioreactor and the clarifier supernatant with average percentage errors (APE) of 4.6% and 10%, respectively. Hydrogen production rates and hydrogen yields predicted by the model were in close agreement with the observed experimental results as reflected by an APE of less than 4%, while the hydrogen content was well correlated with an APE 10%.
- Despite using a different pathway for biohydrogen production, the predicted and measured acetate and butyrate concentrations for the six experimental runs agreed well with APEs of 12% and 14.8%, respectively. These results were validated by the accurate prediction of total volatile fatty acids in the reactor with an APE of 14%.
- The calibrated model confirmed the advantages of decoupling of the solids retention time (SRT) from the hydraulic retention time (HRT) in biohydrogen production, with the average hydrogen yield decreasing from 3.0 mol H_2 / mol glucose to 0.8 mol H_2 / mol glucose upon elimination of the clarifier.

- Steady state simulation for the IBRCS at HRT of 4 h, 2 h and 1 h utilizing 32 g/L of glucose showed that the hydrogen yield decreased drastically from 3.0 mol H_2 / mol glucose at HRT of 8 h to 1.5, 0.7 and 0.3 mol H_2 / mol glucose for HRT of 4 h, 2 h and 1 h, respectively, due to washout as a result of clarifier limitation.

References

1. Cheong, D.Y., and C.L. Hansen. Bacterial stress enrichment enhances anaerobic hydrogen production in cattle manure sludge. Appl Microbiol Biotechnol. 2006; 72: 635–643.
2. Wu, K. J., and J. S. Chang. Batch and continuous fermentative production of hydrogen with anaerobic sludge entrapped in a composite polymeric matrix. Proc Biochem. 2007; 42:279–284.
3. Hafez, H., G. Nakhla, H. El Naggar, E. Elbeshbishy, and B. Baghchehsaraee.. Effect of Organic Loading on a Novel Hydrogen Bioreactor. Int J Hydrogen Energy. 2010; 35: 81-92.
4. Hafez H, Nakhla G, El Naggar H. Integrated system for hydrogen and methane production during landfill leachate treatment. Pending patent, US Patent Application US 61/202,137, 2009.
5. Gadhamshetty V, Arudchelvam Y, Nirmalakhandan N, Johnson DC. Modeling dark fermentation for biohydrogen production: ADM1-based model vs. Gompertz model. Int J of Hydrogen Energy. 2010; 35:479-490.
6. Jianlong W, Wei W. Kinetic models for fermentative hydrogen production: a review. Int J Hydrogen Energy. 2009; 34(8):3313–23.
7. Chen X, Sun Y, Xiu Z, Li X, Zhang D.. Stoichiometric analysis of biological hydrogen production by fermentative bacteria. Int J Hydrogen Energy. 2006; 31(4): 539–549.
8. Chen WH, Chen SY, Kumar KS, Sung S. Kinetic study of biological hydrogen production by anaerobic fermentation. Int. J. Hydrogen Energy. 2006; 31(15):2170–2178.
9. Mu Y, Yu HQ, Wang G. A kinetic approach to anaerobic hydrogen-producing process. Water Res. 2007; 41 (5): 1152–1160.

10. Zheng XJ, Yu HQ. Inhibitory effects of butyrate on biological hydrogen production with mixed anaerobic cultures. J Environ Manage. 2005; 74 (1): 65–70.

11. Lin PY, Whang LM, Wu YR, Ren WJ, Hsiao CJ, Li SL, Chang JS. Biological hydrogen production of the genus Clostridium: metabolic study and mathematical model simulation. Int J Hydrogen Energy. 2007; 32 (12): 1728–1735.

12. Whang LM, Hsiao CJ, Cheng SS. A dual-substrate steady-state model for biological hydrogen production in an anaerobic hydrogen fermentation process. Biotechnol Bioeng. 2006; 95 (3): 492–500.

13. Hafez H, Baghchehsaraee B, Nakhla G, Karamanev D, Margaritis A, El Naggar M.H. Comparative Assessment of Decoupling of Biomass and Hydraulic Retention Times in Hydrogen Production Bioreactors. Int J Hydrogen Energy. 2009;34:7603-7611.

(427e) Nanocomposite Ni-ZrO_2 Catalysts for Ethanol Steam Reforming: Effect of Support Size On the Catalytic Behavior

Xinbin Ma, Shuirong Li, Maoshuai Li, Shengping Wang, **Jun Wang** and **Yongli Sun**, Key Laboratory for Green Chemical Technology of Ministry of Education, School of Chemical Engineering & Technology, Tianjin University, Tianjin, China

With heightened concerns for the clean air and clean environment, hydrogen energy derived from renewable sources in conjunction with fuel cell has recently been under investigation actively. Producing hydrogen from steam reforming of ethanol is considered as a preferred method. Ni-based catalyst is highly recommended in this process while the oxide supports play an important role on its selectivity and stability. Researches in the nanomaterials recently have revealed that catalytic nanoarchitecture with comparably sized metal and oxide domains can retain high activity because of an increase the interfacial contact area and hindering agglomeration of the active metal. It opens a new route in the development of highly active and stable catalysts for the steam reforming process. Here we present primary study on the impacts of zirconia particle size on the catalytic behavior of nanocomposite Ni-ZrO_2 catalysts in the steam reforming of ethanol. Three nanocomposite Ni-ZrO_2 catalysts with particle sizes of zirconia concentrated at 50 nm, 25 nm and 15 nm were carefully prepared by precipitation followed by impregnation method and nominated here as Ni-ZrO_2-50, Ni-ZrO_2-25 and Ni-ZrO_2-15 respectively. The average nickel particle sizes in the first two catalysts were about 62 nm and 23 nm after reduction. To better understand the size effect of zirconia support, the Ni-ZrO_2-15 catalyst underwent specially treated to enlarge the nickel particles to 80 nm in the reduced samples. The impact of weight flow rate on the catalytic activity was investigated at 400 ≡C, ethanol/H_2O = 1:3 for all the three catalysts. The catalysts showed complete conversion of ethanol at W/F higher than 2.0×10-4 (gcat h)/ml but dropped in an order of Ni-ZrO_2-50, Ni-ZrO_2-15 and Ni-ZrO_2-25 catalysts in the reverse direction, indicating different activities of the catalysts. Analysis of the selectivity towards different carbonous compounds in the reactions operating at condition range that incomplete conversion of ethanol was achieved reveals clue in the reaction pathway of different catalysts, in which a main route of ethanol undergoing dehydrogenation to produce acetaldehyde followed by decomposition of acetaldehyde into CO and CH_4 was found for all the three catalysts. Besides, trace of acetone and ethylene were also detected in the product stream of the Ni-ZrO_2-50 and Ni-ZrO_2-15 catalysts while little was found for the Ni-ZrO_2-25 catalyst under the reaction condition range studied here. In ethanol steam reforming, acetaldehyde could also be oxidized into acetic acid and further transform into acetone by ketonization reaction, which could lead to unfavorable coke deposition on the surface of the support. Zirconia is reported to have high surface oxygen mobility especially in tetragonal phase when comparing with monoclinic phase. In the Ni-ZrO_2-15 catalyst with a reduced zirconia particle size under the size limitation, predominant tetragonal phase along with trace of monoclinic phase of zirconia were detected in the XRD analysis while only a single monoclinic phase of zirconia was detected in the reduced Ni-ZrO_2-50 and Ni-ZrO_2-25 catalysts. The increase of oxidation activity due to the reduction of zirconia particle size has led to a higher selectivity towards acetone formation in the ethanol steam reforming for Ni-ZrO_2-15 catalyst. In spite of the improvement in oxidation activity of the catalyst, the reduction in particle size of zirconia would not have significant positive impact on the catalytic activity unless the particle size of the nickel metal is also reduced into a limited range. Although the zirconia and the nickel particles are in a comparable range, 50 nm seems to be too large for the Ni-ZrO_2-50 catalyst to perform a good activity in the steam reforming of ethanol. Reducing the particle of zirconia while keeping the nickel metal particle size in a range of larger than 50 nm, as in the situation of the Ni-ZrO_2-15 catalyst, the improvement in the catalytic activity could only be in a limited range since insufficient C-C bond cleavage activity is supplied by the nickel metal. When both of the particle sizes of the zirconia and the nickel metal are reduced into a suitable range as in the situation of the Ni-ZrO_2-25 catalyst, the increase of metal support interaction as revealed in the H_2-TPR profiles lead to a significant improvement in both the activity and selectivity of the catalyst.

See more of this Session: Renewable Hydrogen Production I
See more of this Group/Topical: Topical 8: Hydrogen Production and Storage

(427f) Hydrogen Production From Ammonia Using a Micro-Reformer System Integrated with a Heat-Recirculating Micro-Combustor

Jung Ho Kim[1], Sang Ik Lee[1], Jae Moon Joo[1] and Oh-chae Kwon[2], (1)School of Mechanical Engineering, Sungkyunkwan University, Suwon, Gyeonggi-do, South Korea, (2)School of Mechanical Engineering, Sungkyunkwan University, Suwon, South Korea

In order to evaluate the potential of reforming ammonia as a carbon-free fuel in production of hydrogen, an annulus-type micro reforming system integrated with a micro-combustor is studied experimentally. The micro-combustor is a simple cylinder with an annular-type shield to apply for a heat-recirculation concept and an expanded exhaust outlet to control ignition, satisfying the primary requirements for designing a micro-combustor as a heat source, i.e., stable burning in a small confinement and uniform distribution of temperature along a wall. The micro-combustor is surrounded with a micro-reformer for converting ammonia gas to hydrogen. The annulus-type micro reforming system is designed to produce 1-10 W (based on lower heating value, LHV) of hydrogen using various catalysts (Ru, Ni/SiO2/Al2O3 or Ir), which satisfies the primary requirements for designing an effective micro reforming system, i.e., maximum heat transfer through the micro-combustor wall and enhanced hydrogen production. The feed rate of ammonia, the micro-combustor inlet velocity of fuel-air mixtures and the catalyst materials substantially affect the performance of the designed micro reforming system. Under optimized design and operating conditions, the micro reforming system using Ru produces 5.4 W (based on LHV) of hydrogen with a conversion rate of 98% and an overall system efficiency of 13.7%. Thus, the present configuration can be applied to practical micro reforming systems, supporting the potential of using ammonia as a clean fuel.

See more of this Session: Renewable Hydrogen Production I
See more of this Group/Topical: Topical 8: Hydrogen Production and Storage

(506a) Hybrid Sulfur Cycle Options for the Next Generation Nuclear Plant

Maximilian B. Gorensek, Computational Sciences Directorate, Savannah River National Laboratory, Aiken, SC and William A. Summers, Energy Security Directorate, Savannah River National Laboratory, Aiken, SC

The hybrid sulfur (HyS) cycle is one of the two high-temperature water-splitting processes under consideration for use with the Next Generation Nuclear Plant (NGNP) demonstration at Idaho National Laboratory. This presentation will describe the reference case HyS flowsheet developed for NGNP and discuss the synergies and challenges of integrating HyS with other potential NGNP applications.

See more of this Session: Advances in Thermochemical Hydrogen Production
See more of this Group/Topical: Topical 8: Hydrogen Production and Storage

Development of Non-Pt Electrocatalysts in HBr Proton Exchange Membrane Electrolyzer for Hydrogen Production

Rui Zhang, John W. Weidner

Center for Electrochemical Engineering, Department of Chemical Engineering, University of South Carolina, Columbia, SC 29208

1. Introduction

Hydrogen could be produced via indirect splitting of water at a lower temperature (800-1000°C) than in the direct thermal dissociation (>2500°C) [1-4], through the promising thermal-chemical reaction of modified hybrid calcium-bromide-based cycle. The whole cycle basically consists of a thermo-chemical reaction to produce O_2 and an electrolytic process to produce H_2. In the electrolytic process, Br_2 is evolved at the anode with the feed of anhydrous HBr to promote mass transfer, while hydrogen evolution reaction occurs at the cathode. The electrode reactions are [1]:

Anode: $$2HBr \rightarrow Br_2 + 2H^+ + 2e^- \qquad E^\circ = 0.58 \text{ V vs. SHE}$$

Cathode: $$2H^+ + 2e^- \rightarrow H_2 \qquad E^\circ = 0\text{V vs. SHE}$$

Pt, as a standard electrode, has been widely investigated for both reactions above [5-10]. However, scarcity and high cost of Pt would impede scaling-up and commercialization of this electrolyzer. What's more, strong adsorption of Br^- on the Pt would significantly decrease its activity towards HER at the cathode side [7-9], when HBr crossover takes place. Therefore, non-Pt catalysts need to be developed for both half reactions to mediate the cost and improve the electrode stability.

In this paper, RuO_2, Vulcan XC 72R carbon black and Niobium doped TiO_2 are prepared onto the anode side of the membrane electrode assembly and tested in the HBr electrolyzer. MoS_2 and IrO_x are prepared by a low temperature method, prepared into a thin film electrode, tested on the rotating disk electrode towards HER.

2. Experimental

HBr PEM Electrolyzer

The HBr PEM electrolyzer uses the same testing system as previous one [10], except the cell hardware. The cell hardware uses the aluminum backing plate of the 10 cm^2 cell purchased from Fuel Cell Technologies, Inc. The flowing fields on both sides were grooved on a composite of graphite and fluoro inert polymer (Diamond F100), with 0.031 inch width and 0.04 inch depth of the groove. A couple of Kynar plates were placed between the aluminum ending plates and flowing fields to prevent the ending plates from corroding by hydrogen bromide. Also copper current collectors are followed between the Kynar plates and backing plates.

MEAs (Membrane electrode assembly) with different electrode materials are applied and usual procedure for assembling the cell was followed. Data points of current and cell potential are measured and collected on HP power supply.

MEA Preparation

MEA (membrane electrode assembly) was prepared by the regular spray and hot-pressing method. The carbon cloth (totally hydrophilic) from ETEK was used as the gas diffusion layer. At the anode, 1.5 mg/cm^2 catalysts were sprayed directly onto the carbon cloth substrate. Nafion 212 (poly [perfluorosulfonic] acid membrane) from Dupont was used as the electrolyte. The nafion membrane was boiled in 1 N H_2SO_4 solution for 90 minutes and rinsed thoroughly with de-ionized water. The commercial 0.5mg/cm^2 Pt/C EDL (ELAT from ETEK) served as the cathode directly. Thus derived sandwich was placed in a Carver hot press (Model # 3851-0) for hot-pressing at 500 tons and 120°C for 2 minutes. Cool the MEA in the air and save it ready to use.

RDE Experiment

Synthesis of MoS_2 and IrO_2/C

MoS_2 is prepared by the solvent-heat method [11] by reacting sulfur (Alfa Aesar) and $Mo(CO)_6$ (Alfa Aesar, 98%) in p-xylene. In a typical synthesis, 30.6 mg sulfur and 126 mg $Mo(CO)_6$ powders are dissolved in 100 mL p-xylene in a three-neck round bottom flask. Nitrogen is purged to protect the solution from air. Then, this solution is heated to 140 °C and held for three hours. The black precipitate (MoS_2) is then separated from the solution via filtering and dried in the vacuum oven at 90 °C for about 4 hours. In the post-treatment experiment, the material is heated to 450 °C and 800 °C under nitrogen in a horizontal furnace.

IrO2 is prepared from EG Method [12]. Measure $RuCl_3$ or $IrCl_3$ into a crucible and incipient wetness with ethylene glycol (EG). Sonicate the mixture for 30 minutes to obtain a homogeneous mixture. Add carbon and stir with a spatula until fully moistened. Sonicate again for 30 minutes. Dry the mixture at 140°C for 12 hours in vacuum oven. Grind dried solids into a fine powder. Heat sample to calcinate and hold for one hour. Samples were calcinated at 300°C, 350°C, and 400°C.

Preparation of the film electrode

The film electrode was made from depositing proper catalyst ink onto the disk electrode from Pine.The catalyst ink is prepared by mixing 6 mg of solid catalysts with 3 mL Isopropyl Alcohol (IPA, >99%, Sigma-Aldrich) and 3 mL water (MilliQ water, $\rho > 18$ MΩcm). The ink was sonicated for 20 minutes and then pipetted onto the disk electrode to gain the loading of 20 μg solid catalysts. To make the catalyst film uniform, a drop of 5 μL catalyst ink at most was pipetted stepwise and 5 μL of binding solution was finally applied and dried after the desired catalyst loading was achieved. The binding solution was prepared by mixing 5% wt. Nafion in IPA solution (Ion-power, Inc) and additional IPA in the volume ratio of 1:20.

Catalyst characterizations

Electrochemical characterizations are performed by cyclic voltammetry (CV) and steady state polarization in a three-electrode setup from Pine Instrument in the aqueous solution of 1.0 M $HClO_4$ as the electrolyte. The reference electrode is Hg/Hg_2SO_4 (-0.68 V vs. NHE) and a

Pt wire was used as the counter electrode. Before each test, the electrolyte was de-aerated with purging N_2 for 10 minutes. Potential is swept at the rate of 50 mV/s under the control of the Princeton Applied Research 263A potentiostat controlled by Corrware (Scribner Associates Inc.). After holding at the open circuit for 20 minutes, the potential starts from the open circuit potential in the positive direction and swept back and forth in the potential range set. Pseudo-Steady state polarization was conducted by potentiodynamic method at 2 $mV \cdot s^{-1}$ at the same potential window. The current densities were all normalized to the geometric area of the working electrode.

Characterization of bulk structure is done on the powder X-ray Diffraction (XRD, Rigaku miniflex II) and the microscopy of the surface morphology is done on High Resolution Transimission Electron Microscopy (HRTEM, JEOL JEM 2100F)

3. Representative Results and Discussion

The primary performance of the vapor phase HBr PEM electroyzer is shown in figure 1. It is seen that HBr electrolytic cell has higher efficiency and productivity than HCl process. Gaseous phase feeding of HBr could improve mass transfer.

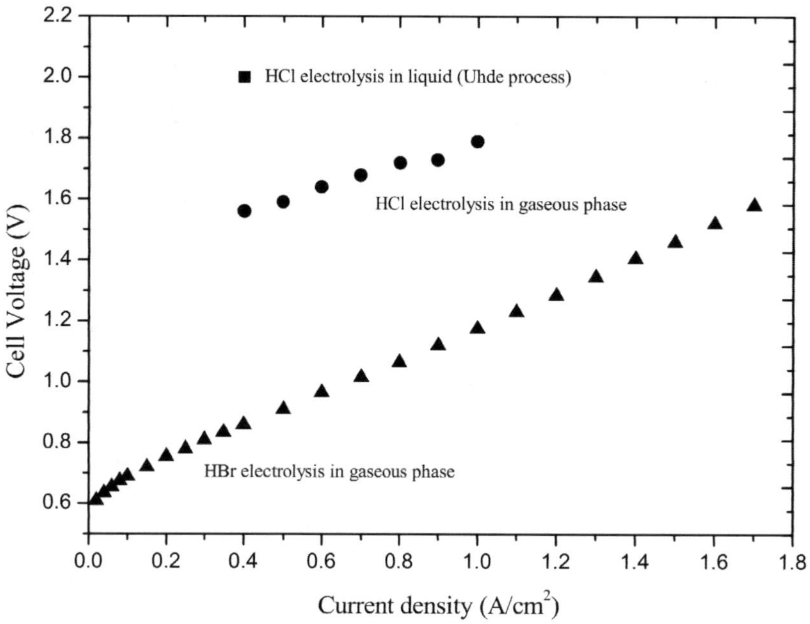

Figure 1: Comparison of the HBr process with HCl process with standard MEA.

Effects of the anode catalysts are in figure 2. It is seen that the non Pt catalyst of RuO_2, Vulcane XC 72R and TiO_2-Nb have comparable performance with the commercial MEA at the anode side.

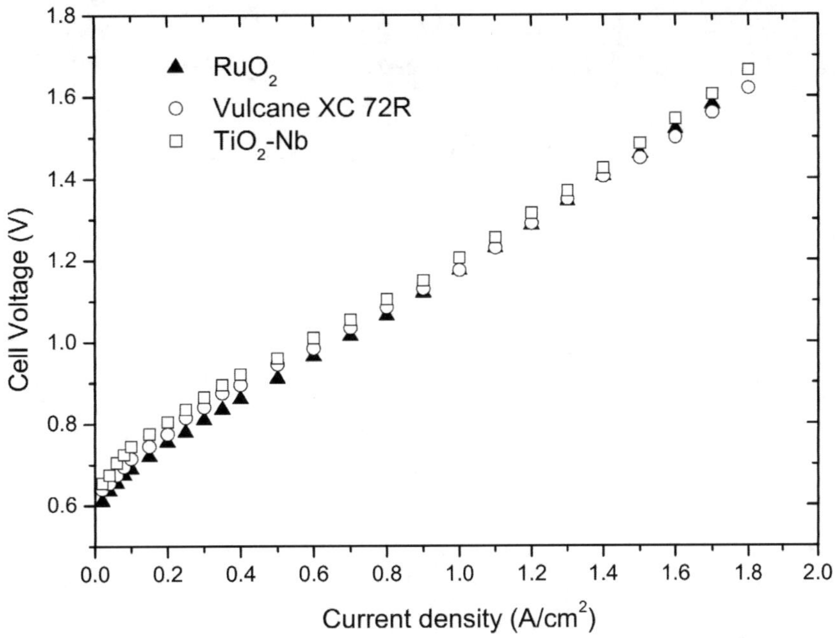

Figure 2: Effects of MEAs with different anode catalysts

From the previous results, it is seen that the overpotential for the HER catalyst takes a small part in the total cell voltage loss. Also, Pt is well known to be poisoned by the strong adsorption of Br^- and Br_2. Therefore, it is desired to develop non Pt electrocatalyst on the MEA. In this study, MoS_2, RuO_2, and IrO_2 are investigated, with respect to the synthesis procedures, screening the activity on the RDE and characterization. The results show that the low temperature synthesis of MoS_2 can significantly increase its activity. Also, the catalyst of IrO_2/C prepared from the EG method has a very high activity, which is even close to Pt. Characterization shows that the carbon support and the oxidation state of Ir probably play the key roles in this catalyst system.

4. Conclusion

The gas phase electrolysis has been successfully demonstrated to be used to produce hydrogen. HBr electrolysis is expected to have facile kinetics and thus the catalyst is not crucial, which enables us to use less expensive material at the anode side. Cathodic overpotential is relatively small than other voltage losses and thus we can use non Pt material to lower the cost and avoid Br^- and Br_2 poisoning.

Reference:

1. P. Sivasubramanian, R. P. Ramasamy, F. J. Freire, C. E. Holland and J. W. Weidner, *Int. J. Hydrogen Energy*, 32, 463 (2007).

2. W. Kondo, S. Mizuta, Y. Oosawa, T. Kumagai and K. Fujii, *Bull. Chem. Soc. Jpn.*, 56, 2504 (1983).

3. W. Kondo, Y. Shimizu, N. Miura and N. Yamazoe, *Int. J. Hydrogen Energy,* 13, 345 (1988).

4. W. Kondo, Y. Shimizu, N. Miura, N. Yamazoe, CN. Wauters and J. Winnick, *AIChE* J., 44, 2144 (1998).

5. S. Ferro and A. DE. Battisiti, *Journal of Applied Electrochemistry,* 34, 981 (2004).

6. S. Ferro, C. Orsan and A. DE. Battisti, *Journal of Applied Electrochemistry,* 35, 273 (2005).

7. J. M. Orts, R. Gomez and J. M. Feliu, *Journal of Applied Electrochemistry,* 467, 11 (1999).

8. J. M. Orts, R. Gomez, J. M. Feliu and A. Aldaz, *J. Phys. Chem.,* 100, 2334 (1996).

9. J. M. Orts, R. Gomez, J. M. Feliu and A. Aldaz. *Langmuir*, 13, 3016 (1997).

10. P. Sivasubranmanian, R. P. Ramasamy, F. J. Freire, C. E. Holland and J. W. Weidner, *Int. J. Hydrogen Energy*, **32**, 463 (2007).

11. D. Duphil et al. *Nanotechnology.,* 15, 828, (2004).

(506c) Hydrogen Production Via Two - Step High Temperature Thermochemical Process Using Porous $MnFe_2O_4$ Material

Michael K. Opoku, Rajesh Shende and Jan A. Puszynski, Chemical and Biological Engineering, South Dakota School of Mines and Technology, Rapid City, SD

Porous $MnFe_2O_4$ material was synthesized and investigated as a potential candidate for thermochemical water splitting. The basic idea is that manganese-ferrite can split water molecules by extracting oxygen atoms and reversibly incorporating them into its lattices. $MnFe_2O_4$ was produced from manganese acetate, iron citrate, and sucrose precursors via a novel microwave assisted synthesis method which resulted in a pure crystalline material. DSC/TGA redox experiments indicated that synthesized $MnFe_2O_4$ can gain up to 3.34wt% in oxygen atmosphere at 850 °C and 1 atm. The results have also showed that these materials can be fully regenerated in nitrogen atmosphere at 950 °C and 1 atm. Based on the TGA analyses of oxidation and reduction in air and nitrogen respectively, it was estimated that 47.7 STPml/g-solid of H_2 might be generated in each cycle. Hydrogen production in multiple cycles via a two-step thermochemical process using packed-bed reactor filled with porous $MnFe_2O_4$ material will be presented. In addition, details of synthesis and characterization of this promising material will be presented.

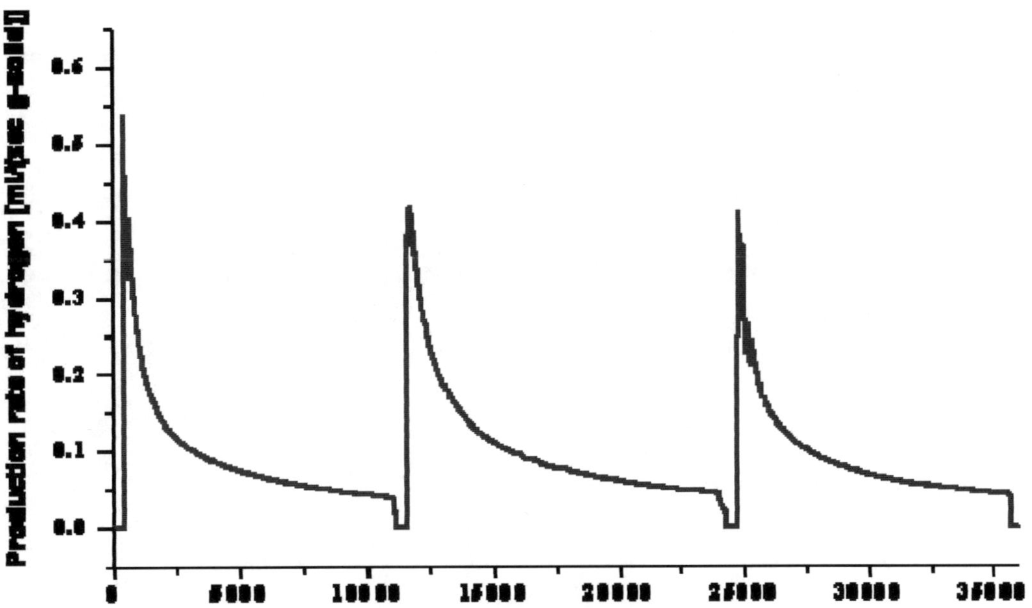

Figure 1. Production rate of hydrogen at 800 °C and atmospheric pressure using porous $MnFe_2O_4$ material.

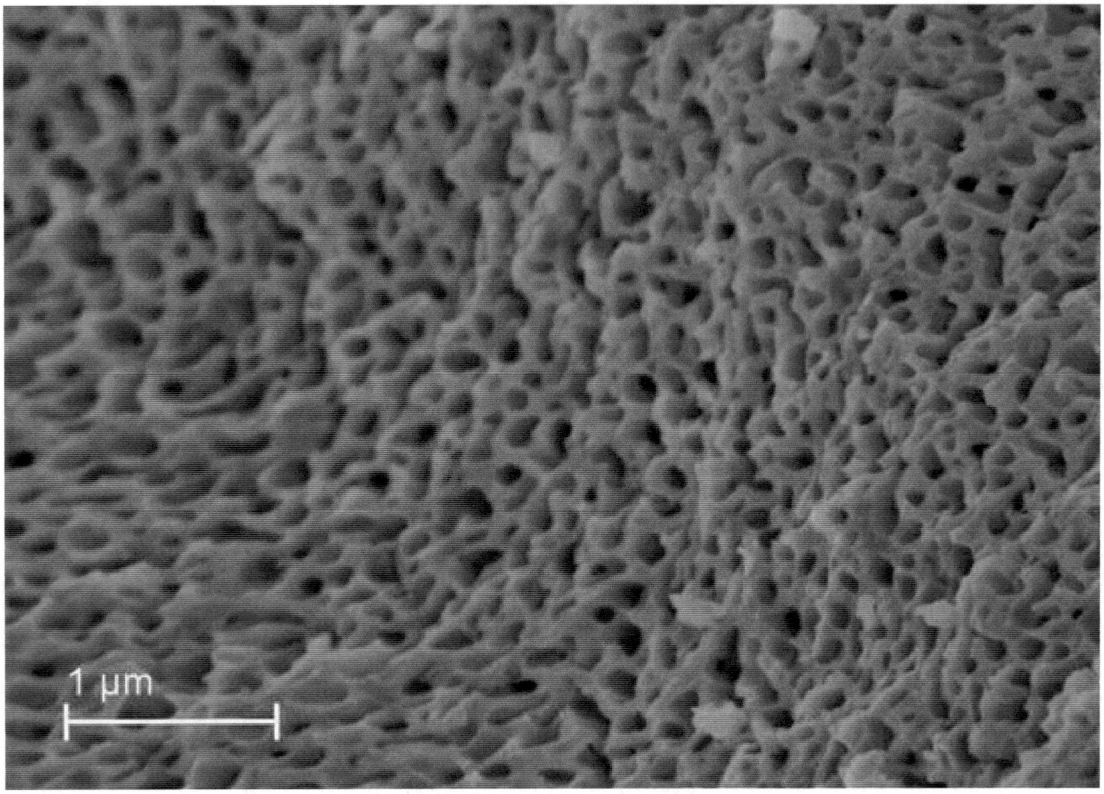

Figure 2. SEM image of synthesized $MnFe_2O_4$ materials.

See more of this Session: Advances in Thermochemical Hydrogen Production
See more of this Group/Topical: Topical 8: Hydrogen Production and Storage

H_2 GENERATION FROM THERMOCHEMICAL WATER-SPLITTING USING SOL-GEL SYNTHESIZED FERRITES ($M_xFe_yO_z$, M = Ni, Zn, Sn, Co, Mn, Ce)

Rahul R. Bhosale, Rajesh V. Shende and Jan A. Puszynski,

Department of Chemical and Biological Engineering,
South Dakota School of Mines & Technology,
Rapid City, SD 57701 USA

Abstract

This paper reports the sol-gel synthesis of Zn-ferrite with nanosized or porous morphology. This method involves the addition of chloride salts of Zn and Fe in ethanol followed by gelation using propylene oxide (PO). As-prepared gels were aged, dried and heated rapidly upto different temperatures of 600°C, 800°C and 1000°C and quenched in air or N_2 environment. Powder X-ray diffraction (XRD), scanning electron microscopy (SEM) and BET surface area analysis were performed to characterize the calcined ferrite powders. To achieve nanosize particles and higher specific surface area (SSA), the gels were rapidly calcined upto 600°C and quenched in air. The calcined Zn-ferrite powder thus obtained was then placed in an Inconel packed bed reactor and the H_2 generation and regeneration abilities were investigated by performing four consecutive thermochemical cycles where water-splitting and regeneration were carried out at 800°C and 1050°C, respectively.

Introduction

The energy density of H_2 (143 MJ/kg) is relatively higher than that of the fossil fuels such as oil (46.4 MJ/kg), gas (53.6 MJ/kg) and coal (32.5 MJ/kg), which makes it as an attractive and potential energy carrier. H_2 can be stored, delivered and converted to electricity by fuel cell technologies and used directly as a fuel without emitting any polluting gases such as SO_2 or carbon CO_2 [1]. Industrially H_2 is produced by gasification and reforming of fossil fuels, pyrolysis and reforming of biomass, and solar thermal processes [2]. However, by far the most attractive reaction for the production of H_2 is the thermal decomposition of water yielding pure H_2 [3]. Thermolysis of water (water-splitting in absence of chemical compounds) is extremely challenging as the direct thermal dissociation of water requires very high temperatures (>2500°C) and a unique high temperature membrane technology for separation of the H_2 and O_2 [4]. In contrast to this, thermochemical water-splitting cycles have huge potential as an environmentally and thermodynamically feasible approach for producing H_2 [5]. A metal oxide based thermochemical water-splitting is a two step process in which the first step belongs to the reduction of the metal oxide by releasing O_2 at higher temperatures; whereas, the second step corresponds to the oxidation of the reduced metal oxide by taking O_2 from water and producing H_2 via water-splitting at lower temperatures [2]. The combination of one regeneration / activation step (endothermic) and one water-splitting step (exothermic) is termed as one thermochemical cycle. The two-step thermochemical process eliminates the formation of explosive H_2-O_2 mixture, avoiding the need for high temperature separation of O_2 and H_2 [6].

Among the several redox materials investigated so far for the water-splitting application, ferrites such as $(A_xB_y)Fe_2O_4$ (doped spinel) or $A_xB_yFe_{1-x-y}O$ (doped wustite) or their mixtures with elements A and B being bivalent metal cations such as Mn, Ni, Zn, Co or Cu were reported to be most active materials for H_2 generation [4]. These ferrites have been synthesized by several synthesis methods such as solid-state mixing, oxidation of aqueous metal hydroxide suspension, self-propagation high temperature synthesis (SHS) and aerosol spray pyrolysis method [7 - 10]. Although beneficial, there have been reports indicating that these methods sometimes results into chemical, heterogeneity, non-stoichiometry, coarser particles and lower specific surface area (SSA) [12]. It appears that coarser and non-porous particles may pose the hydrodynamic limitations such as diffusional and mass-transfer resistances if used in packed bed reactors. In addition to this, during the thermochemical cycles the redox materials undergo thermal cycling which might results into grain growth, thermal stress and structural disintegration such as spalling of the redox material coating. This may further clog the reactor and the process lines, which may results into increase in the pressure drop and thereby creating unsafe processing conditions [12].

We believe that the use of ferrite with porous nanoparticle morphology with higher SSA will lower down the water-splitting and regeneration temperatures. Additionally, it will help overcoming the diffusional limitations and mass-transfer resistances. The packed bed reactors loaded with porous ferrites with nanoparticle morphology will provide lower pressure drop and minimize resistance for gas transport. Ferrite with porous nanoparticle morphology can be synthesized using sol-gel synthesis method which involves molecular or atomic level mixing and networking of chemical components. By using the sol-gel method significant improvements in chemical homogeneity, stoichiometry, phase purity and powder characteristics, for instance, ultrafine powder and narrow particle size distribution are often realized [2]. The powders produced by the sol-gel synthesis method usually have higher SSA, which may provide higher number of active sites for water-splitting reactions. In our previous investigations we have reported the H_2 generation and regeneration abilities of sol-gel derived Ni-ferrite and Sn-doped Ni-ferrite at lower water-splitting and regeneration temperatures of 700°C and 900°C, respectively [2, 12]. The temperature 900°C is the lowest ever reported for the regeneration of the Ni-ferrite in thermochemical water-splitting application.

In this investigation, we report the synthesis of Zn-ferrite using sol-gel approch. As synthesized gels were aged, dried and calcined in air or N_2 environment. The effect of PO concentration and calcination temperature and environment on the compositional purity, crystallite size (C_s), SSA, pore volume (P_v) and pore diameter (P_d) were investigated in detail. The H_2 generation and regeneration ability of the calcined Zn-ferrite were investigated by performing four thermochemical cycles where water-splitting and regeneration were performed at 800°C and 1050°C, respectively.

Experimental Methods and Characterization

Zn-ferrite Synthesis

In this synthesis, anhydrous zinc chloride ($ZnCl_2$) and iron chloride tetrahydrate ($FeCl_2.4H_2O$) [1:0.95 (w/w) ratio] were dissolved in ethanol (C_2H_5OH) to obtain a clear solution. To this clear solution, a predetermined amount of propylene oxide (C_3H_6O) i.e. PO was added and the solution was set aside undisturbed at room temperature until a gel was observed. As-synthesized gels were aged for 24 h at room temperature, dried at 100°C for 1 h, and then heated rapidly (ramp of 40°C/min) upto various temperatures and quenched in air or N_2 environment.

Zn-ferrite Characterization

The compositional purity and C_s of the calcined Zn-ferrite powders at different temperatures were determined using Rigaku Ultima-Plus X-ray diffractometer (XRD) provided with Cu$K\alpha$ radiation (λ = 1.5406 Å, 40 kV, 40 mA) and a graphite monochromator. The Zeiss Supra 40VP field-emission scanning electron microscopy (SEM) was used to analyze the morphology of the calcined powders. BET surface area analyzer, Gemini II-2375 from Micromeritics, was used to determine the SSA, P_d and P_v of the calcined powders from the adsorption isotherm obtained after degassing the material at 200°C.

H_2 Generation Set-up

The water-splitting and regeneration ability of the sol-gel synthesized Zn-ferrite was investigated using the experimental set-up shown in Fig. 1. It consists of a tubular Inconel packed bed reactor enclosed in a vertical split furnace (Carbolite Inc., USA). The temperature of the furnace was controlled and regulated precisely with the in-built PID controller. Accurately weighed 5.0 g of the Zn-ferrite was packed in the middle section of the tubular reactor using raschig rings and quartz wool as support materials. A horizontal tube furnace was provided with a stainless steel tube and it was used as a water vaporizing/preheating set-up. Distilled water from the reservoir was fed to the vaporizing/preheating furnace using a metering pump (Fluid Metering Inc., USA) at the flow rate of 1 ml/min. Water vaporized and the preheated steam obtained was fed to the packed bed reactor with N_2 as a carrier gas. A mass flow meter for N_2 (AALBORG Inc., USA) was mounted on the feed line to control the carrier gas flow rate; a flow rate of about 100 SCCM/min was randomly selected. The exit gas stream from the packed bed reactor was first cooled down in a water bath and finally fed to a moisture adsorption column containing Drierite (anhydrous calcium sulfate), purchased from W.A. Hammond Drierite Company Inc., OH, USA. The exit gas stream (mixture of N_2 and H_2) from the adsorption column was continuously monitored using an online H_2 sensor, HY-OPTIMA 730 from H_2 Scan Corp., USA. The sensor provided H_2 concentration in vol % at an interval of 1 sec. For each water-splitting cycle, the water supply was shut down when the online H_2 sensor indicated a minimum of 0.5 (vol %) concentration level. Finally, the effluent gas stream was burned continuously.

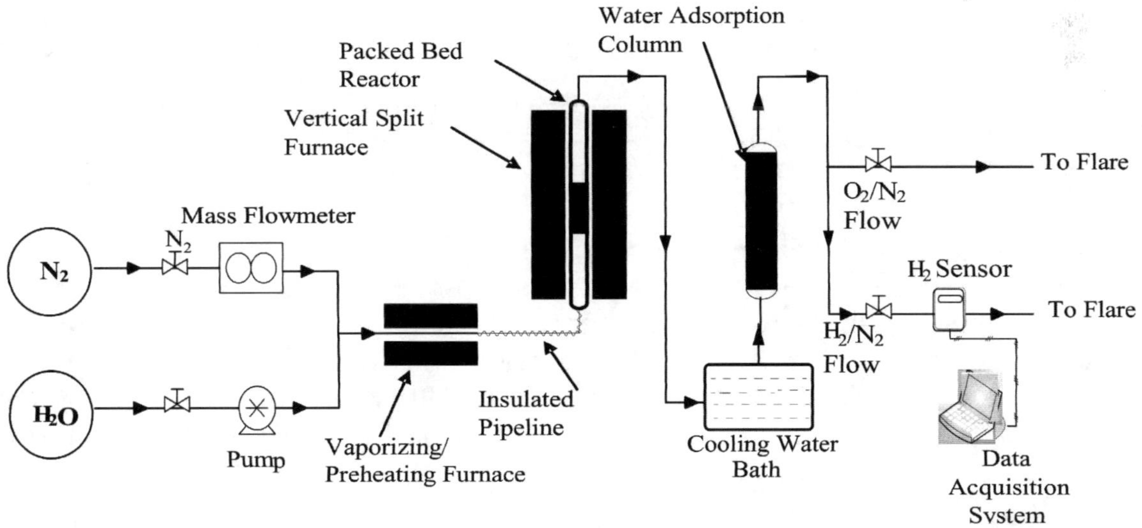

Fig. 1 Thermochemical water-splitting reactor set-up for H_2 generation [2].

Results and Discussion

Effect of Sol-gel Reaction Parameters and Calcination Temperature and Environment

In sol-gel synthesis method, addition of a proton scavenger such as PO helps in freezing the molecular chains and reduces the time required for the gel formation [2]. A set of experiments were performed to study the effect of PO concentration on gel time. To synthesize Zinc-ferrite, 10 g of the respective precursors (the weight ratios are mentioned in previous section) were added in 20 mL of ethanol and sonicated to obtain a clear solution. Different amounts of PO were added to achieve the gel formation. Table 1 presents the gel time with respect to PO addition. It was observed that as the concentration of PO increases, the time required for gel formation decreases. As-synthesized gels were aged for 24 h, dried at 100°C and heated rapidly upto 700°C and quenched in air or N_2 environment. The C_s, SSA, P_v and P_d of the calcined powders were determined and are presented in Table 1. These results indicate that the SSA and P_v of the calcined powders increases as the concentration of PO increases, whereas the C_s and P_d remain constant. The phase compositions of the calcined Zn-ferrite powders were observed to be unchanged irrespective of the PO concentration used.

Table 1. Effect of PO concentration on gel time, SSA, P_v, P_d and C_s of the sol-gel synthesized Zn-ferrite.

PO (mL)	Gel Time (sec)	Air				N_2			
		SSA (m^2/g)	P_v (cm^3/g)	P_d (nm)	C_s (nm)	SSA (m^2/g)	P_v (cm^3/g)	P_d (nm)	C_s (nm)
5	399	6.66	0.0089	5.393	30.33	5.49	0.00516	2.659	42.32
10	312	13.12	0.0201	5.424	29.83	10.45	0.00698	2.646	41.66
15	240	19.96	0.0278	5.432	29.44	16.33	0.01152	2.655	42.12
20	150	20.11	0.0296	5.387	29.41	16.98	0.01193	2.639	43.19

To study the effect of calcination temperature and environment, as-prepared gels with 20 mL of PO were aged for 24 h, dried at 100°C and heated rapidly upto 600°C, 800°C and 1000°C and quenched in air or N_2 environment. The compositional puritiy of the calcined powders were determined using X-ray diffractometer and the XRD patterns obtained are shown in Fig. 2a and 2b. These patterns indicate that the gels calcined in air at different temperatures resulted in a mixed composition containing $ZnFe_2O_4$, Fe_2O_3 and ZnO (Fig. 2a). Whereas, the gels calcined in N_2 environment resulted in mixed composition of $ZnFe_2O_4$, FeO and ZnO (Fig. 2b).

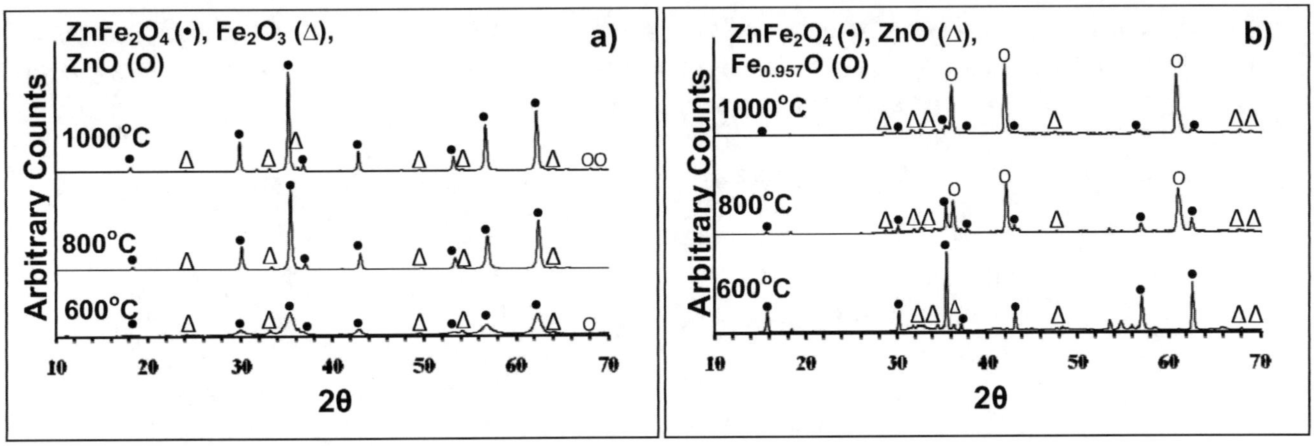

Fig. 2 XRD patterns of Zn-ferrites calcined at different temperatures in a) air and b) N_2 environments.

The BET SSA analysis results presented in Table 2. Indicate that the SSA, P_v, and P_d decrease and C_s increases with increase in calcination temperature. It can be observed that the samples calcined in air have higher SSA, P_v, and P_d as compared to that of calcined in N_2 environment. It was believed that the calcinations in N_2 environment might have induced the grain growth and decreased the SSA, P_v, and $P_{d,}$ which required further validation using SEM analysis. The SEM images of calcined powders obtained after rapid calcination of the dried gels upto 600°C in air or N_2 environment are shown in Fig. 3a and 3b. The SEM image of the Zn-ferrite calcined in air indicated agglomeration of plate-like nanoparticles stacked together (Fig. 3a); whereas, the gels calcined upto 600°C and quenched in N_2 resulted into a foam- like

Table 2. BET and XRD analysis of the Zn-ferrites calcined at different temperatures in air or N_2 environment.

Calc. Temp. (°C)	Air					N_2				
	Phase Composition	**SSA (m^2/g)**	**P_V (cm^3/g)**	**P_D (nm)**	**C_S (nm)**	**Phase Composition**	**SSA (m^2/g)**	**P_V (cm^3/g)**	**P_D (nm)**	**C_S (nm)**
600	$ZnFe_2O_4$ (87%), Fe_2O_3 (12%), ZnO (1%)	36.43	0.0527	6.349	20.37	$ZnFe_2O_4$ (81%), FeO (6%), ZnO (13%)	21.33	0.0134	2.679	36.05
800	$ZnFe_2O_4$ (98%), Fe_2O_3 (2%)	13.66	0.0188	5.391	35.75	$ZnFe_2O_4$ (25%), $Fe_{0.957}O$ (68%), ZnO (7%)	11.78	0.0103	2.724	45.76
1000	$ZnFe_2O_4$ (95%), Fe_2O_3 (3%), ZnO (2%)	8.98	0.0115	5.098	40.53	$ZnFe_2O_4$ (6%), $Fe_{0.957}O$ (89%), ZnO (5%)	6.53	0.0058	2.623	50.92

morphology with few faceted larger particles of smaller (<1μm) and larger (>2μm) sizes (Fig. 3b). It indicates that the calcination in N_2 environment has produced few additional larger faceted grains, which were absent in the samples calcined in air.

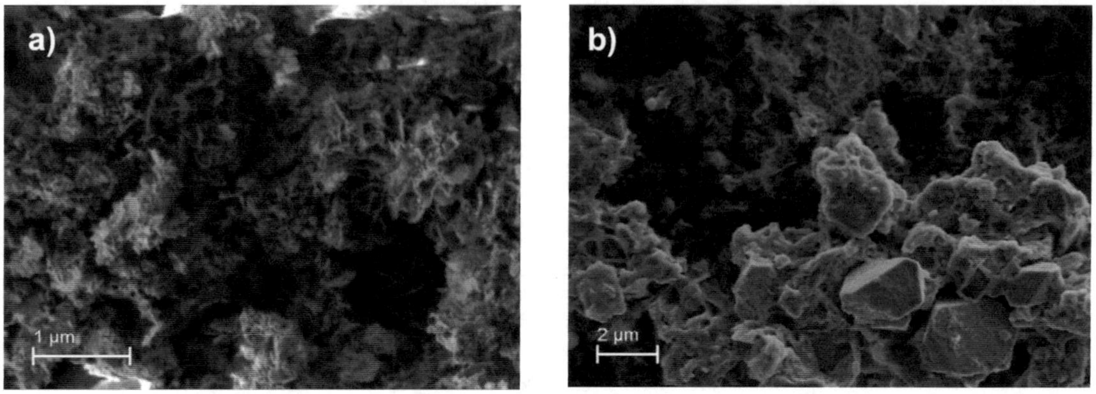

Fig. 3 SEM images of Zn-ferrite gels rapidly heated upto 600°C and quenched in a) air and b) N_2 environments.

H_2 Generation from Thermochemical Water-Splitting

Accurately weighed 5.0 g of Zn-ferrite was packed in the middle section of the Inconel tubular packed bed reactor and the temperature of the packed bed was increased upto the activation temperature of 1050°C. After activating the material at 1050°C for 4 h, the temperature of the packed bed was brought down to 800°C to perform water-splitting reaction. The preheated steam was fed to the packed bed reactor and the H_2 evolution was recorded continuously using the online H_2 analyzer. After water-splitting step, the material was regenerated at 1050°C for 4h in N_2 environment and the reactor again cooled down to 800°C to perform subsequent water-splitting cycle. Four consecutive thermochemical cycles were performed and the transient H_2 profiles obtained are shown in Fig. 4a. The results indicate that this material has regenerated itself at 1050°C, which is relatively lower regeneration temperature as compared with those that reported by other investigators. However, the H_2 levels were found to decrease in subsequent water-splitting cycle. From the H_2 profiles presented in Fig. 4a, the volume can be worked out using numerical integration method.

As the regeneration of the Zn-ferrite was performed at higher temperatures of 1050°C, Zn-ferrite might undergo thermal fatigue and that could eventually alter the powder characteristics. Therefore, after performing four thermochemical cycles, the ferrite material was removed from the reactor and analyzed using XRD, BET surface area analyzer and SEM. The XRD pattern of the reacted Zn-ferrite indicated a phase pure $ZnFe_2O_4$ (100wt %) indicating a good chemical stability during thermochemical cycling operation (Fig. 4b). The SSA of the original calcined Zn-ferrite before performing the water-splitting step was 36.43 m^2/g; whereas, for the material obtained after performing four thermochemical cycles it was decreased to 1.77 m^2/g. In addition to this, the P_v and P_d values of the reacted material were observed to be significantly lower as compared to that of original unreacted material. The SEM image shown in Fig. 4c support these results as the particle size was increased from nanosize to micron size with absence of any porous morphology. The investigation is in progress to examine the H_2 generation ability of Zn-ferrite material calcined in N_2 environment using thermochemical water-splitting.

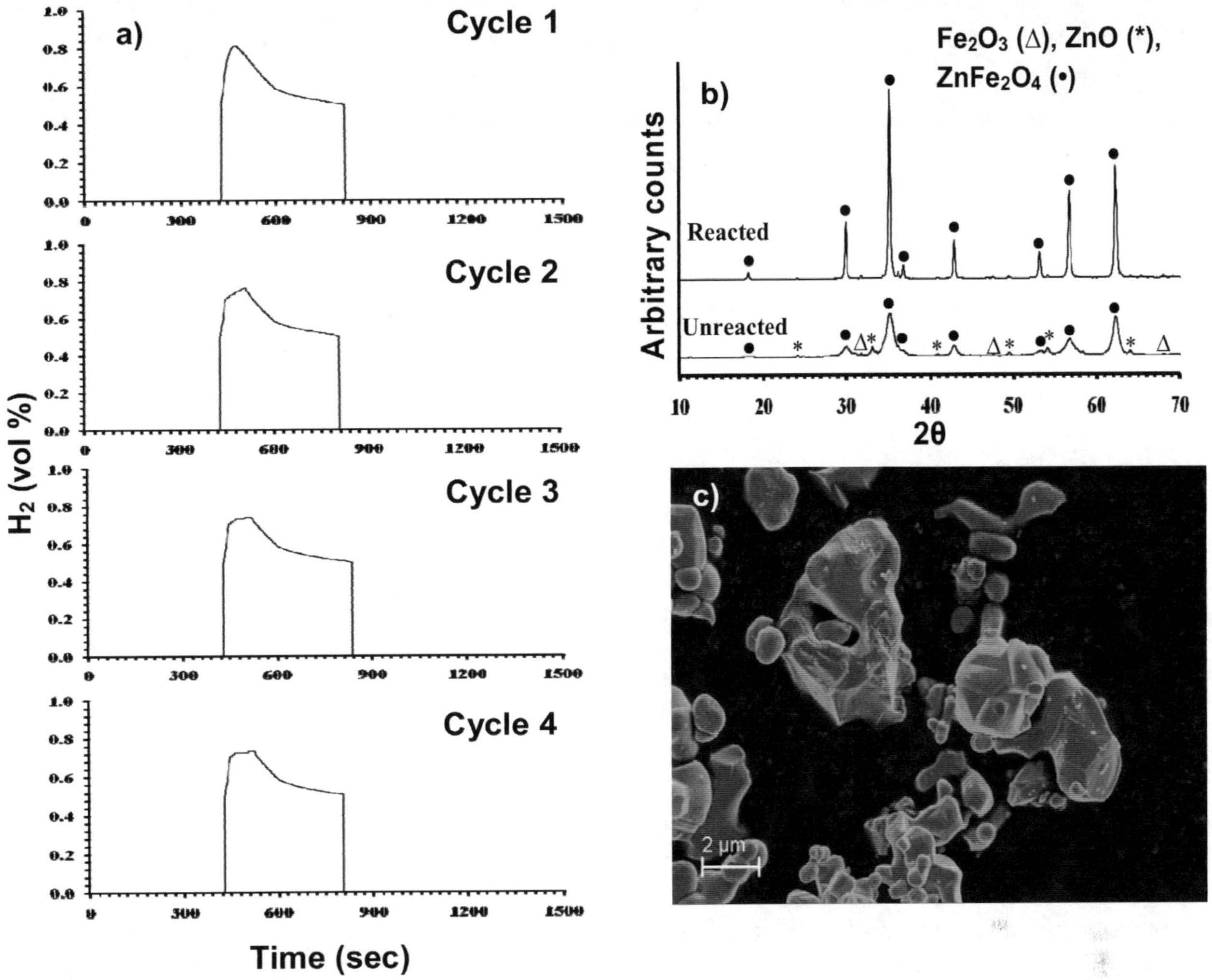

Fig. 4 a) Transient H_2 profiles obtained during four thermochemical cycles where water-splitting and regeneration were performed at 800°C and 1050°C, b) XRD patterns of the as-prepared Zn-ferrite and Zn-ferrite after four thermochemical cycles, c) SEM image of the Zn-ferrite after four thermochemical cycles .

Conclusions

Zn-ferrite material with nanosize or porous morphology was successfully synthesized using sol-gel method. The Effect of PO concentration and calcination temperature and environment on the compositional purity, SSA, C_s, P_v and P_d were studied in detail. Higher concentration of PO was found to reduce the gel time and increase the SSA of the Zn-ferrite. The results also indicated that the calcination temperature and environment have significant effects on the compositional purity and particle morphology. The Zn-ferrite gel rapidly calcined upto 600°C in air indicated a mixed phase composition containing the major phase as $ZnFe_2O_4$. The calcined Zn-ferrite powder was used for H_2 generation in four consecutive thermochemical cycles where water-splitting was performed at 800°C. The sol-gel synthesized Zn-ferrite material was observed to have potential to produce H_2 during water-splitting step at

800°C; however, its regeneration was not possible below 1050°C. The x-ray diffraction analysis of the reacted material revealed that the Zn-ferrite is chemically stable during thermochemical cycling operation. The BET SSA analysis results indicated that the SSA of the Zn-ferrite after performing four thermochemical cycles was significantly reduced. In addition to this, the SEM analysis of the reacted material showed that after four consecutive thermochemical cycles the ferrite results into non-porous micron size crystals.

Acknowledgement

The authors gratefully acknowledge the financial support by the National Science Foundation, grant award CBET-0756214.

References

1. Kang, K., Kim, C., Cho, W., Bae, K., Kim, S., and Park, C. (2009), "Novel two-step thermochemical cycle for hydrogen production from water using germanium oxide: KIER 4 thermochemical cycle," *International Journal of Hydrogen Energy*, **34**, pp. 4283-4290.
2. Bhosale, R. R., Shende, R. V. and Puszynski, J. A. (2010), "H_2 generation from thermochemical water-splitting using sol-gel derived Ni-ferrite" *Journal of Energy and Power Engineering*, **4**, pp. 27-38.
3. Roeb, M., Sattler, C., Kluser, R., Monnerie, N., Oliveira, L., Konstandopoulos, A., Agrafiotis, C., Zapalis, V., Nalbandian, L., Steele, A., and Stobbe, P. (2006), "Solar hydrogen production by a two-step cycle based on mixed iron oxides" *Journal of Solar Energy Engineering*, **128**, pp. 125-133.
4. Shende, R. V., Puszynski, J. A., Opoku, M. K., and Bhosale, R. R. (2009), "Synthesis of novel-ferrite foam material for water-splitting application," *Nanotech ISBN 978-1-4398-1782-7*, **1**, pp. 201-204.
5. Seralessandri, L., Varsano, F., La Barbera, A., and Padella, F. (2006), "On the oxygen-releasing step in the water-splitting thermochemical cycle by $MnFe_2O_4$ – Na_2CO_3 system," *Scripta Materialia*, **55**, pp. 875-877.
6. Bhosale, R. R., Shende, R. V., Opoku, M. K., and Puszynski, J. A. (2009), "H_2 generation from water-splitting using novel ferrite foam materials," *AIChE Conference Proceedings*.
7. Kojima, M., Sano, T., Wada, Y., Yamamoto, T., Tsuji, M., and Tamaura, Y. (1996). "Thermochemical decomposition of H_2O to H_2 on cation-excess ferrite," *J. Phys. Chem. Solids*, **57**, pp. 1757-1763.
8. Tamaura, Y., Hasegawa, N., Kojima, M., Ueda, Y., Amano, H., and Tsuji, M. (1998). "Water - splitting with the Mn(II)-ferrite-CaO-H_2O system at 1273K," *Energy*, **23**, pp. 879 – 886.
9. Agrafiotis, C., Pagkoura, C., Lorentzou, S., Kostoglou, M., and Konstandopoulos, A. (2007). "Hydrogen production in solar reactors," *Catalysis Today*, **127**, pp. 265 – 277.
10. Steinfeld, A. (2005). "Solar hydrogen production via a two-step water-splitting thermochemical cycle based on Zn / ZnO redox reactions," *International Journal of Hydrogen Energy*, **27**, pp. 611 – 619.
11. Bhosale, R. R., Shende, R. V., and Puszynski, J. A. (2010), "Sol-gel synthesis of ferrite foam materials for H_2 generation from water-splitting reaction," *Nanotech ISBN 978-1-4398-3401-5*, **1**, pp. 368 - 371.

Canada's Experimental Program on Nuclear Hydrogen Production with the Cu-Cl Cycle

G. F. Naterer
Faculty of Engineering and Applied Science
University of Ontario Institute of Technology (UOIT)
Oshawa, Ontario, Canada, L1H 7K4

ABSTRACT

This paper presents recent advances in Canada's program on nuclear hydrogen production with the thermochemical copper-chlorine (Cu-Cl) cycle. Enabling technologies for the Cu-Cl cycle are being developed by a Canadian consortium, as part of the Generation IV International Forum (GIF) for hydrogen production with the next generation of nuclear reactors. Recent advances towards an integrated lab-scale Cu-Cl cycle are discussed in this paper, including experimentation on the hydrolysis, spray drying, oxygen production and CuCl / HCl electrolysis processes.

1. INTRODUCTION

The ominous signs of climate change and air pollution are all around us: poorer air quality, retreating glaciers, illness and disease, stronger hurricanes, rising sea levels, among others. Hydrogen is a potentially major solution to the problem of climate change. But a key future challenge for hydrogen as a clean energy carrier is a sustainable, low-cost method of producing it in large capacities. Most of the world's hydrogen (about 97%) is currently derived from fossil fuels through some type of reforming process (such as steam-methane reforming; SMR). This paper presents recent advances in Canada's program of sustainable hydrogen production from nuclear energy, for a carbon-free source of hydrogen production.

Hydrogen is becoming an increasingly important energy carrier. It can be stored and used to generate electricity, either onboard vehicles or stationary power systems [1]. Hydrogen is also used widely by petrochemical, agricultural (e.g., ammonia for fertilizers), manufacturing, food processing, electronics, plastics, metallurgical, aerospace and other industries. For example, in Alberta, Canada, the oil sands sector needs vast amounts of hydrogen to upgrade bitumen to synthetic crude oil and remove impurities. Hydrogen has recently made significant commercial penetration with fuel cells in the lift-truck market, as well as back-up power systems for cell phone communication towers. The envisioned future hydrogen economy in the transportation sector will require significant increases to the world's current capacity to generate hydrogen, particularly in a clean, sustainable manner without relying on fossil fuels.

Thermochemical water decomposition is an emerging technology for sustainable large-scale production of hydrogen. Typically using two or more intermediate compounds, a sequence of chemical and physical processes split water into hydrogen and oxygen, without releasing any pollutants externally to the atmosphere. These intermediate compounds are recycled internally within a closed loop. Naterer et al. [2] presented an overview of Canada's program on nuclear hydrogen production with the thermochemical Cu-Cl cycle. Previous studies have identified over 200 possible thermochemical cycles [3, 4]. The sulfur-iodine (S-I) and hybrid sulfur cycles are prominent cycles that have been developed extensively by several countries. These include the USA (General Atomics, Savannah River National Laboratory, among others), Japan (Japan Atomic Energy Agency, JAEA), France (CEA), Italy and others [5, 6]. The following seven cycles (in addition to the S-I and hybrid sulfur cycles) were identified in a Nuclear Hydrogen Initiative [4] as the most promising cycles: copper-chlorine (Cu-Cl) [7], cerium-chlorine (Ce-Cl) [8], iron-chlorine (Fe-Cl) [8], vanadium-chlorine (V-Cl) [8], copper-sulfate (Cu-SO4) [8] and hybrid chlorine [8].

Experimental work has been conducted for processes in these cycles to demonstrate their scientific feasibility. However, most of these cycles require heat at temperatures over 800 C from very high temperature (Generation IV) reactors, which are not currently available and entail major design and material challenges. The Cu-Cl cycle has a significant advantage over these other cycles, due to lower temperature requirements around 530 C and lower. This paper focuses on recent experimental developments in Canada on the Cu-Cl cycle. Particular emphasis is given to hydrogen production with Canada's Super-Critical Water Reactor, SCWR.

2. THERMOCHEMICAL CU-CL CYCLE

The Cu-Cl cycle performs a sequence of reactions to achieve the overall splitting of water into hydrogen and oxygen. Using intermediate copper chloride compounds, the cycle decomposes water into hydrogen and oxygen, in a closed internal loop that re-cycles all chemicals on a continuous basis. Process steps in the Cu-Cl cycle and a schematic realization of the cycle are shown in Figure 1.

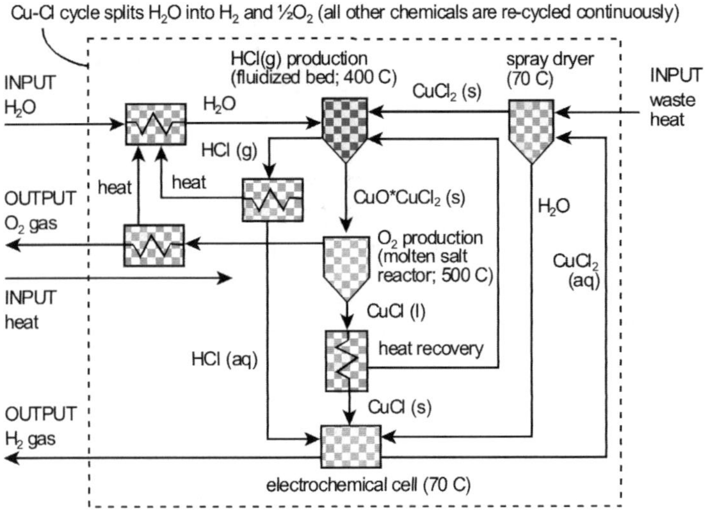

Figure 1: Schematic of the Cu-Cl cycle

There are three key variations of the Cu-Cl cycle: 5-step, 4-step and 3-step cycles. In the 5-step cycle, copper is produced electrolytically, moved to an exothermic thermochemical hydrogen reactor and then reacted with HCl gas to produce hydrogen gas and molten CuCl. The 4-step cycle combines these processes together to eliminate the intermediate production and handling of copper solids, through a CuCl / HCl electrolyzer that produces hydrogen electrolytically and aqueous Cu(II) chloride. Afterwards, the aqueous product is dried to generate Cu(II) chloride particles; then fed to a hydrolysis reactor to produce copper oxychloride. The 3-step cycle further combines these processes by supplying aqueous Cu(II) chloride directly into the hydrolysis chamber, such as spraying the solution with co-flowing steam to produce the same copper oxychloride product. This paper will focus on the 4-step cycle (Figure 1), since separation of the hydrolysis and drying processes can significantly improve the cycle's thermal efficiency.

3. CUCL / HCL ELECTROLYSIS FOR HYDROGEN PRODUCTION (STEP 1)

Oxidation of Cu(I) chloride (CuCl) during an electrochemical reaction occurs in the presence of hydrochloric acid (HCl) to generate hydrogen. The Cu(I) ion is oxidized to Cu(II) at the anode, and the hydrogen ion is reduced at the cathode. While the majority of the Cu species present is expected to be anionic (which would not permeate through the membrane), an appreciable concentration of cationic or neutral Cu species still exists and can permeate through the membrane and enter the cathode compartment where they are reduced to metallic copper. Metallic copper at the cathode acts as a poison, reducing the activity of the Pt catalyst towards the hydrogen evolution reaction (HER). This could result in a decrease in cell performance, so active research is underway to eliminate any possible copper crossover.

Atomic Energy of Canada Limited (AECL) has successfully demonstrated experimentally the production of hydrogen over a period of several days with a CuCl / HCl electrolyser. Hydrogen is produced electrolytically at the cathode and Cu(I) chloride is oxidized to Cu(II) chloride at the anode. The overall cell reaction can be expressed as

$$2CuCl + 2HCl \rightarrow 2CuCl_2 + H_2 \quad (1)$$

At the anode, several Cu(I) and Cu(II) chloride complexation species can form, via the following reactions:

$$Cu^+(aq) + n\,Cl^-(aq) \leftrightarrow CuCl_n^{1-n} \quad (2)$$

$$Cu^{2+}(aq) + n\ Cl^{-}(aq) \leftrightarrow CuCl_n^{2-n} \quad (3)$$

In chloride media, Cu(I) can exist as several different anionic species (e.g.$CuCl_2^-$, $CuCl_3^{2-}$). The relative ratios of each species are dependent on the concentration of chloride ions. Regardless of the exact species present, the anodic reaction proceeds by the electro-oxidation of Cu(I) species into Cu(II) species.

Recent studies have shown that the rate of the CuCl / HCl electrolysis reaction increases with temperature and CuCl concentration. A variety of electrode materials have been tested and hydrogen production was consistently observed at potentials as low as 0.5 V. However, approximately 0.65 V was typically required to achieve good current density. Promising results have successfully demonstrated that the reaction is feasible at low potentials and with relatively inexpensive electrode materials.

4. DRYING OF AQUEOUS CU(II) CHLORIDE (STEP 2)

Step 2 of the Cu-Cl cycle is the drying step as expressed by: $2CuCl_2\ (aq) \rightarrow 2CuCl_2\ (s)$. Aqueous $CuCl_2$ exiting from the previous electrolysis cell is dried to produce solid $CuCl_2(s)$, which is required for a subsequent hydrolysis step that produces copper oxychloride ($CuO*CuCl_2$) and HCl gas. The drying process is an energy-intensive step within the Cu-Cl cycle. Although the amount of heat required for the drying step is much higher than other steps in the cycle, it occurs at a lower temperature (lower quality) and therefore with heat that is more readily available. Within the drying step, the energy requirement increases from 1 to 5 times higher for slurry feed to solution, respectively, depending on the $CuCl_2$ concentration. The required heat can be obtained from low-grade "waste" heat from nuclear or other thermal power plants to improve the cycle efficiency. Spray drying is an efficient method of water removal due to the relatively large surface area available for heat and mass transfer, provided the liquid atomizes into sufficiently small droplets (on the order of a few hundred microns).

Experimental spray drying studies have been performed to produce solid Cu(II) chloride particles from an aqueous solution. Spray drying is a well established industrial process that allows powder characteristics and properties to be controlled and maintained nearly constant throughout a drying operation. Current research is performing experimental studies at successively lower inlet air temperatures, in order to eventually utilize waste heat from thermal power plants or other industrial sources for the spray drying heat requirements. Numerical studies have also been conducted to analyse the process of spray drying of aqueous Cu(II) chloride droplets. The results indicate that evaporative drying is possible down to low temperatures as low as 35°C, although such low temperatures may limit the product throughput rate.

5. HYDROLYSIS REACTION (STEP 3)

The following hydrolysis reaction occurs within the Cu-Cl cycle:

$$H_2O(g) + 2CuCl_2(s) \rightarrow Cu_2OCl_2(s) + 2HCl(g) \quad (4)$$

The reaction is an endothermic non-catalytic gas-solid reaction that operates between 350 and 400 C. The solid feed to the hydrolysis reaction is Cu(II) chloride, from the dried $CuCl_2$ product of step 2. Aqueous Cu(II) chloride is dried in step 2 to produce $CuCl_2$ particles, which are then moved to the hydrolysis unit and reacted with superheated steam to produce copper oxychloride solid and hydrochloric gas. This section describes recent developments in both experimental and numerical studies of transport phenomena associated with the non-catalytic gas-solid reaction of hydrolysis of Cu(II) chloride.

Lewis et al. [9] have demonstrated experimentally the scientific practicality of the hydrolysis reaction to produce HCl gas and solid copper oxychloride. Experimental results and simulations showed a large excess of water was needed for complete conversion (up to 98%) of $CuCl_2$. Byproducts other than Cu_2OCl_2 may appear, i.e., CuCl and CuOHCl, although ongoing research is investigating how and why these byproducts appear.

For an efficient and low capital cost process, it is important to reduce the water consumption. Aspen Plus simulations predict that 100% yield can be obtained with steam to $CuCl_2$ molar ratio of 17 at 370 C.

Unlike a spray reactor (nebulizer) used by the Argonne National Laboratory for the hydrolysis reaction, UOIT experiments have separated the drying and reaction processes, thereby using separate steps 2 (drying) and 3 (gas-solid reaction in a fluidized bed for hydrolysis). At UOIT, high temperature fluidized bed experiments with actual Cu-Cl compounds are being investigated, as well as low temperature experiments with simulated working fluids (air and spherical beads) to examine fluid dynamics and heat transfer phenomena in the fluidized bed.

The experimental unit was designed similar to the high temperature hydrolysis reactor, both geometrically and dynamically, by matching the Archimedes number, Froude number, density ratio, particle to column diameter and particle size distribution. The solid to gas density ratio is matched for accurate scaling of the hydrodynamics in a bubbling or turbulent bed. The experimental data indicates the importance of the pressure drop on the fluidization quality at high temperatures. Variations of particle properties at high temperatures reduce the fluidization quality.

Additional numerical and experimental studies have been performed to investigate the transport phenomena associated with the gas-solid reaction of particles in the fluidized bed during hydrolysis of Cu(II) chloride [2]. The conversion of solids depends on the rate of reaction and residence time of a particle.

A shrinking-core model has been developed to estimate the reaction rate of the gas-solid reaction. When the diffusion of gaseous reactant into a particle is much faster than the chemical reaction, the solid reactant is consumed nearly uniformly throughout the particle. In this situation, a uniform-reaction model can be used. On the other hand, when the diffusion of gaseous reactant is much slower and it restricts the reaction zone to a thin layer that advances from the outer surface into the particle, then the shrinking-core model is adopted. As the mole fraction of steam in the gas increases, the time of solid conversion decreases. Numerical simulations indicated that the fraction of reacted core volume to the total particle volume increases over time, regardless of the resistance mode for mixed flow of solids.

6. OXYGEN PRODUCTION BY COPPER OXYCHLORIDE DECOMPOSITION (STEP 4)

The oxygen production step receives solid feed of copper oxychloride ($CuO*CuCl_2$) and produces O_2 gas and molten Cu(I) chloride. The decomposition reaction is given by $CuO*CuCl_2(s)$ = 2CuCl (molten) + $0.5O_2$ (gas) at 530 C. Copper oxychloride particles decompose into molten salt and oxygen gas bubbles. The reactant particles absorb decomposition heat from the surrounding molten bath. Gas species leaving the oxygen reactor include oxygen gas and potentially impurities of products from side reactions, such as CuCl vapor, chlorine gas, HCl gas (trace amount) and H_2O vapor (trace amount). The substances exiting the reactor are molten CuCl, potentially unreacted solid $CuCl_2$ from the upstream hydrolysis reaction, due to the incomplete decomposition of $CuCl_2$, as well as reactant particles entrained by the flow of molten CuCl.

Recent Canadian developments have demonstrated experimentally the processes in a large engineering scale reactor, with flow capacities of 2 kg/day of oxygen production, by decomposing about 21 kg/day of solid $CuO*CuCl_2$ feed that enters the reactor. A number of experimental and operational issues with the unit reactor are currently being investigated, such as aggregation into solid blocks of $CuO*CuCl_2$ when conveying and feeding solid particles. The aggregation may choke or clog the feeder and cause sudden spouting of particles. Also, embedded particles of $CuCl_2$ from the upstream hydrolysis reactor in $CuO*CuCl_2$ may lead to undesirable products and side reactions, such as $CuCl_2$ decomposing to CuCl and Cl_2 gas. If particles enter the reactor at a temperature lower than 430 C, a challenge arises with the presence of bubbles in the molten salt. Some CuCl vapor might condense and molten CuCl can solidify around the $CuO*CuCl_2$ particles. If an aggregation develops with particles, molten salt and bubbles, the contact area between a reactant particle and heating medium (molten CuCl) will decrease and the aggregations float along the surface of the molten salt. This deters the decomposition of reactant particles and potentially leads to choking of the reactor (a major safety concern). These issues and others are currently under investigation for the copper oxychloride decomposition experiments.

Additional studies have also been performed to effectively recover heat from molten CuCl exiting the copper oxychloride decomposer. A configuration with convective heat transfer between molten CuCl droplets and air in a counter-current spray flow heat exchanger has been developed. Recovering heat from molten CuCl is challenging due to the phase transformations of molten CuCl, as it cools from liquid to two different solid states. Various configurations of heat recovery from molten CuCl have been examined.

7. CONCLUSIONS

This paper has presented an overview of Canada's experimental program on nuclear hydrogen production with the thermochemical copper-chlorine (Cu-Cl) cycle. The Cu-Cl cycle was identified by Atomic Energy of Canada Limited, AECL (CRL; Chalk River Laboratories), as the most promising cycle for thermochemical hydrogen production with the Generation IV nuclear reactor, SCWR (Super-Critical Water Reactor). Objectives of Canada's nuclear hydrogen program are to develop commercially viable processes for producing hydrogen based on the thermochemical Cu-Cl cycle and water electrolysis. The Cu-Cl cycle has attractive features that should help to meet these goals, particularly a 550 C maximum temperature that reduces demands on materials, compared to higher temperature cycles. Also, the Cu-Cl cycle couples well with various heat sources, such as the SCWR, solar power tower or Na-cooled fast reactor. The yields are nearly 100% in the hydrolysis and copper oxychloride decomposition processes, without catalysts and no recycle streams in these reactions. Also, conceptual process designs use commercially practiced processes in industry.

ACKNOWLEDGEMENTS

Support of this research and assistance from Atomic Energy of Canada Limited, Ontario Research Excellence Fund, Natural Sciences and Engineering Research Council of Canada (NSERC), University Network of Excellence in Nuclear Engineering (UNENE) and the Canada Research Chairs (CRC) program are gratefully acknowledged.

REFERENCES

[1] Stevens MB, Fowler MW, Elkamel A, Elhedhli S. Macro-level optimized deployment of an electrolyser-based hydrogen refuelling infrastructure with demand growth. Engineering Optimization 2008; 40 (10): 955 - 967

[2] Naterer, G. F., Suppiah, S., Lewis, M., Gabriel, K., Dincer, I., Rosen, M. A., Fowler, M., Rizvi, G., Easton, E. B., Ikeda, B. M., Kaye, M. H., Lu, L., Pioro, I., Spekkens, P., Tremaine, P., Mostaghimi, J., Avsec, A., Jiang, J., "Recent Canadian Advances in Nuclear-Based Hydrogen Production and the Thermochemical Cu-Cl Cycle", International Journal of Hydrogen Energy, vol. 34, pp. 2901 – 2917, 2009

[3] McQuillan BW, Brown LC, Besenbruch GE, Tolman R, Cramer T, Russ BE, Vermillion BA, Earl B, Hsieh HT, Chen Y, Kwan K, Diver R, Siegal N, Weimer A, Perkins C, Lewandowski A. High Efficiency Generation of Hydrogen Fuels Using Solar Thermochemical Splitting of Water: Annual Report, GA-A24972, General Atomics, San Diego, CA, 2002

[4] Lewis M, Taylor A. High Temperature Thermochemical Processes, DOE Hydrogen Program, Annual Progress Report, Washington, DC, 2006, p 182 - 185

[5] Sakurai M, Nakajima H, Amir R, Onuki K, Shimizu S. Experimental Study on Side-Reaction Occurrence Condition in the Iodine-Sulfur Thermochemical Hydrogen Production Process. International Journal of Hydrogen Energy 2000; 23: 613 – 619

[6] Schultz K, Thermochemical Production of Hydrogen from Solar and Nuclear Energy, Technical Report for the Stanford Global Climate and Energy Project, General Atomics, San Diego, CA, 2003

[7] Sadhankar RR, Li J, Li H, Ryland D, Suppiah S, Hydrogen Generation Using High-Temperature Nuclear Reactors, 55th Canadian Chemical Engineering Conference, Toronto, Ontario, October, 2005

[8] Carty RH, Mazumder M, Schreider JD, Panborn JB. Thermochemical Hydrogen Production, vol. 1, Gas Research Institute for the Institute of Gas Technology, GRI Report 80-0023, Chicago, IL 60616, 1981

[9] Lewis MA, Masin JG, Vilim RB, Serban M. Development of the Low Temperature Cu-Cl Thermochemical Cycle, International Congress on Advances in Nuclear Power Plants, May 15 - 19, 2005, Seoul, Korea

USING *iMASS* TO SIMULATE THE TRACKING/MOVEMENT OF SPECIAL NUCLEAR MATERIAL

Miltiadis Alamaniotis[1], Shanjie Xiao[2], Jason Young[1], Rong Gao[3], Lefteri Tsoukalas[1], Dong-Ok Choe[2], Tatjana Jervremovic[2]
[1]Purdue University, W. Lafayette, IN 47907
*[2]Utah Nuclear Engineering Program **(UNEP)**, The University of Utah, Salt Lake City, UT 84112*
[3]ARETE Company, W. Lafayette, IN 47906

Abstract

The **iMASS** (intelligent Model-Assisted Sensing System) is software modeling toll for fast and reliable detection of SNM developed jointly by The University of Utah, Purdue University and Wright State University. The **iMASS** is a new theoretical paradigm developed as a new computational methodology consisting of a number of advanced tools applicable to a wide range of interrogation techniques of interest in nuclear material detection and numerical monitoring of the materials of interest. The **iMASS** was developed using the Nuclear Resonance Fluorescence (NRF) as a test bed, and demonstrated that the software is widely applicable to any kind of nuclear signals or the fusion of multiple signals from different types of detectors. In this paper, we present a different application of the **iMASS** tool; we use it to simulate the effect of the movements/tracking/trafficking of SNM in some predefined environment. The effects of the movements/tracking/trafficking of SNM are analyzed in terms of radiation level change in predefined environment. Thus, instead of using the NRF signal for which the **iMASS** tool has been widely verified, we use the radiation level signal as a diagnostic "marker" in finding the movement/trafficking of the SNM from the known location in predefined environment.

Introduction

Special Nuclear Material (SNM) detection has been one of the identified areas of importance and as such it has been a very active research topic in recent years. Significant progress has been made during this period. An impressive variety of detectors is developed and still new and better detectors are under constant revision for their efficient use for safety and security. Detection of such materials is of great interest for homeland security and various methods have been proposed [1-4].

When a facility owns any type of special nuclear material, it is important to be able to track these materials in the interest of safety and security for the facility and the surrounding area. It is crucial to make sure that a facility can track these SMNs both in transport between facilities (as the NRC tracks) and to track within the facility itself. Part of this would be tracking how moving these SNMs could affect the people in the area during and after the transport of the material. According to Nuclear Regulatory Commission, "*Special nuclear material is only mildly radioactive, but it includes some fissile materials-- uranium-233, uranium-235, and plutonium-239 -- that, in concentrated form, can be the primary ingredients of nuclear explosives. These materials, in amounts greater than formula quantities*[1]*, are defined as "strategic special nuclear material" (SSNM). The uranium-235 content of low-enriched uranium can be concentrated (i.e., enriched) to make highly enriched uranium, the primary ingredient of an atomic bomb.*"

There are two significant parameters for developing efficient tracking processes: *accuracy* and *speed.* Thus new techniques and methods should maximize tracking accuracy while minimizing decision time. The *tracking accuracy* is defined as a main performance criterion; it is desirable to develop a methodology that will keep record of the SNM movement while keeping the number of false track paths as low as possible. In the latter, it is significant that methodology is not confused by movement of any but SNM materials. The *detection speed* is crucial; time consuming or delayed tracking of SNM at a facility might affect negatively its security, slow-down the efficiency of movement monitoring and bring unacceptable delays in taking safety measures. Therefore, the tracking process should be accurate and quick.

In this paper, an application of the **iMASS** [5-7] tool is presented. Specifically, it is used for simulation of the effect of the movements/tracking of special nuclear material in some predefined environment. In the next two sections **iMASS** is described and some preliminary results are shown while the last section concludes the paper.

iMASS (intelligent Model-Assisted Sensing System)

The iMASS is a novel system that is composed from several parts that work synergistically in order to provide quick and accurate detection of SNM [6]. Its main goal is to be used in ports and a country's entry points for scanning incoming cargo containers. The **iMASS** was developed using the Nuclear Resonance Fluorescence (NRF) as a test bed, and demonstrated that the software is widely applicable to any kind of nuclear signals or the fusion of multiple signals from different types of detectors. The architecture of the proposed system is presented in **Fig. 1**. Specifically, iMASS is comprised of the following parts:

- **Radiography/CT unit.** This is the first part of the iMASS project and based on the incoming data from the CT sensors builds a virtual cargo model. In order to the model to be built Monte Carlo methods are used.
- **NRF processing unit.** This component has as an input the actual signal arriving from the NRF detectors. Then the signal is processed using the Local-Global Graph (LG Graph) method [8] and possible peaks are marked.
- **Decision-Making unit.** This part gets the results of the processing of the spectra (Monte Carlo and NRF) and compares them. The main purpose of this unit is to via the evaluation of the findings from the spectra to decide whether the cargo merits further examination (see **Fig. 2**).

At this point it should be mentioned that In order to accelerate the Geant4 simulation it was developed a specialized to iMASS grid computing software named LightGrid. LightGrid [9] is an optimal solution for putting together in one frame software technologies such as PHP, MySQL, Apache, and the Network File System to create a fast, grid computing system for Geant4.

---

[1] Strategic special nuclear material in any combination in a quantity of 5000 grams or more computed by the formula, grams = (grams contained U-235) + 2.5 (grams U-233 + grams plutonium). This class of material is sometimes referred to as a Category I quantity of material (10CFR70.4). [http://www.nrc.gov/reading-rm/basic-ref/glossary/formula-quantity.html]

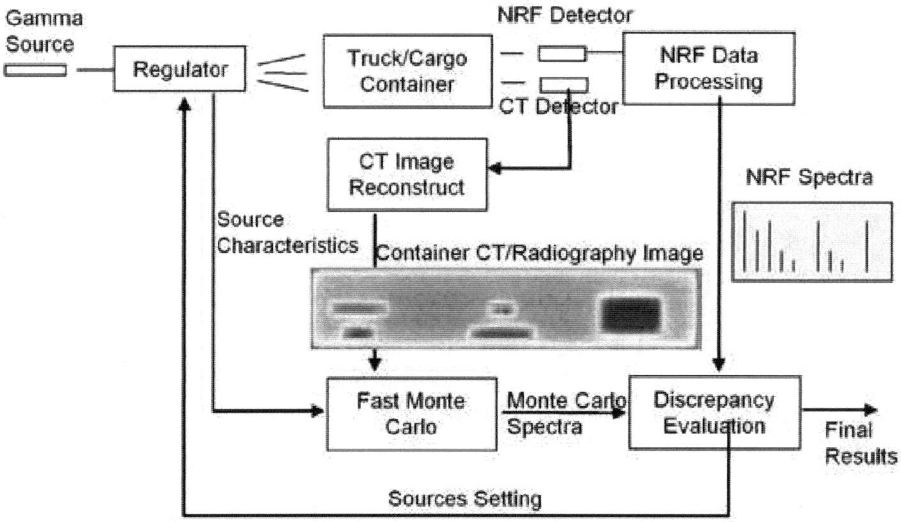

Figure 1. *intelligent Model Assisted Sensing System (iMASS) architecture as shown in [6].*

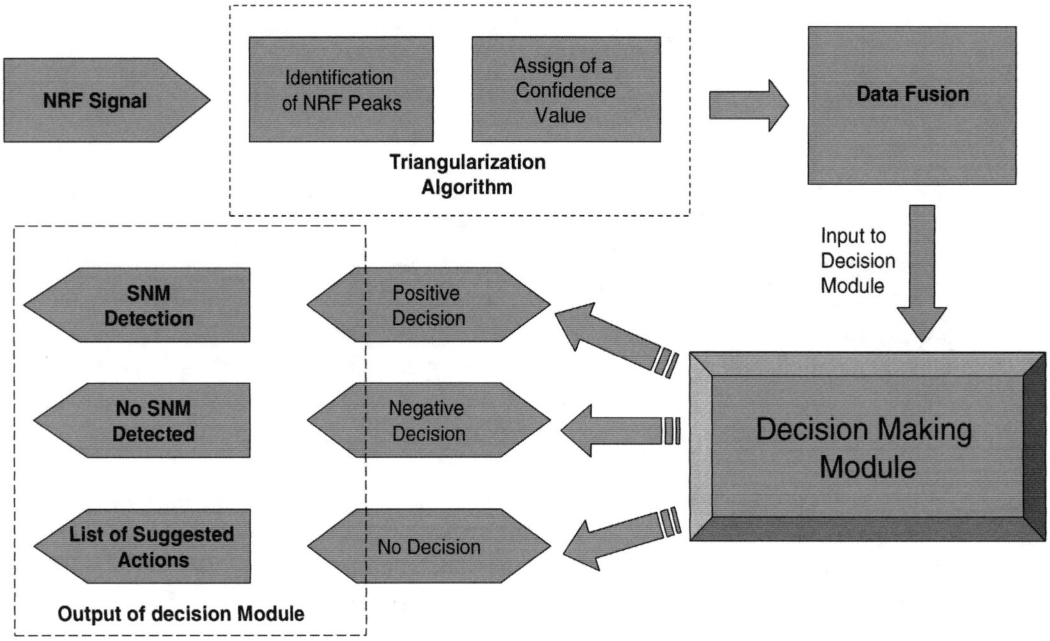

Figure 2. *Flow of signal processing and decision making in iMASS.*

Demonstration Case

For the tracking purposes, we use the radiation level signal as a diagnostic "marker" in finding the movement/trafficking of the SNM from the known location in predefined environment. We will consider certain amounts of typical SNMs that move (scenario of insiders intending to take these SNMs out from the predefined environment) and show the spatial change of the radiation levels changing in time in regard to the known background.

The **iMASS** tool gives plots as the one shown in **Fig.2**, i.e. will indicate the movements of the radiation peaks in time and space, thus giving the information of the unexpected activities in predefined environment. The challenge is to read very often a low signal imbedded into the background noise and be able to make a confident decision. To validate the simulation data, we create a simple experiment involving the movement of some radioactive material.

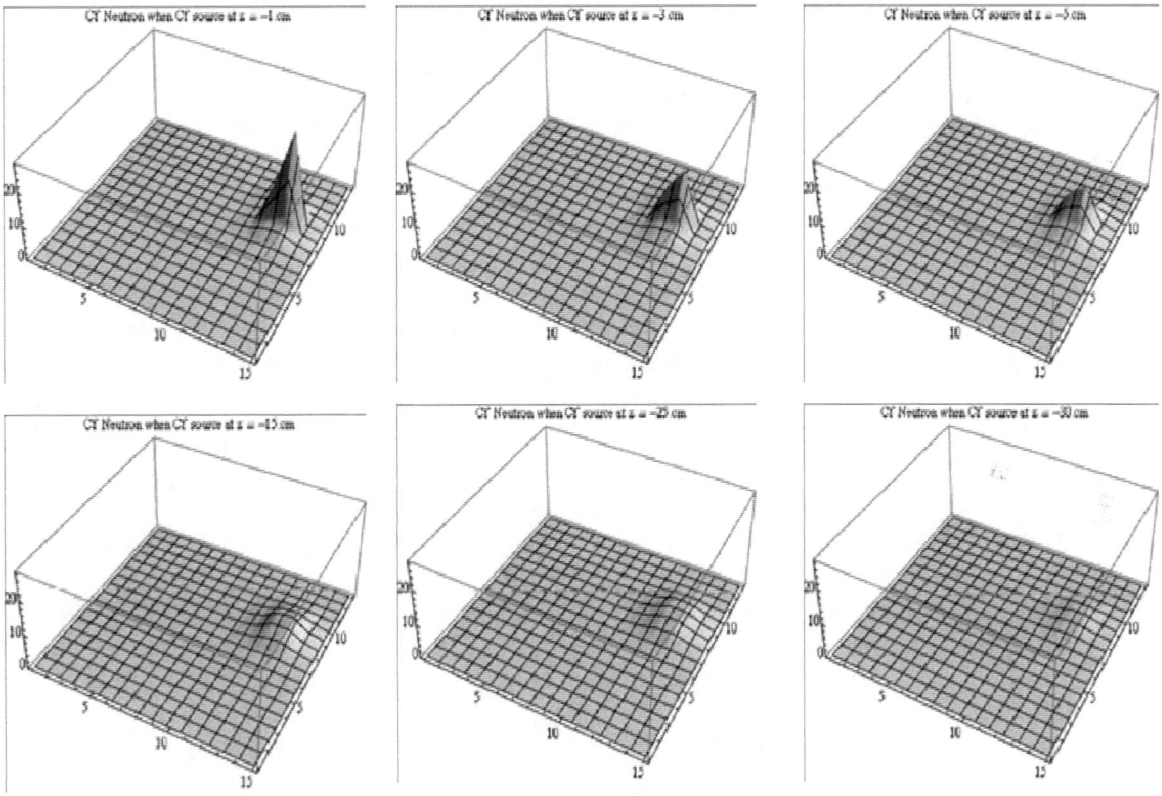

Figure 2. *Radiation field perturbation caused by the movement of the radioactive material [simulation result based on GEANT4]; this research was presented at 1st National Conference: Advancing Tools and Solutions for Nuclear Material Detection, March 10-11, 2010: "Dynamic Geant4 Simulation of UNEP Radiation Source Room for Nuclear Safeguarding", Samantha Winkle, Olga Baykova, Shanjie Xiao, Dong-OK Choe and Tatjana Jevremovic]*

Detection of SNM movement is mainly based on the processing and decision making module of the iMASS. Specifically, the processing unit extracts the useful information from the obtained signal and the decision making uses fuzzy rules to infer over the movement of the material. In the case shown in Fig. 2, the goal is to detect similar series of patterns taken from a sensor located in a different place. In case the latter occurs, movement of the material is identified and movement mapping is possible. At this point, it should be emphasized that iMASS using its Monte Carlo module can simulate the expected signal in case material properties are known. In that case the decision module takes also into consideration possible discrepancies between expected signal and measurement.

Conclusions

In this paper, a simple approach for tracking of SNM movement was presented. Specifically, the recently developed iMASS system was used to simulate the effect of the movements/tracking/trafficking of SNM in some predefined environment. The novel iMASS system for cargo inspection was briefly described. To be more specific, its functionality and its individual components that comprise the system were identified and their functionality explained shortly. Lastly, a demonstration case on radiation field perturbation caused by the movement of the radioactive material was exhibited. Future work will focus on extension of the demonstration case and application of iMASS on various paths for SNM movement.

References

1. W. Bertozzi, E.S. Korbly, J.R. Ledoux and W. Park,(2007). Nuclear resonance fluorescence and effective Z determination applied to detection and imaging of special nuclear material, exlosives, toxic substances and contraband, Nuclear Instruments and Methods in Physics Research A, Elsevier: Amsterdam,vol. 261, pp. 331-336.
2. M.J. Hall, S. Asztalos, P. Biltoft, J. Church, A.M. Descalle, T. Luu, D. Manatt, G. Mauger, E. Norman, D. Petersen, J. Pruet, S. Prussin and D. Slaughter,(2007). The Nuclear Car Wash: Neutron interrogation of cargo containers to detect hidden SNM, Nuclear Instruments and Methods in Physics Research B, Elsevier: Amsterdam, vol. 261, pp. 337-340.
3. A.A. Diaz, B.J. Burghard, J.R. Skorpik, C.L. Shepard, T.J. Samuel and R.A. Pappas (2003). Non-invasive ultrasonic examination technology in support of counter-terrorism and drug interdiction activities – the acoustic inspection device (AID), Proc. of SPIE, vol. 5048, pp. 1275-1280.
4. Perry, J., Xiao, S., & Jevremovic, T. (2008). iMASS: Computational NRF Spectra Signal from Geant4. In 2008 20th IEEE International Conference on Tools with Artificial Intelligence (pp. 541-546). Dayton, OH: IEEE. doi: 10.1109/ICTAI.2008.94
5. Alamaniotis, M., Young, J., Perry, J., Xiao, S., Agarwal, V., Forsberg, P., et al. (2009). Engineering Solution to Nuclear Material Detection at Ports: Introducing the Novel iMASS Paradigm. In Proceedings of 21st IEEE International Conference on Tools with Artificial Intelligence. Newark, NJ.
6. Perry, J., Xiao, S., & Jevremovic, T. (2009). iMASS: Evolved NRF Simulations for More Accurate Detection of Nuclear Threats. In Proceedings of the 17th International Conference on Nuclear Engineering. Brussels, Belgium

7. Bourbakis, N., Pantelopoulos, A., & Kannavara, R.(2008). A NRF Spectra Signature Detection Model Using Local Global Graphs", November, 2008, 20th IEEE Int'l Conference on Tools with Artificial Intelligence, pp.547-550.
8. Young, J., Perry, J., & Jevremovic, T. (2010). LightGrid - An Agile Distributed Computing Architecture for Geant4. Nuclear Instruments and Methods in Physics Research Section A: Accelerators, Spectrometers, Detectors and Associated Equipment, 614(1), 154-158.)

GEANT4 NRF Simulation for Material Detection

Shanjie Xiao and Tatjana Jevremovic
Utah Nuclear Engineering Program, The University of Utah, Salt Lake City, UT

Introduction

Nuclear resonance fluorescence (NRF) is similar to X-ray fluorescence, except that nucleus and not atom absorbs a photon; in the subsequent de-excitation of the excited nucleus by re-emission of one or more photons a spectrum of emitted photons is obtained that is related to the nuclear structure of that nuclei and is thus its unique *signature*. The NRF therefore can be applied as an excellent nondestructive method for material detection to probe different objects in finding specific NRF spectrum signatures, therefore isotopes of interest. This NRF signature that create a unique thumbprint is used to identify isotopes and elements in a bulk of materials [1].

GEANT4 [2] is a set of particle physics Monte Carlo simulation toolkits. It provides a diverse but cohesive set of software components which can be employed in any scientific and engineering areas where particle interactions in matter is investigated. GEANT4 provides flexible geometry definition, physics process simulation from standard electromagnetic interactions to quark interactions, and in wide energy range. GEANT4 is an open-source code and its C++ object-oriented programming model makes it easy to be fully customized by the application developer. GEANT4 provides various physics models based on theoretical equations and experimental data. The developer can select the model fitting the specific application, or even create his/her own physics model.

The NRF simulation is not existent in most of the available Monte Carlo codes, as well as in the original GEANT4 package. GEANT4 NRF module, called G4NRF, was developed by Pacific Northwest National Labs (PNNL) [3] to simulate the NRF. This paper will first review the NRF simulation implemented in the G4NRF module, and then introduce the method to manage the required nuclear data to simulate the NRF spectrum for nuclear materials, such as for example U-235. One test-example with hidden nuclear materials in a cargo container is described to demonstrate the application and validity of GEANT4 NRF simulation.

GEANT4 NRF Simulation

G4NRF Module

In GEANT4, the transport of a particle is simulated step by step. At the beginning of the first step, the interaction type length (mean free path) of each individual physics process assigned to the particle is calculated based on the cross section and target number density. Therefore there are several different lengths, plus the distance to the next volume boundary. The interaction type with the shortest length is chosen to occur (it may just cross the volume boundary when the distance to the boundary is shorter than the length of all other processes). After the process occurs, the interaction length of the process, which has just occurred must be re-calculated, and the processes, which did not occur must have the previously occurred length subtracted from the interaction length. With all the refreshed lengths, still the process with the shortest length will occur. It continues until the particle is terminated. The process is sketched in Fig. 1, for instance three physics processes are taken.

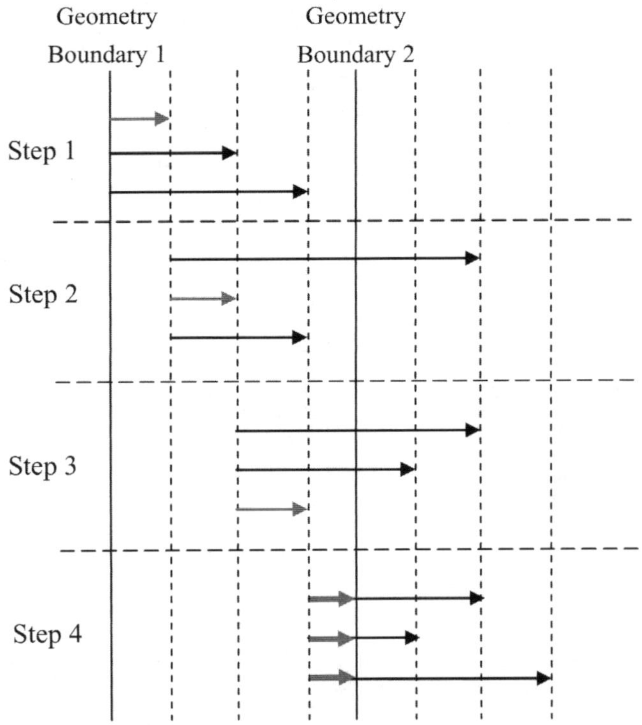

Figure 1 Transport of particles in GEANT4

To calculate the interaction length, the corresponding cross section must be known. In the absence of Doppler broadening, the NRF resonance cross section has a Breit-Wigner shape [4]:

$$\sigma(E)=\pi\left(\frac{\lambda}{2\pi}\right)^2\frac{2J_1+1}{2J_0+1}\frac{\Gamma_0(\Gamma/2)}{(E-E_r)^2+(\Gamma/2)^2},$$

where λ is the wavelength of the incident photon of energy E, E_r is the resonance energy, J_0 and J_1 are the nuclear spins of the ground state and excited state, Γ is the total decay width of the excited state, and Γ_0 is the partial width of EM de-excitation to the ground state.

Considering the Doppler broadening, the NRF cross section of a photon of energy E is [4]

$$\sigma(E)=\pi^{3/2}\left(\frac{\lambda}{2\pi}\right)^2\frac{2J_1+1}{2J_0+1}\frac{\Gamma_0}{\Delta}\exp\left[-\left(\frac{E-E_r}{\Delta}\right)^2\right].$$

The Doppler width is given by

$$\Delta=E\left(\frac{2kT}{M_c^2}\right),$$

where M is the mass of the nucleus, k is Boltzmann's constant, and T is the absolute temperature of the material.

Nuclear Data

In the current GEANT4 NRF simulation module, the nuclear level parameters J_0, J_1 and Γ_0 are extracted from a set of database files derived from the ENSDF (Evaluated Nuclear Structure Data File) database. However, in the original version of GEANT4 NRF errors were detected when using nuclear materials, such as U-235 and U-238, causing the simulation to stop; the reason for errors were a lack of necessary nuclear data. In addition, the GEANT4 simulation requires also the photon evaporation data. What we have found was that the nuclear levels in the two sets of data files cannot match with each other for some of the isotopes.

In order to perform the NRF simulations for all isotopes, it is required to match nuclear levels. Further analysis showed that nuclear levels in the NRF database was identical with the latest ENSDF database available online (last updated March, 2006) [5]. The photon evaporation data are claimed to come from ENSDF, but the levels are not exactly the same as in ENSDF; for instance, with the tolerance of 50 eV, there are seven (7) different levels for U-235 and 53 different levels for U-238. The problem was resolved by developing an additional subroutine that converts the ENSDF database to a format of GEANT4 photon evaporation data files. The new files replaced the original photon evaporation data, so as to perform NRF simulations for all isotopes.

Table 1. Nuclear levels related to NRF process of U-235 and Pu-239 [6]

Isotope	Transition (keV)	Statistical significance	Cross section (eV·b)
^{235}U	1656.23(80)	5.8	4.1(13)
	1687.26(33)†	10.2	6.1(11)
	1733.60(22)†	56.4	29.8(39)
	1769.16(28)‡	9.3	4.4(10)
	1815.31(22)‡	19.9	9.7(17)
	1827.54(23)	13.3	6.7(12)
	1862.31(20)	20.1	9.6(17)
	2003.32(25)	14.5	9.7(17)
	2006.19(31)	7.2	4.7(16)
^{239}Pu	2040.25(21)	5.8	8(2)
	2046.89(31)	4.2	5(2)
	2135.00(37)*	3.5	4(2)
	2143.56(13)*	9.7	13(2)
	2150.98(31)*	4.2	5(2)
	2289.02(25)	6.2	8(2)
	2423.48(22)**	7.2	10(2)
	2431.66(25)**	6.3	9(3)
	2454.37(26)	6.2	9(3)
	2460.46(37)	4.7	6(4)
	2464.60(30)	5.7	8(4)
	2471.07(34)	4.6	6(2)

Another problem of the current ENSDF database hat was accounted was a lack of information regarding nuclear levels. Table 1 lists the main nuclear levels related to NRF in U-235 and Pu-239, but none of them is part of the ENSDF database. Since the data insufficiency problem was not completed at this stage, only U-238 was used to demonstrate the efficiency of the revised GEANT4 NRF in detecting nuclear materials.

GEANT4 NRF Simulation Example

To verify the NRF module of GEANT4, two scenarios are created. The first scenario is very simple, called the fundamental model as shown in Figure 2: the Bremsstrahlung photon beam from the particle gun hits perpendicularly the small target plate; the ring detector collect photons at the back scattering angle of about 135 degree. Figure 3 (a) shows the experimental NRF spectrum of U-238 obtained at MIT HVRL [7], and Figure 3 (b) shows the GEANT4 NRF simulated NRF spectrum of U-238, i.e. the material of the target in Figure 2 is pure U-238. All significant peaks in a measured spectrum appear in the simulated NRF spectrum as well. The difference in the peak heights and background counts is because: 1) the fundamental GEANT4 NRF model is very simplified and thus is not fully comparable with the experiment; and 2) the intrinsic error of current G4NRF module [3].

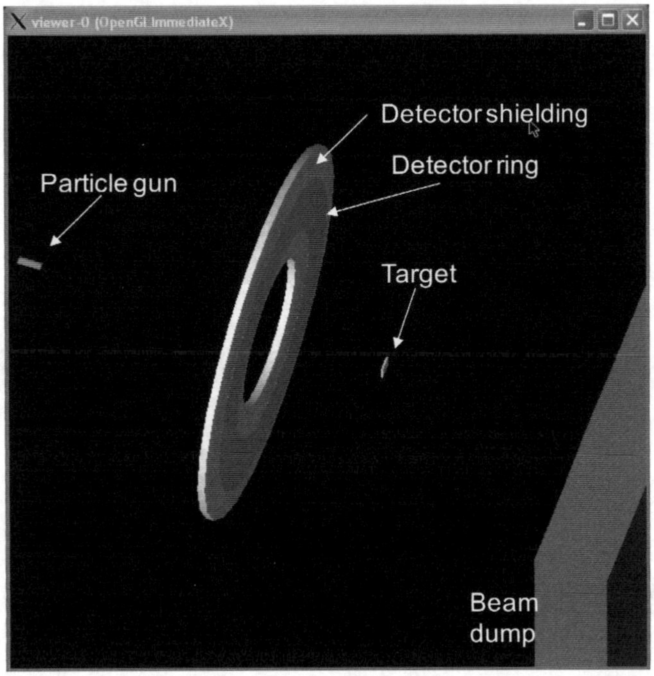

Figure 2 The fundamental GEANT4 NRF simulation model

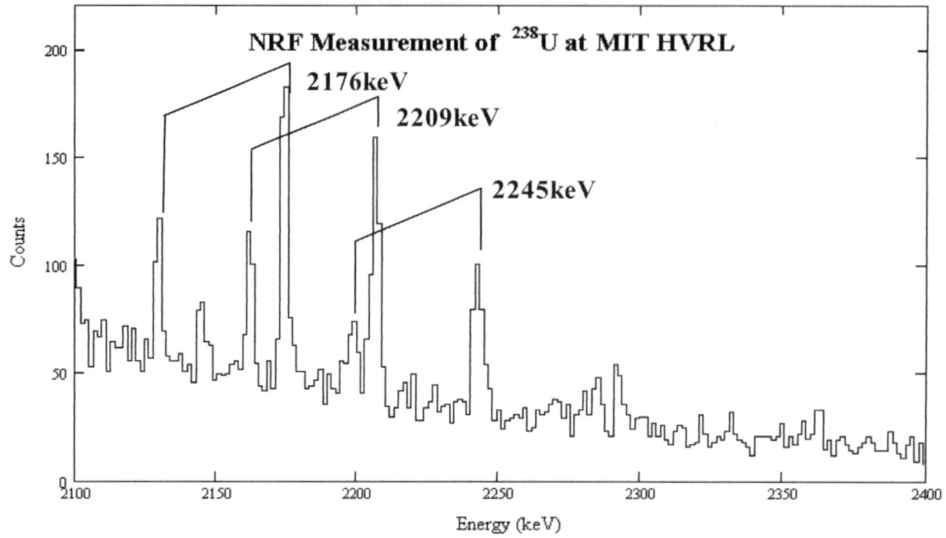

(a) NRF measurement spectrum of U238 [7]

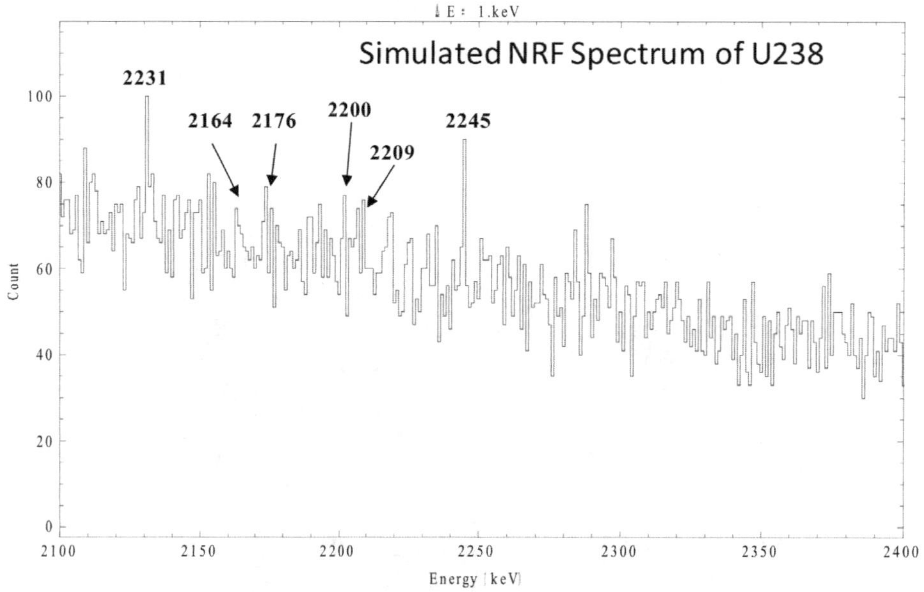

(b) GEANT4 simulated NRF spectrum of U-238

Figure 3 Comparison of NRF spectrums of U-238 from experiment and GEANT4 NRF simulation

The second scenario is a model of a cargo container with two cars; in the battery of one of the two cars there is U-238 hidden inside, as shown in Figure 4. The car battery consists of lead plates, which provide a good shielding for easy detection of uranium. Figure 5 shows the spectrum obtained with Bremsstrahlung photon beam in detecting the battery region of the two cars. The spectrum from car 1 has some peaks corresponding to U-238 while no significant peaks in the spectrum from car 2.

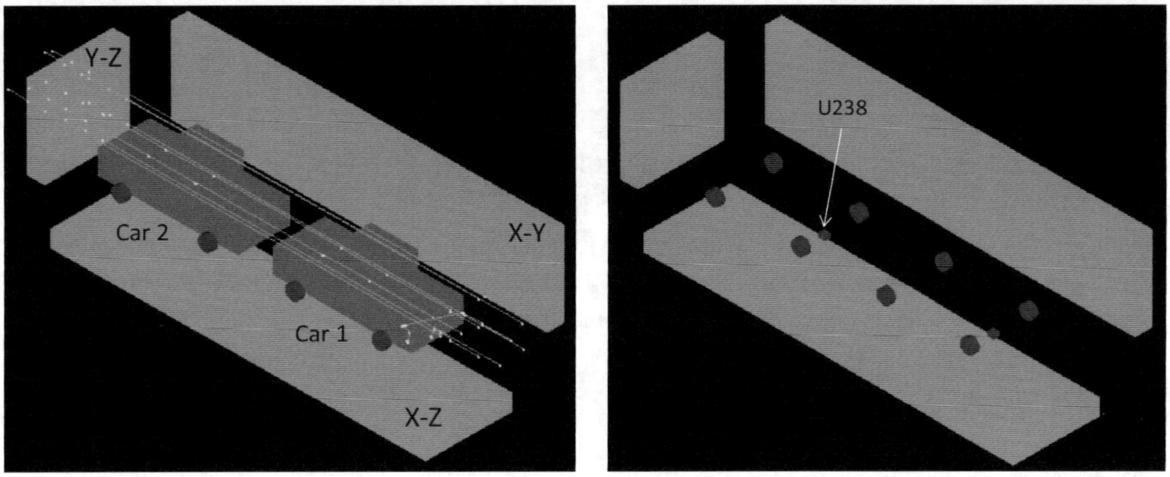

Figure 4 GEANT4 model of two cars with hidden U-238 in a battery of one car

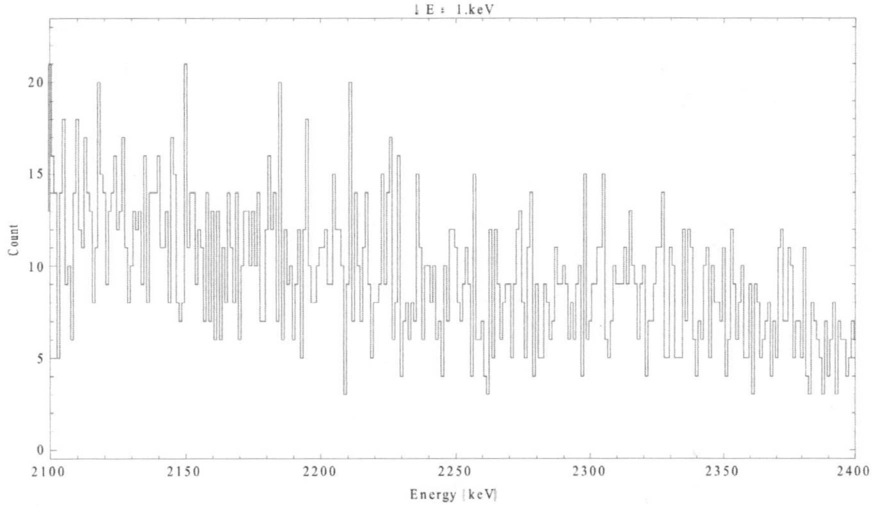

(a) Simulated NRF spectrum from Car 1

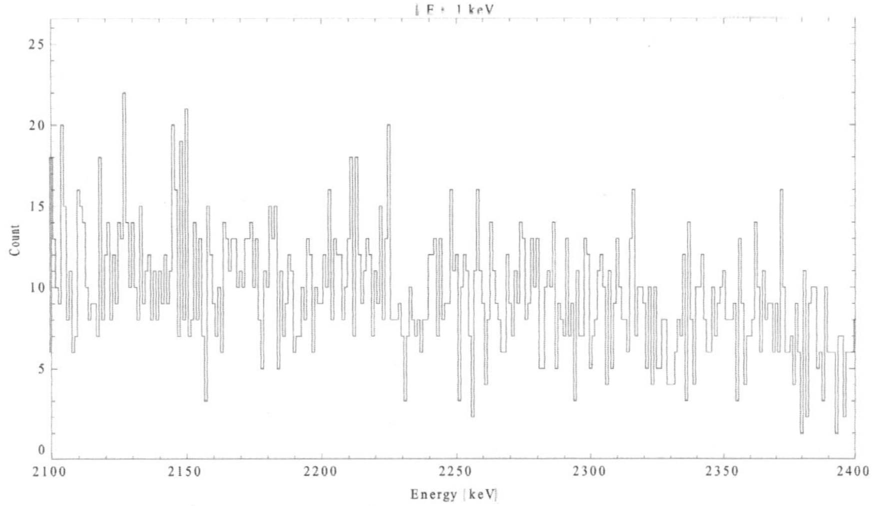

(b) Simulated NRF spectrum from Car 2

Figure 5 Simulated NRF spectrums from Car 1 and Car 2 in the range of U-238 peaks

Conclusion

NRF is a good method for nuclear material detections, because it has special signature for each isotope and the high energy of the signal (NRF peaks are in the range of MeV) has good penetration from inside of hidden objects. GEANT4 is a powerful Monte Carlo particle physics simulation toolkit. The GEANT4 NRF module can simulate NRF process and the simulation can be performed for all isotopes after adjusting the database issues as we identified. Thus the GEANT4 NRF module can be a useful tool for the design and optimization of material detection systems applying the NRF technique.

References

[1] G.A. Warren, J.A. Caggiano, H.K. Walter, E. Lepel, S. Pratt, W. Bertozzi, S.E. Korbly, R.J. Ledoux, and W.H. Park, "Nuclear Resonance Fluorescence of 235U," *Nuclear Science Symposium Conference Record*, 2006. IEEE, vol. 2, 2006, pp. 914-917.

[2] S. Agostinelli, et al. "Geant4—a simulation toolkit," *Nuclear Instruments and Methods in Physics Research Section A: Accelerators, Spectrometers, Detectors and Associated Equipment*, vol. 506, 2003, pp. 250-303.

[3] D.V. Jordan and G.A. Warren, "Simulation of nuclear resonance fluorescence in Geant4," *Proc. NSS'07, IEEE*, 2007, pp. 1185-1190.

[4] F.R. Metzger, "Resonance Fluorescence in Nuclei", in *Progress in Nuclear Physics*, vol. 7, New York: Pergamon, 1959, pp. 53-88.

[5] "Evaluated Nuclear Structure Data File (ENSDF) Database", March, 2006.

[6] W. Bertozz, etc., "*Nuclear resonance fluorescence excitations near 2 MeV in 235U and 239Pu*", Phys. Rev. C, 2008.

[7] W. Bertozzi and R. J. Ledoux, "*Nuclear Resonance Fluorescence Imaging in Non-Intrusive Cargo Inspection*", CAARI 2004, 2004.

Pulse Shape Study in High Purity Germanium Detectors

Haori Yang, Canberra Industries, Meriden, CT

Introduction

In the area of radiation detection and measurement, semiconductor radiation detectors have become the most significant tools over the past decade. This class of detectors has unique characteristics and provides superior performance over other candidates in many respects. In a radiation detection system, generally speaking, the energy resolution is related to the number of equivalent signal carriers produced by unit energy deposition when Poisson statistics noise dominates. This is the case for most popular radiation detectors, for example, gas-filled detectors, scintillation detectors coupled with PMT, and semiconductor detectors. Because of the small band gap, the energy deposition required to produce an electron-hole pair inside a semiconductor crystal (~ 3 eV for germanium) is much less than what is required to generate a signal carrier in other competing technologies (~100s eV for a photoelectron in scintillators-PMT combination and ~30 eV for an ion pair in gas-filled detectors). As a result, semiconductor-based detectors provide energy resolution that is superior to that of other technologies.

In addition to the superior energy resolution, which is directly related to pulse amplitude of the output signal from a semiconductor detector, the pulse shape contains a lot of useful information as well. In a semiconductor-based detector, incident photons usually deposit their energies via three major types of interactions: photoelectric absorption, Compton scattering, and pair production. Energetic electrons generated during these interactions further produce electron-hole pairs through ionization. By applying an electric field across the crystal, these electron-hole pairs can be collected on electrodes and an electric pulse is generated as the output signal. The amplitude of this pulse is proportional to the energy deposition within the active area of the crystal, while the shape of the pulse contains other information like interaction position, material properties and etc. By tailoring the electric field and the potential distribution within a semiconductor crystal, one can obtain capabilities that no other technologies can offer. For example, by using pixelated anodes, He and etc. developed a 4π Compton camera using 3D position sensitive CZT detectors [1]; through segmentation, GRETINA detectors are able to track gamma rays at a very high count rate [2].

Pulse shape of the output signal is primarily determined through charge collection process. In order to better understand the process of charge collection inside a semiconductor, a study approach is developed and described in this work. First, a Monte Carlo model is created based on dimensional details of the detector. This model can then be benchmarked against a set of well controlled measurements. Using this model, energy deposition process of incident photons inside the crystal can be very reliably simulated using Monte Carlo method. The electric field distribution is calculated using commercially available analysis software, utilizing the knowledge of the geometric dimensions and material properties of the detector. A customized code package then integrates all the information and calculates the pulse shape of the output signal. This approach can be used to study position sensitive detector response, the effect of material impurity level, electrode design and etc. Although the approach shown in this paper is applicable to other semiconductor materials, the current study is focusing on High Purity Germanium (HPGe) detectors only.

Monte Carlo Model and Detector Simulation

Monte Carlo method is often used to study radiation transportation. It relies on repeatedly random sampling, validated physics models and reliable data libraries to simulate the interaction of radiation with various materials. Monte Carlo method is only practical when sufficient computing power is available. Along with the development of electric computers, Monte Carlo method quickly finds various applications including space mission, radiotherapy, nuclear criticality and etc.

Currently, two most commonly used Monte Carlo codes in radiation transportation are MCNP/MCNPX and GEANT4. MCNP is a general purpose Monte Carlo code developed by LANL that can be used for radiation transportation involving neuron, photon, electron, and coupled problems [3]. To represent real geometric configuration, MCNP creates three-dimensional cells filled with materials. It utilizes available cross section data or physics models to simulate radiation interactions with cell materials. The simulation results are provided through a rich set of flexible tallies, including surface current and flux, volume averaged flux, particle heating, pulse height tally and more. Compared with other options, MCNP is versatile and easy to use, due to its powerful source term definition ability, extensive data collection and sophistic variance reduction technique. MCNPX extends the capabilities of MCNP to nearly all particles, nearly all energies. GEANT4 is a toolkit managed by the international GEANT4 collaboration [4]. GEANT4 is the first Monte Carlo simulation toolkit to use Object oriented programming (C++). It provides a set of tools to cover all areas in radiation transportation, including geometry, particle tracking, detector response, source particle generation and etc. GEANT4 uses a different approach to model real geometric configuration from MCNP. Instead of defining the surfaces that bound the volumetric cells, objects are directly defined as solids in GEANT4. Compared with MCNP/MCNPX, GEANT4 is much more flexible, but it does require some level of C++ programming skill. The GEANT4 source code is freely available, while a license is required to run or modify MCNP/MCNPX.

For the purpose of this work, either MCNP/MCNPX or GEANT4 can be used to create the Monte Carlo model. In order to build a model that can well represent the real geometric configuration, one would need to know the dimensions of the semiconductor crystal and the packaging components. Aside from the geometric dimensions, the composition of the materials involved is also needed for the Monte Carlo model. Once the Monte Carlo model of the semiconductor detector is created, it can be benchmarked against real measurements under well controlled counting conditions. Both MCNP/MCNPX and GEANT4 provide a rich set of tools for source definition. The model can then be adjusted to match the measurement results. Using the benchmarked detector model and source model, Monte Carlo simulation can be performed. In MCNP/MCNPX simulation, one can use the PTRAC card to track the history of individual incident photons. In GEANT4 simulation, one would need to write his own functions to get the necessary information out. In either case, the goal is to extract the series of (interaction position, energy deposition) doublets along the track of each individual incident photon. This information is needed for later pulse shape calculation.

Field Calculation and Pulse Shape Calculation

As discussed above, the output signal from a semiconductor detector is caused by the charge induced on electrodes during the collection of electron-hole pairs in an electric field caused by external bias. In order to correctly calculate the output signal pulse shape, two field distributions need to be determined ahead: the actual electric field within the crystal due to externally applied high voltage, and the weighting potential field. In semiconductor detectors, charge collection track is determined by the actual electric field, while the amount of the induction charge is related to the change in weighting potential between the start and stop points of the track. The calculation of these two fields will be discussed in more details below.

In the calculation of the actual electric field in a semiconductor crystal, there are a few factors that need to be taken into account. First of all, the geometric configuration of the crystal and the electrodes needs to be carefully modeled. The dimensions of the crystal can be physically measured. However, it is hard to accurately model electrodes or contacts that are created through chemical process or ion plantation. Estimation will have to be made based on other measurements or experience. Fortunately, this only affects the field distribution in the very vicinity of the electrode, where the weighting potential should be very close to the value on the electrode anyway. The error on the overall calculation result caused by this is then negligible. Secondly, the material properties need to be determined. The values of electron and hole mobility are material specific. Charge carriers' drift velocity is proportional to the value of their mobility in a particular field. In most semiconductor materials, the mobility value is not a constant in all directions. This anisotropy effect of the material needs to be included in the model. The distribution of impurity concentration can also alter the actual electric field, especially in the depleted region. In the depleted region, impurities act as space charge and create an electric field on top of the field generated by the external voltage bias. Sometimes, one has to take into account the defect in the crystal lattice created by radiation damage as well. However, this is out of the scope of this work.

The way that weighting potential is calculated was discussed by Shockley and Ramo [5-6]. The weighting potential of an electrode is a mathematical approach to simplify the calculation of the induced charge, not a physical parameter. A weighting potential field can be calculated for each electrode. Weighting potential is a dimensionless quantity. Its numerical value can be calculated as following. First, one needs to remove all space charge. Then, assign positive unit voltage to the interested electrode and 0 volt to all other electrodes. The numerical value of weighting potential field is equal to the electrostatic potential generated under this condition. The same value for geometric configuration and material properties should be used as the above calculation of actual electric field.

As shown above, both actual operating electric field and the weighting potential field within a semiconductor crystal can be obtained via electrostatic field calculation. Quite some software packages are available for this purpose. In this work, a commercially available software tool, called QuickField, is used to calculate the field distribution. QuickField is a very efficient finite element analysis package for electromagnetic, thermal, and stress design simulation. The DC magnetic package handles electrostatic field calculation. One limitation of QuickField is that it can only model axial symmetric geometry. Using a graphic user interface, one can define the two dimensional cross section of the model, specify material properties including impurity concentration distribution, boundary conditions and manually create finite element mesh. The results can be output on a grid with user-specified density.

A customized toolkit is developed in MATLAB for signal calculation. Given the actual electric field distribution within the semiconductor crystal, the trajectory of the charge carriers can be constructed. In solid semiconductor materials, the charge carriers are constantly undergoing collisions with surrounding atoms. As a result, these charge carriers follow the electric field line while being collected by electrodes. The trajectory of charge carriers overlap with the electric field lines passing through the position of the production of electron-hole pairs. The drift velocity of charge carriers at any given location equals the product of their mobility and the strength of the operating electric field. The change in the induced charge on a particular electrode due to movement of a charge Q is give by the product of charge Q and the difference in weighting potential between final and initial location of the charge Q. Knowing the trajectory and the drift velocity, one can then construct the output signal due to induced charge on the electrode.

In addition to creating the pulse shape caused by charge carriers produced at a specific position, the toolkit can also read in the Monte Carlo simulation results. For each incident photon in the simulation, the position and energy deposition of each interaction are used to construct an output pulse. This signal can then be analyzed using various digital shaping techniques, such as trapezoidal shaper.

Sample Applications

The method to calculate the output signal described above has many applications, such as to provide basis for signal decomposition algorithm, to give predictive results for new detector design, to help explaining measured results and etc.. In this section, a few examples of the application of this signal calculation method are shown.

1. Pulse shape study for generic p-type point contact detector

As a proof of concept, the method was first used on a generic p-type point contact HPGe detector. The detector crystal has a diameter of 100 mm and a length of 25 mm. In the model, the germanium crystal is mounted on a copper holder and wrapped in isolation materials. The detector endcap is made of aluminum. At the front end of the endcap, there is a thin carbon window to improve low energy gamma sensitivity. The Monte Carlo model of this detector is shown in Figure 1. The module is then used to simulate the energy deposition inside the crystal by gamma rays using GEANT4. For each photon event, the position and energy deposition of each interaction are recorded during the simulation. Figure 2 shows the operating electric field, when +4000v bias was applied on the front and side electrodes, which are n^{+} contacts. It can be observed that the electric field is much stronger near the point contact. Figure 3 shows the weighting potential field inside the crystal. It can be observed that the difference in weighting potential is much larger around the point contact. Thus, majority of the signal comes from collection of holes. Figure 4 is the user interface of the customized toolkit, showing the charge carrier collection trajectories, pre-amplifier output signal and the output signal after a trapezoidal shaper.

Figure 1. Monte Carlo model of a generic p-type point contact HPGe detector

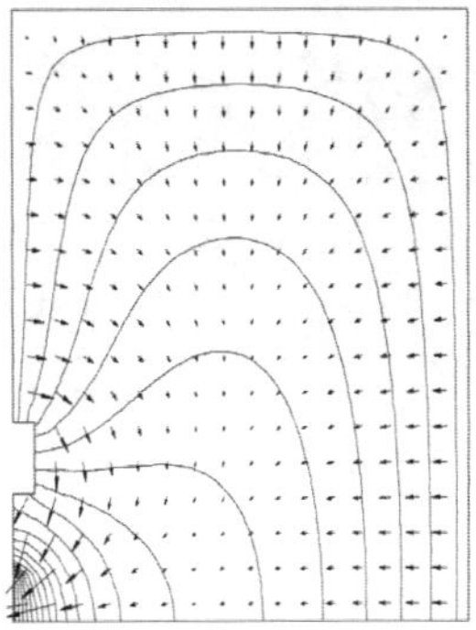

Figure 2. Operating electric field inside the germanium crystal

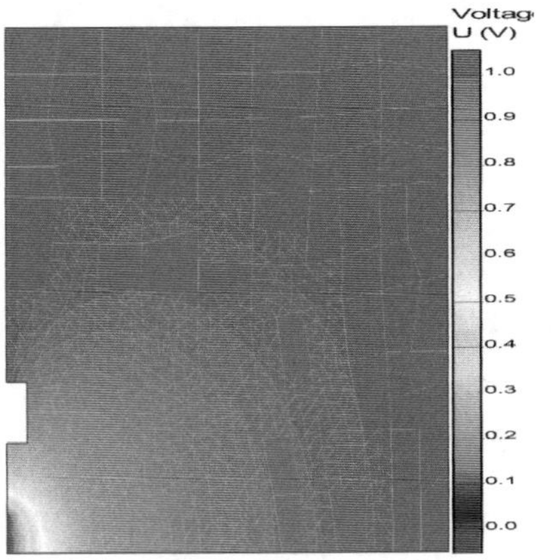

Figure 3. Weighting potential field inside the germanium crystal

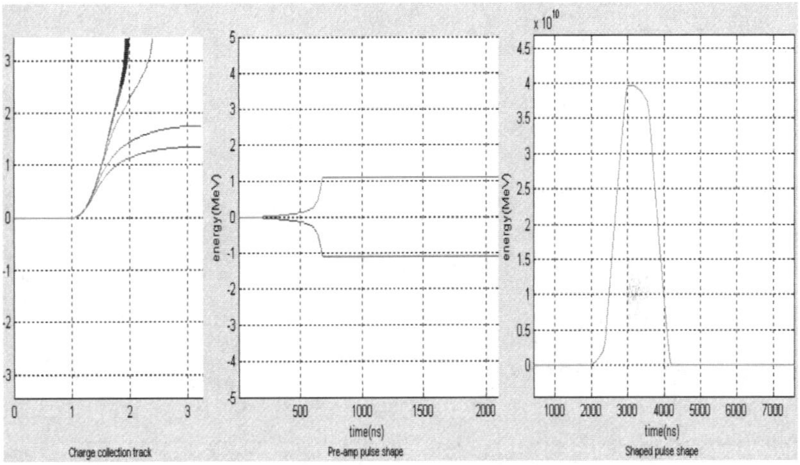

Figure 4. Charge collection trajectories, pre-amp signal and shaping amplifier output

2. Signal amplitude loss due to surface channels

The charge collection near edges of semiconductor detectors may be dramatically disturbed by surface channels. This has been previously discussed by Malm and Dinger in their paper [7]. It was shown by scanning measurements with a collimated low energy gamma ray beam that the loss of signal amplitude is caused by the loss of one charge carrier through surface channel. In this process, one carrier in the electron-hole pair is attracted to the surface by the electric field and does not contribute to the pulse height. The method described above has been used to investigate this surface channel effect. The surface channels are approximated using a step function of charge distribution in this preliminary study, where $|N_D - N_A| \sim 10^{11}\ cm^{-3}$. The operating electric field inside a planar germanium crystal is shown in Figure 5. Constructed using the method described above, the charge collection trajectories and the output signals are shown in Figure 6. The loss in pulse height due to the loss of one charge carrier can be clearly observed.

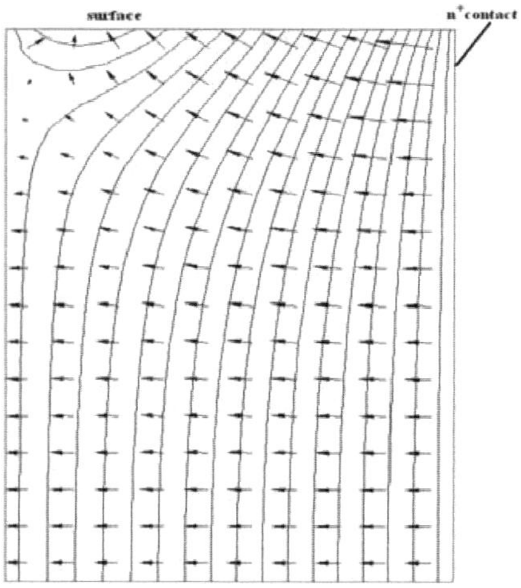

Figure 5. Electric filed inside a planar crystal with surface channels

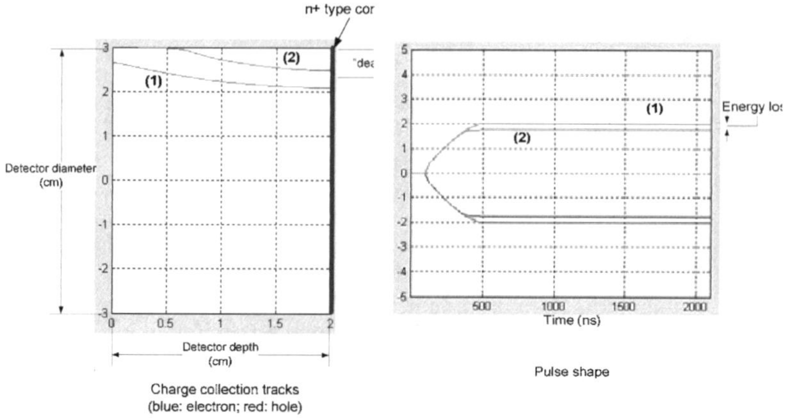

Figure 6. loss of one charge of carrier and the subsequent energy loss due to surface channels

3. Pulse shape study for segmented p-type point contact HPGe detector

By carefully tailoring the electric field inside a semiconductor detector, one can extract position information from a single crystal. At LBNL, Amman and etc. have developed a segmented p-type point contact HPGe detector that can provide two-dimension position information in τ-θ space [8]. The cross section of their design is shown in Figure 7. There are three contacts in this design: the point contact, the front contact and the side contact. Assuming a constant impurity level in the crystal, the operating electric field was calculated and shown in Figure 8. The weighting potential fields for the point contact and the front contact are shown in Figure 9. In Figure 10, the measured signals presented by Amman and the signal calculated using the method described in previous sections are compared. It is demonstrated that a signal calculation method as described here can reliably predict the output signal pulse shape. This is very useful to produce a decomposition pulse shape library for template matching technique.

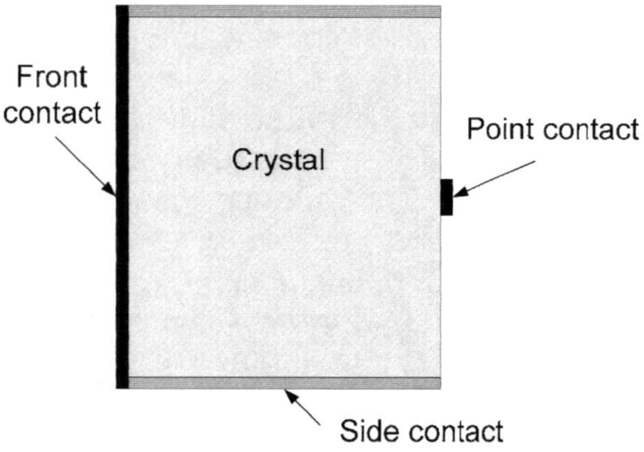

Figure 7. Cross section of LBNL segmented p-type point contact HPGe

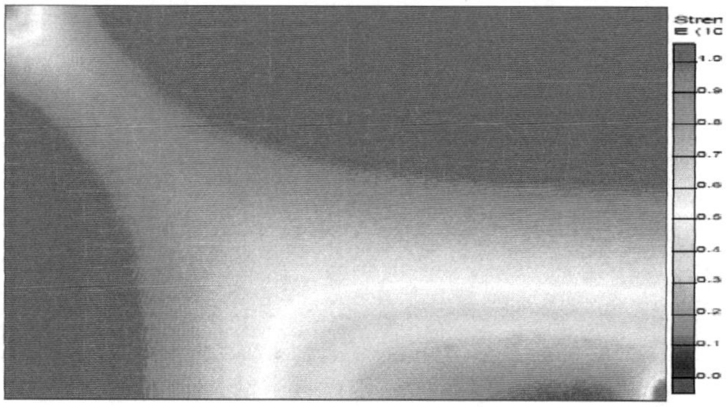

Figure 8. Calculated operating electric field

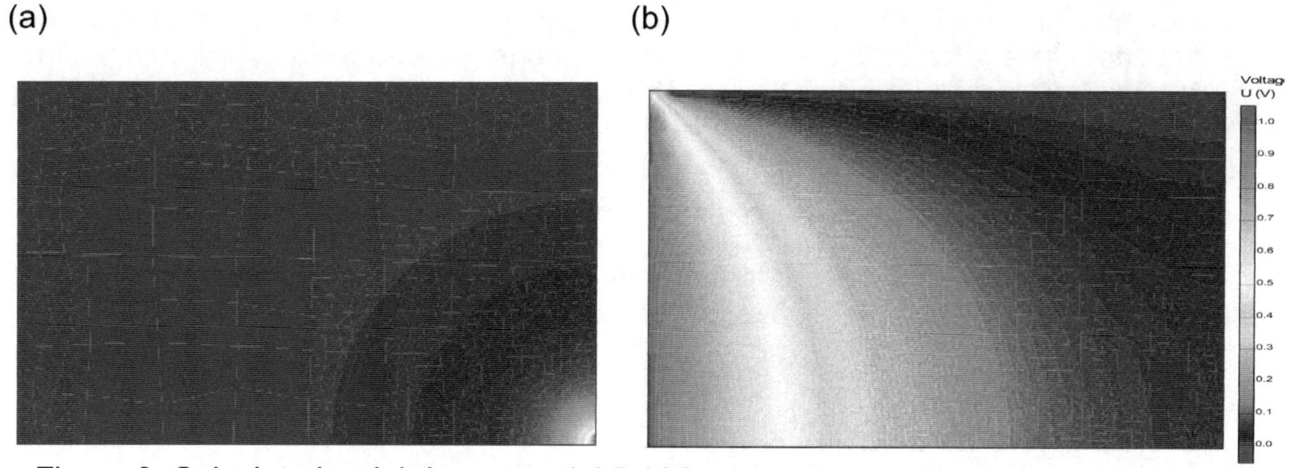

Figure 9. Calculated weighting potential field for (a) point contact and (b) front contact

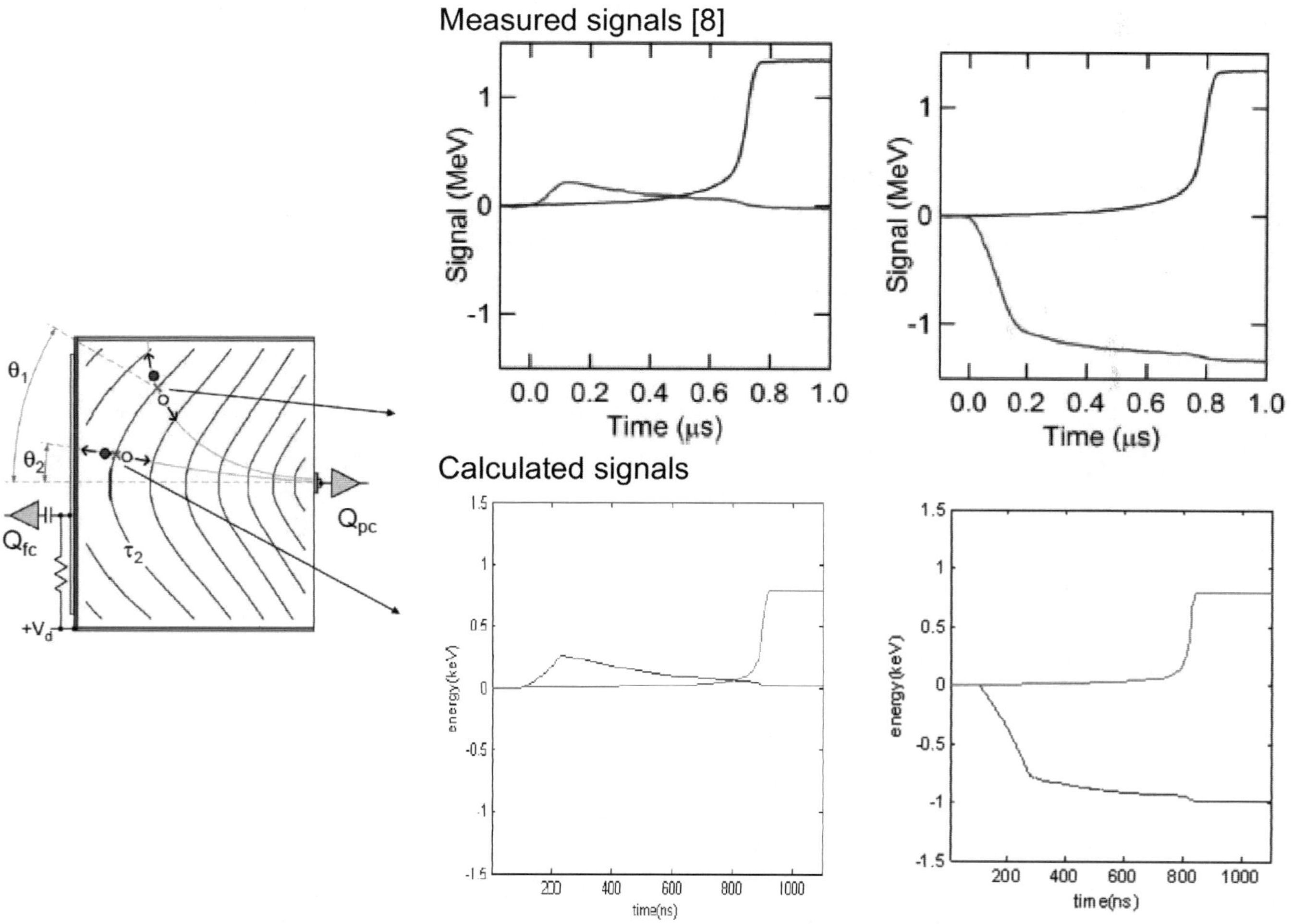

Figure 10. Comparison between measured and calculated signals

Conclusions

A calculation method is shown here to study pulse shape of the output signal from semiconductor detectors. For each incident photon, energy deposition and interactions positions are simulated using Monte Carlo method. Operating electric field and weighting potential field within the semiconductor crystal are calculated using finite element method. A customized toolkit has been developed to construct charge collection trajectory, calculate changes in weighting potential. Pulse shape of the output signal is then calculated based on Shockley-Ramo theorem. Several samples of the potential applications of this method are shown in this paper. The method can be further improved by implementing finite range of the primary electron and finite size of the electron cloud during the ionization process.

References

[1] Z. He, "Review of the Shockley-Ramo theorem and its applications in semiconductor gamma-ray detectors," *Nuclear Instruments and Methods*, A463 (2001) 250-267

[2] I. Y. Lee, "Gamma-ray tracking detectors", *Nuclear Instruments and Methods*, A422 (1999) 195-200

[3] T. E. Booth, F.B. Brown, J.S. Bull, et al, "MCNP5 1.50 Release Notes", (2008).

[4] S. Agostinelli, et al, "G4-a simulation toolkit", *Nuclear Instruments and Methods in Physics Research* A 506 (2003) 250-303

[5] W. Shockley, "Currents to conductors induced by a moving point charge, " *J. Appl. Phys.*, vol.9, pp. 635-636, 1938

[6] S. Ramo, "Currents Induced by electron motion," *Proc. I.R.E.*, vol. 27, pp. 584-585, 1939

[7] H. L. Malm, R. J. Dinger, "Charge collection in surface channels on high-purity Ge detectors, " *IEEE Trans. on Nucl. Sci.*, vol. NS-23, No. 1, 1976

[8] M. Amman, P. Luke, Y. Chan, K. Lesko, and the Majorana group, "P-type point contact germanium detectors, " *IEEE Nuclear Science Symposium* 2009

Modeling of a System of Equations to Determine the Concentrations of U^{235} and Pu^{239} in a Sample

Dongok Choe
Utah Nuclear Engineering Program (UNEP)
The University of Utah
50 S. Central Campus Dr. Room MEB 1206
Salt Lake City, UT 84112

ABSTRACT

Environmental Pu^{239} arises from the development, production and testing of nuclear weapons. Fall out of this weapon debris has produced an average worldwide Pu^{239} level of about 1 ppb. Traditional chemical separation methods are available to measure Pu^{239} at fCi levels, but these methods require significant works on chemical separation process and are unable to detect concentration of Pu^{239} at environmental levels (~100aCi).

Fissile samples are exposed to distinct neutron spectra to achieve quantitative analysis of each fissile isotope using track registration method with minimum chemical processing. U^{235} and Pu^{239} have unique fission cross-sections for different neutron energies. It is possible to differentiate between the isotopes based on the number of fission tracks produced. The fission tracks can be generated using a neutron source such as a nuclear reactor, high-energy gamma source or a neutron generator. In this research, a Monte Carlo method (MCNP5) was used to generate theoretical number of fissions and from this data a system of equations was developed. MCNP5 will generate two or three distinct neutron spectra that can be obtained from a nuclear reactor. These distinct energy spectra are cold, thermal and epithermal.

The proposed method can measure or predict the amount of Th^{232}, U^{235}, and Pu^{239} in the sample simultaneously. Theoretically, the detection limit for these isotopes is ~100 aCi/gram. The calculated result from this study must be compared with an experimental measurement. The uncertainties associated with this theoretical approach stem from the differences in theoretical and actual neutron flux in the irradiators. By providing more realistic neutron spectra and neutron filtering system, these uncertainties can be reduced.

Introduction

After World War II, scientists such as Fleischer and Price measured Pu^{239} in the environment by fission track analysis[1]. During the same time period, Silk and Barnes[2] discovered fission tracks caused from fissile nuclei such as Th^{232}, U^{235}, and Pu^{239}. The 1939 publication by Hahn and Strassmann[3] is generally considered to be the initial document that identifies and describes the fission process with the first theoretical explanation for the fission of a nucleus presented shortly after by Bohr and Wheeler[4].

In 1988, Van Assche suggests that the first fission process was found much earlier in 1925 when Noddack identified the first fission product, Tc^{99} [5]. In 1963, Fleischer and Price measured fission tracks in a mica detector using a Cf^{252} source[6]. Fleischer and Price used a high-energy argon ion beam to generate fission tracks incrystalline detectors. Originally, Fission Track Analysis (FTA) was used in the field of archeological age dating. Later, it was used to measure high-energy cosmic rays in space, and to detect some isotopes, especially Pu^{239} and U^{235}, in the environment[7]. The production of Pu^{239} is mainly from the following nuclear reaction:

$$U_{92}^{238} \quad n_0^1 \quad U_{92}^{239} \qquad Np_{93}^{239} \qquad Pu_{94}^{239} \tag{1}$$

Interest in measuring Pu^{239} increased with the development and testing of nuclear weapons. The risk to the health of the human population from low level exposure of Pu is unknown.

The recent promulgation of ultra-conservative regulatory limits on the allowable concentrations of selected actinides in the environment (i.e., drinking water) and the protection from exposure to plutonium at active and discontinued processing sites (i.e., Rocky Flats) suggest that the radiotoxicity may be overestimated. The cost of implementing overzealous safety and radiation protection (ALARA) programs may require a financial burden that does not correlate with protection from significant risk to the occupational worker. An overly conservative regulation slows the progress of environmental restoration and remedial safety programs. Ultra conservatism, while appearing prudent, may only reduce the number of concerns that can be addressed without realizing a significant safety benefit. Delaying other remedial tasks to later dates may allow for further deterioration of containment systems or barriers resulting in a possible of increase exposure risk. Regulations concerning exposure limits should be scientifically based, thus allowing the proper levels of protection to be applied.

Two well-characterized neutron energy spectra were used to generate fission tracks that can be compared with the calculated model in this study. Variations in the nucleonics and interaction parameters (i.e. cross-sections) will provide an effective means to quantify the presence of both fissile radioisotopes. We implemented a multiple fluence exposure that enabled the distinction between fission tracks caused by Pu^{239} and U^{235}.

The theoretical basis for the technique and mathematical details of the system of equations are discussed for the University of Utah TRIGA reactor. The system of equations was derived to calculate the concentration of Th^{232}, Pu^{239}, and uranium in a sample using a multi group neutron model. Neutron spectra for two different positions were obtained using the Monte Carlo computer code MCNP5[8].

Fission Track Analysis: Physics and Applications

There are two different kinds of track registration processes: track formation in crystalline materials, and track formation in polymers. Damage to a crystalline lattice, due to the ionizing fission fragments, is a result of the mutual repulsion of like ions that are adjacent to each other. When high-energy fission fragments pass through crystalline material, the atoms in the material ionize and move from their original locations leaving vacant lattice sites. This theory, also known as lattice instability model, suggests that all materials possess some threshold ionization density below which a particle will not leave a fission track. The second process is the registration of fission tracks in polymers such as polycarbonate. The fission fragments break the polymer chain to create chemically reactive sites[9].

In general, two detector types are used to register fission tracks: mica and polycarbonate (Lexan). These two detectors produce different fission track shapes. The shape of fission tracks in the mica detector is elliptical or a distorted circle. Alternatively, fission tracks in the polycarbonate detectors have a thin cone shape. The mica detector requires more etching time than a polycarbonate detector[9]. Both U^{235} and Pu^{239} are α emitters but α particles or any particle with a mass smaller than 20 amu will not register tracks on a polycarbonate detector[9]. Hepburn and Windle[4] have listed a few of more common solid state track detectors and have cataloged their sensitivity to ions by the lightest ion that will leave an etchable track. From their data, the lightest detectable ion for polycarbonates, such as Lexan and Makrofol, is carbon. Hepburn and Windle acknowledge polycarbonates produced by different manufacturing processes will have different sensitivities. Even if some variations in the sensitivity of different polycarbonates exist, this data would indicate that polycarbonate is more than sensitive enough to register tracks from fission fragments.

Neutron Fission Cross-Section and Flux Parameters

The fission cross-sections for fissile isotopes such as U^{235} and Pu^{239} follow the *1/v* law in the thermal neutron energy region. The total cross-section in the thermal energy range is given by:

$$(v) \quad \frac{1}{h^4}|M|^2 \frac{P_b^{\ 2}}{v_b v}(2I_b \quad 1) \tag{2}$$

where *b* is the emergent particle and has a kinetic energy on the order of 1 MeV, *M* is the nuclear matrix element, and h is Planck's constant. If the ratio P_b^2/v_b and matrix element *M* are independent of *v*, then the above equation can be reduced to:

$$_{total}(v) \quad \frac{_0 v_0}{v} \tag{3}$$

where σ_0 is the measured neutron fission cross-section at thermal neutron velocity v_0=2200 m/sec. The total cross-section $\sigma(v)$ does not change exactly as *1/v* because there exist several resonance regions nearby. These resonance peaks can vary from 0~1000 barns. For the pure Maxwellian neutron spectrum, we can calculate the neutron cross-section at temperature T_n as follows :

$$\frac{{}_0\sqrt{\frac{2}{m}} \ {}_0 \ (E)E\exp(\ \frac{E_n}{kT_n})dE}{\sqrt{E}\exp(\ \frac{E_n}{kT_n})dE \ {}_0} \qquad g(T_n) \ {}_0 \tag{4}$$

The Westcott *g*-factor or non *1/v* factor $g(T_n)$ is defined as:

$$g(T) \quad {}_0 \ \frac{(E_n)\sqrt{E_n}M(E_n,T)}{{}^{th}\sqrt{E_n^{\ th}}}dE_n \tag{5}$$

where

$$M(E_n,T) \quad \frac{4n_0}{\bar{v}_n^0\sqrt{\ }}v_n^2\exp\ \frac{v_n^2}{(\bar{v}_n^0)^2} \ ,$$

$\bar{v}_n^{\ 0}$ is the most probable velocity, $v_n^{\ 0}$ is the velocity at temperature T_n = 293 °K, and $E_n^{\ th}$ is 0.0253 eV. The *g*-factors for U^{235} and Pu^{239} at T_n = 293 °K are 0.9780 and 1.0723, respectively[10].

The neutron fission cross-sections for the three isotopes, Th^{232}, U^{235}, and Pu^{239} have approximately the same shape in the thermal energy region. In general, the fission cross-section at the thermal neutron energy is given by:

$$^{th}_{n,f} \quad 2.57 \quad 10^9 \frac{{}_n \ {}_f}{(E_n^0\ )^2 \ \ (\frac{\ }{2})^2}(barns) \tag{6}$$

where Γ is the reduced neutron widths.
The fission cross-section in fast neutron energy varies smoothly with neutron energy. The average fission cross-section for fast neutron can be given by:

$$\sigma_{nf} \approx \frac{1}{\Delta E_n} \int_{E_n - \frac{\Delta E_n}{2}}^{E_n + \frac{\Delta E_n}{2}} \sigma_{nf}(E_n')\,dE_n' \qquad (7)$$

where ΔE_n is the averaged neutron energy interval and should be large enough to diffuse possible structural effects.

In the neutron energy range 1 eV ~ 100 eV, the neutron fission cross-section is composed of many sharp and closely-packed resonance peaks. In this energy range, the cross-section is very sensitive to neutron energy. When the incident neutron energy is comparable to that of the lowest energy level in the target nucleus, the incident neutron has more of a chance to react with the target nucleus[11]. The interaction of slow neutrons with the target nucleus proceeds essentially through out the formation of compound nucleus states.

Th^{232} has a threshold fission cross-section. Below the neutron energy of 2 MeV, the fission cross-section is not well known and is considered to be 0 barns. Some experiments[12] have shown that Th^{232} has a fission cross-section of a few micro-barns for thermal energies. Unfortunately, the uncertainties associated with this cross-section preclude its use in this study. The fission cross-section goes up to ~10 barns at a neutron energy of 3 MeV and varies smoothly with neutron energies. The typical thermal research reactor like the University of Utah TRIGA has a small portion of fast neutrons in the thermal irradiator.

U^{235} and Pu^{239} have very similar patterns of fission cross-sections throughout the resonance and fast region. In fact, in the resonance area, it is impossible to differentiate between the two cross-section curves. In the fast neutron energy region, the fission cross-section varies smoothly with neutron energies in both cases. Furthermore, U^{235} and Pu^{239} have very similar characteristics in their masses and cross-sections. Since their masses and resultant fission fragments are very similar, it is almost impossible to differentiate each isotope from the shape of the fission tracks.

In the thermal region, there is a difference in fission cross-sections between U^{235} and Pu^{239}. Where U^{235} has a fission cross-section of 582 barns at $E_n = 0.025$eV, Pu^{239} has a fission cross-section of 750 barns at the same neutron energy. This difference can also be seen in the epithermal energy region. To take the advantage of fission cross-section differences, the spectrum of the neutron flux needs to be skewed to the thermal and epithermal region. The neutron flux in which the epithermal region is dominant can be generated using several combinations of materials such as cadmium and boron. Further, a silicon crystal can be used as a fast neutron absorber if it is cooled with liquid nitrogen. Utilizing the fission cross-section differences and two well-known neutron spectra, we can define a mathematical model to calculate the amounts of two or three isotopes.

In the database ENDF/B-V[12], the fission cross-section for U^{238} at neutron energies below 1.6 MeV was considered to be 0 barn. According to the ENDF/B-VI database, the fission cross-section of U^{238} has several resonance peaks between 1 eV and 0.1 MeV, then varies smoothly through the scattering dominant neutron energy region. The fission cross-section for U^{238} follows the 1/v law in the thermal energy region and has a maximum value of 15 mbarns at $E_n = 0.025$ eV. Since U^{238} is more abundant in nature than U^{235}, this small fission cross-section corresponds to ~0.15 mbarns of U^{235} fission cross-section.

The neutron flux and fission cross-section are energy dependent quantities. The whole neutron spectrum from the nuclear reactor core can be divided into three distinct regions; thermal, epithermal and fast neutron or scattering dominant regions. The thermal neutron spectrum has Maxwell-Boltzmann distribution and has a peak value at the neutron energy $E = 0.025$ eV. The neutron spectrum in thermal region can be approximated by:

$$_{Th}(E) \quad CE^{\frac{1}{2}}\exp(\quad \frac{E}{kT}) \tag{8}$$

where E is neutron energy, C is a constant and k is the Boltzmann constant.

Development of Theoretical Basis Model

The total number of fission fragments entering the detector is:

$$\begin{aligned} R_{m,total} & \quad N_i \;\; _{i,f}(E) \;\; (E)\,dE\,dt \\ & _{t,E} \\ & \quad tN_i \;\; ^{E}_{0} \;\; _{i,f}(E) \;\; (E)\,dE \end{aligned} \tag{9}$$

The actual number of fission fragments entering the detector for a steady state neutron flux is:

$$\begin{aligned} R_{m,fission,actual} & \quad tr_0 N_i \;\; _{i,f}(E) \;\; (E)\,dE\,dA \\ & _{A,E} \\ & \quad Ar_0 tN_i \;\; ^{E}_{0} \;\; (E) \;\; _{i,f}(E)\,dE \end{aligned} \tag{10}$$

where t is irradiation time, A is detector area, and r_0 is the maximum range of the fission fragment in a specific material. The normalized number of fission tracks FT can be calculated using equation (10).

$$\begin{aligned} FT & \quad \frac{R_{m,fission,actual}}{At} \\ & \quad (E)N_i \end{aligned} \tag{11}$$

where

$$(E) \quad r_0 \;\; ^{E}_{0} \;\; (E) \;\; _{i,f}(E)\,dE$$

The variable r includes the efficiency factor e and the range of the fission fragment r_0 together. If we consider four isotopes; Th^{232}, U^{235}, U^{238}, and Pu^{239}, equation (11) becomes:

$$FT \quad _{Th}N_{Th} \quad _{U}N_{U} \quad _{Pu}N_{Pu} \tag{12}$$

where α is the coefficient to be determined. To solve equation (12), three unique irradiation positions are required, one of which should have ample fast neutrons to produce sufficient Th^{232} fission reactions to count. In general, the number of fission tracks for a mixed sample in the single energy neutron model can be written as:

$$FT \quad _{i}N_i \;\; _{i} \tag{13}$$

where

$$_{Th} \quad r_{02} \;\; (E) \;\; _{02,f}(E)\,dE \;\; _{E}$$

$$_{U} \quad r_{25} \underset{E}{} \ (E) \ {}_{25,f}(E)dE \quad r_{28} \underset{E}{} \ (E) \ {}_{28,f}(E)dE$$

$$r_{U} \underset{E}{} \ (E)[\ {}_{25,f}(E) \quad {}_{28,f}(E)]dE$$, and

$$_{Pu} \quad r_{49} \underset{E}{} \ (E) \ {}_{49,f}(E)dE$$

In equation (13), *FT* is a function of the fission cross-section, therefore this equation has limited applications. For example, if a suitable filter system is designed, equation (13) can be applied for the Maxwellian energy distribution of cold neutrons with small errors. Conversely, the flux in a reactor core has a broad energy spectrum ranging from 0 ~ 20 MeV. Therefore, to approximate a real system, a multi-group, energy dependent system of equations needs to be developed. For three different irradiation positions such as thermal – fast – cold, or thermal – cold – thermal with cadmium, equation (13) becomes:

$$\begin{matrix} FT_1 & {}_{Th,1}N_{Th} & {}_{U,1}N_{U} & {}_{Pu,1}N_{Pu} \\ FT_2 & {}_{Th,2}N_{Th} & {}_{U,2}N_{U} & {}_{Pu,2}N_{Pu} \\ FT_3 & {}_{Th,3}N_{Th} & {}_{U,3}N_{U} & {}_{Pu,3}N_{Pu} \end{matrix} \qquad (14)$$

Equations (14) can be solved using N_{Th}, N_U, and N_{Pu} and the following matrix relationships:

$$\begin{matrix} FT_1 \\ FT_2 \\ FT_3 \end{matrix} \quad \begin{matrix} {}_{Th,1} & {}_{U,1} & {}_{Pu,1} \\ {}_{Th,2} & {}_{U,2} & {}_{Pu,2} \\ {}_{Th,3} & {}_{U,3} & {}_{Pu,3} \end{matrix} \quad \begin{matrix} N_{Th} \\ N_{U} \\ N_{Pu} \end{matrix} \qquad (15)$$

After some mathematical manipulations, we obtain the concentration of the three isotopes as follows:

$$N_{Th} \quad \frac{1}{M}[(\ {}_{U,2}\ {}_{Pu,3} \quad {}_{U,3}\ {}_{Pu,2})FT_1 \quad (\ {}_{Th,3}\ {}_{Pu,2} \quad {}_{Th,2}\ {}_{Pu,3})FT_2$$
$$(\ {}_{Th,2}\ {}_{U,3} \quad {}_{Th,3}\ {}_{U,2})FT_3] \qquad (16)$$

$$N_{U} \quad \frac{1}{M}[(\ {}_{U,3}\ {}_{Pu,1} \quad {}_{U,1}\ {}_{Pu,3})FT_1 \quad (\ {}_{Th,1}\ {}_{Pu,3} \quad {}_{Th,3}\ {}_{Pu,1})FT_2$$
$$(\ {}_{Th,3}\ {}_{U,1} \quad {}_{Th,1}\ {}_{U,3})FT_3] \qquad (17)$$

$$N_{Pu} \quad \frac{1}{M}[(\ {}_{U,1}\ {}_{Pu,2} \quad {}_{U,2}\ {}_{Pu,1})FT_1 \quad (\ {}_{Th,2}\ {}_{Pu,1} \quad {}_{Th,1}\ {}_{Pu,2})FT_2$$
$$(\ {}_{Th,1}\ {}_{U,2} \quad {}_{Th,2}\ {}_{U,1})FT_3] \qquad (18)$$

where subscript 1, 2, and 3 stand for 3 unique irradiation positions.

The thermal neutron fission cross-section for Th^{232} is approximately 40 mbarns[12] and increases approximately 4 barns in the scattering dominant neutron energy region. For this reason, the Th^{232} contribution in equations (16) through (18) is insignificant. Therefore, considering only natural uranium and Pu^{239}, the above three equations can be reduced to the following two equations:

$$N_{U} \quad \frac{{}_{Pu,2}FT_1 \quad {}_{Pu,1}FT_2}{{}_{U,1}\ {}_{Pu,2} \quad {}_{U,2}\ {}_{Pu,1}} \qquad (19)$$

$$N_{Pu} \quad \frac{{}_{U,1}FT_2 \qquad {}_{U,2}FT_1}{{}_{U,1} \quad {}_{Pu,2} \qquad {}_{U,2} \quad {}_{Pu,1}} \tag{20}$$

Development of Experimental Facilities and Procedures

Two different reactor-core positions that possess significantly different characteristics were used to generate two distinct sets of equation. The samples were prepared using a direct deposit method. Samples containing known amounts of Pu^{239}, natural uranium (U^{235} and U^{238} only), and two mixed isotopes were deposited on the polycarbonate Lexan detectors. After irradiating these samples for a suitable time (10 minutes at 90 kW) in the reactor core, the fission tracks were enhanced by a chemical etching process and counted using an optical microscope. These results were then inserted into a mathematical model that calculates the amount of Pu and U in the sample. Table 1 shows two sets of calculated α values for the thermal irradiator and the central irradiator with cadmium foil.

Table 1. Calculated a coefficients for the thermal and central irradiator with and without cadmium foil

Coefficients	Irradiation positions	Calculated Values
$\alpha_{U,1}$	TI	$(6.2479 \pm 0.17)\text{x}10^{-13}$
$\alpha_{U,2}$	CI with Cadmium Foil	$(1.1304 \pm 0.23)\text{x}10^{-13}$
$\alpha_{Pu,1}$	TI	$(8.1976 \pm 0.16)\text{x}10^{-11}$
$\alpha_{Pu,2}$	CI with cadmium foil	$(1.1492 \pm 0.23)\text{x}10^{-11}$

Polycarbonate (Lexan) slides were used as the detectors to register the fission tracks. The dimension of the detectors were 26 mm x 56 mm with a thickness of 0.35 mm and each detector has four circles for sample deposit and two small circles for Lexan blanks. To minimize the background effects, these Lexan slides were cleaned for 30 minutes in the 8 normal HCl solution and another 30 minutes in the 0.1 normal HCl solution. Both HCl solutions were pretreated with anion exchange columns and sub-boiling distillation to lower fissile background. After cleaning with the HCl solution, the detectors were cleaned with EtOH solution and dried in a class 100 clean room. Natural uranium and Pu^{239} solutions were directly deposited on the Lexan slides. Different amounts of U^{235}, U^{238}, Pu^{239}, and mixed natural uranium and Pu^{239} were deposited in the individual circles. The samples were dried then under the infrared lights. Spacers with the same thickness as the Lexan detectors were placed between each detector. For each irradiation, three sets of samples (low, medium, and high concentrations of natural uranium, natural uranium + Pu^{239}, and Pu^{239}) were provided. The concentration of U^{238} in these samples is at least two order of magnitude higher than those of U^{235}. Since the fission cross-section of U^{238} in thermal and epithermal regions is very small (less than a mbarn) compared to those of U^{235}, the occurrence of fission tracks due to U^{238} is negligible. The fission cross-section for U^{238} was taken as 0 below a neutron energy of 1.2 MeV as reported in ENDF/B-V. To obtain a different neutron spectrum, each set of Lexan detectors was covered with cadmium foil and then placed in the central irradiator. The detectors were stacked in the polyethylene

TRIGA holder. Samples were deposited in a very small area which allowed the sample to be treated as a point source. All samples were irradiated for 10 minutes, which corresponds to a total neutron fluence of 2.166 x 10^{14} neutrons/cm^2 for a thermal irradiator.

The University of Utah's 100 kW TRIGA reactor was used to obtain two different neutron spectra: thermal and epithermal. To obtain a modified or epithermal neutron spectrum, a 0.4 mm cadmium foil was placed around the sample to reduce the thermal segment of the whole neutron spectrum. The absorption cross-section of cadmium is ~2800 barns at a neutron energy of 0.025 eV and drops to almost 0 near the neutron energy of 0.4 eV. Table 2 shows the number of fission tracks and calculated concentration from equation (20).

Table 2. A comparison of the concentrations of U^{235} and Pu^{239} with deposited concentrations using Equation (20)

Sample ID		Number of Fission tracks	Deposited concentrations ($nuclei/cm^3$)	Concentrations from equation (20) ($nuclei/cm^3$)
T-3 (TI)	U-235		3.919E13	2.3426E13
	Pu-239	650	1.978E09	6.5875E09
	U-235		2.864E13	5.4358E13
	Pu-239	502	1.266E09	2.3247E09
	U-235		2.562E13	N/A
	Pu-239	498	1.187E09	N/A
	U-235		3.920E13	1.3098E13
	Pu-239	609	1.978E09	6.8462E09
T-4 (TI + cd)	U-235		3.769E13	N/A
	Pu-239	9	1.899E09	N/A
	U-235		3.919E13	2.3426E13
	Pu-239	16	1.981E09	6.5875E09
	U-235		3.618E13	N/A
	Pu-239	8	1.028E09	N/A
	U-235		3.920E13	1.3098E13
	Pu-239	14	1.980E09	6.8462E09
T-7 (CI +cd)	U-235		1.809E13	N/A
	Pu-239	187	9.493E09	N/A
	U-235		2.862E13	5.4358E13
	Pu-239	73	1.266E09	2.3247E09
	U-235		3.919E13	1.3098E13
	Pu-239	92	1.976E09	6.8462E09
	U-235		5.728E13	2.3426E13
	Pu-239	103	2.373E09	6.5875E09

Results and Discussion

The largest contribution to errors associated with a coefficients was with fission track counting, estimated neutron spectra, and measured flux. Further variance will be introduced in calculating these coefficients by using the average values of the cross-sections in the resonance region. With these coefficients substituted into Equations (19) and (20), the concentrations of natural uranium and Pu^{239} can be calculated.

$$N_U = -2.7540 \times 10^{11}\, T_1 + 1.9654 \times 10^{12}\, T_2$$
$$N_{Pu} = 3.3861 \times 10^{7}\, T_1 - 1.4973 \times 10^{8}\, T_2 \qquad (20)$$

T_1 and T_2 represent the number of fission tracks in irradiation position 1 (thermal irradiator) and irradiation position 2 (central irradiator with cadmium foil), respectively. The results from Equation (20) possess significant errors in the predicted concentrations of both isotopes because the modified (epithermal) neutron flux in the thermal irradiator has a large variance in the thermal neutron energy region.

Conclusion

The purpose of this study is to develop the theoretical basis of the technique in which the detection limit can be maintained or possibly lowered without using complicated and expensive chemical processing. MCNP5 and activation experiments were performed that produced reasonably accurate neutron spectra for two configurations in the core of the 100 kW TRIGA nuclear reactor. Mathematical correlations were produced from two unique neutron spectra:

$$N_U = -2.7540 \times 10^{11}\, T_1 + 1.9654 \times 10^{12}\, T_2$$

$$N_{Pu} = 3.3861 \times 10^{7}\, T_1 - 1.4973 \times 10^{8}\, T_2$$

T_1 and T_2 are the number of tracks counted from samples exposed to a known neutron spectra and flux (thermal irradiator and the central irradiator with cadmium foil) and the coefficients were experimentally determined from NIST traceable standards.

The predicted concentrations of natural uranium and Pu^{239} in a sample were in relatively good agreement. The predicted amounts ranged from one to three times from the known quantities of Pu and U deposited on the polycarbonate detectors. The investigation suggests that the detection limit of ~120 aCi/g can easily be achieved without chemical processing. The detection limit can be greatly reduced by exposing the sample detector to higher fluences (i.e.,10^{15} neutron/cm^2,10^{16} neutrons/cm^2, etc.) which generates a larger number of tracks. The uncertainties associated with this method result from the differences in theoretical and actual neutron spectra and flux in the irradiators, the inefficiencies with etching process and errors with the track counting activity.

It is clear from this work that creating a distinctly different and well-characterized neutron spectra can assist in calculating more effective model coefficients that reduce the detection limit further. Minimizing errors requires a combination of different materials such as boron, gadolinium, and hafnium, which effectively absorb or scatter the undesirable portion of the neutron spectrum. For example, to reduce the population of fast neutrons, we can use an aluminum or silicon cooled with liquid nitrogen which have comparably larger absorption cross-sections. Also, processes not considered in this research, such as sample deposition, detector selection, etching, and track counting can be further refined to better optimize the technique. In general, the work confirms the theoretical basis for eliminating the need to chemically separate other fissile atoms (e.g., Uranium or Thorium) from a sample in order to accurately determine the Pu content.

References

1. R. L. Fleischer, P. B. price, and Walker, "A Simple Method of measuring Low Uranium Concentrations in Natural Crystal," AppliedPhysics letters, Vol. 2, June (1963).
2. E. C. H. Silk, and R. S. Barnes, "Examination of Fission Fragment Tracks with an Electron Microscope," Philosophical magazine, Vol. 4, July (1959).
3. O. Hahn, and F. Strassmann, "Uber den Nachweis Und das Verhalten der bei der Bestrahlung des Urans Mitlels Neutronen Entstehenden Erdalkalimetalle," Der Naturwissens Chaffen, Vol. 22, January (1939).
4. N. Bohr, and J. A. Wheeler, Phys. Rev. **56**, 426 (1939).
5. I. Z. Noddak, Angew, Chemie, **37**, 653 (1925).
6. R. L. Fleischer, and P. B. Price, "Tracks of Charged Particles in High Polymer," Science 140, 1221-1222 (1968).
7. R. L. Flischer, P. B. Price, and R. M. Walker, "Solid State Track Detectors: Application to Nuclear Science and Geophysics," Ann. Rev. Nuc. Sci., 15 (1965).
8. "MCNP-A General Monte Carlo N-Particle Transport Code, Version 5," LA-UR-03-1987, Los Alamos National Laboratory (2003).
9. R. L. Fleischer, P. B. Price, and R. M. Walker, "Nuclear Tracks in Solids," University of California Press (1975).
10. J. R. Lamarsh, Introduction to Nuclear Reactor Theory," Addison-Wesley Publishing Co., Massachusetts (1966).
11. A. M. Weinberg, and E. P. Wigner, "The Physical Theory of Neutron Chain Reactions," The University of Chicago Press (1958).
12. D. J. Hughes, et al., "Neutron Cross Section," 2nd Edition and Supplements, BNL-325, 1958, 1960, 1964 (1965).

(558e) Efficient Adaption of Simulated Annealing and Genetic Algorithms to Atmospheric Inverse-Source Problems

Bhagirath Addepalli, Department of Mechanical Engineering, University of Utah, Salt Lake City, UT and Christopher Sikorski, School of Computing, University of Utah, Salt Lake City, UT

In the present work, simulated annealing and genetic algorithms are examined to determine their most efficient implementations in atmospheric inverse-source problems. The inverse problem is comprised of retrieving the spatial coordinates (x, y, and z), source strength, wind speed, and the wind direction at the source, given certain sensor locations, and concentration values at these sensor locations. The solution methodologies developed work for arbitrary atmospheric inverse-source problems, given a forward model and the limits of estimation of the sensors. In the current work, the Gaussian plume model (GPM) is adopted as the forward model due to its theoretical and computational simplicity. Seven out of the eight model parameters in the GPM were retrieved, thus making the inverse problem seven dimensional (7D).

Atmospheric inverse-source problems suffer from sparse number of sensor measurements in general, and very few non-zero measurements in particular. Therefore, the first step in the solution procedure was the development of a misfit function that takes into account both zero and non-zero measurements, and treats both of them equally. The new misfit function developed is based on the reciprocal of the L1-norm of an indicator function, with the indicator function taking the value one when the base 10 logarithm of the observed and predicted data at a certain receptor falls within certain (derived) bounds. The inverse-source problem is cast as a minimization problem and the final source parameters correspond to the model parameters for which the misfit function is minimized. Since the misfit function for atmospheric source inversion problems is characterized by several local minima, stochastic search algorithms such as SA and GA are employed.

In the current work, SA with the inhomogeneous Metropolis algorithm is implemented. In theory, even though SA is an optimization algorithm (asymptotically converges to set of global minima with probability one), in any implementation, heuristic choices need to be made. This is because the necessary and sufficient conditions for convergence are computationally intractable for large state spaces and hence cannot be implemented. Consequently, heuristic versions of SA were developed which do not necessarily satisfy the conditions for weak ergodicity. One of the objectives of the present work is the identification of a feasible annealing algorithm for atmospheric inverse-source problems. Therefore, the performance of several variants of SA was examined. The variants considered include classical SA, Cauchy annealing, very fast SA, adaptive SA, and generalized SA with Tsallis statistics (GSA-TS). These variants were implemented with several neighborhood generation mechanisms and temperature reduction schedules. Since the atmospheric inverse-problem is a formulated as a continuous minimization problem, a continuous GA (CGA) was implemented. In the CGA, the effect of the various parameters such as the initial population size, selection ratio, and mutation rate, and the mechanisms that can be used in the selection, mating, crossover, and mutation procedures, and their effect on the final solution obtained is investigated. At the conference, the performance of the various versions of SA and GA will be presented. Based on the results, the most feasible version will be identified.

See more of this Session: Role of Chemical Engineering in Nuclear Forensics
See more of this Group/Topical: Nuclear Engineering Division

THE UNIVERSITY OF UTAH TRIGA REACTOR AS NEUTRON SOURCE FOR NEUTRON ACTIVATION ANALYSIS OF VARIOUS DAILY PRODUCTS

Todd H Sherman, the University of Utah, Salt Lake City, UT
Nader Satvat, the University of Utah, Salt Lake City, UT
Dong-OK Choe, the University of Utah, Salt Lake City, UT
Tatjana Jevremovic, the University of Utah, Salt Lake City, UT

Abstract

Neutron activation analysis (NAA) can be used to determine the concentration of different elements. In order to perform such experiment a neutron source is required for the activation process. Various neutron sources are available such as a nuclear reactor, an actinide such as californium, and D-T fusion reaction. Using a gamma detector (semiconductor, scintillation, or gas ionization type) various peaks are recorded from the NAA sample. These peaks are compared with available data for various elements to identify the content of the examined samples. The University of Utah TRIGA reactor built in 1975 operates at 100 kW thermal power; it contains several irradiation facilities for fast and thermal neutron spectrums as used for NAA analysis. This paper will present the capabilities of using the University of Utah research reactor, TRIGA, in performing a number of NAA experiments; we will mainly focus at a number of various daily products to examine the chemical compositions and analyze the environmental quality of the Salt Lake. We will thus generate a library of data indicating the contents of tobacco (various cigarettes brands), bullets, in addition to examining the chemical content of Salt Lake water and shores.

Introduction

Since the 1936, when Hevesy and Levi discovered the Neutron Activation Analysis (NAA), the technique has been used in many areas of science and technology. Its precision has made NAA the standardized way of identifying traces of elements. Recently, at the University of Utah Nuclear Engineering Program the NAA has been place back into a routine use in identifying traces of elements in various samples thus by providing services to local businesses; in addition the NAA demonstration became an integral part of our renewed nuclear engineering curriculum.

Nuclear reactors as sources of neutrons, as well other neutron sources have been fundamental tools to provide valuable and accurate analysis in a myriad of applications. For example, the TRIGA reactor at the Johannes Gutenberg-University of Mainz, Institute of Nuclear Chemistry has been used to perform NAA for thousand samples yearly; some of these were to identify paint compositions from works of art; several reverse glass paintings, from the Middle Ages, were analyzed via NAA to determine their origin and estimate the time period in which they were created (Conejos Sanchez, 2009). Another famous examples are, detailed analysis of dietary mineral and toxicology levels in hair, teeth, bone, and tissue samples from many regions of Brazil, which became a regular function by the Sao Paulo Nuclear and Energy Research Institute (IPEN) to determine the nutrition and potential health risks associated with mal-nutrition. These NAA analysis are related to finding concentrations of

K, P, Na, Fe, Cr, Se, Zn, Ca, and many other elements using the IPEN IEA-R1 reactor (Nuclear Science & Technology, 2008).

NAA is also commonly used to determine mineral profiles of geological samples. Petroleum reservoirs have been explored using NAA and Instrumental Neutron Activation Analysis (INAA) in Western Cuba. The carbonaceous character of the rocks has been determined via INAA to adequately evaluate the compositions of the various petroleum reservoirs (Montero Cabrera, 2000).

Neutron Activation Analysis: Physics and Applications

Detection of unknown elements in a sample of known material is very useful for many applications. One of the methods based on nuclear processes is Neutron Activation Analysis (NAA), which is used to determine elemental composition of the sample. NAA enables isotope identification and their concentrations without considering the chemical form of the sample. NAA is based on activation of material, which requires a neutron source. Neutron source can be any of the followings:

- A nuclear fission reactor
- Natural neutron emitters through spontaneous fission (californium neutron source)
- An alpha emitter mixed with beryllium which generates neutrons (α, ^{12}C+n)
- D-T fusion reaction which emits 14.1 MeV neutrons

Most common is the use of research or test nuclear reactors to irradiate the samples for NAA; samples are placed in a container and depending on the type of irradiation required, several ports are usually available at the reactor core; examples are: fast neutron irradiation facility, or thermal irradiation facility. In some cases several filters are used to create epithermal neutron irradiation available through absorbing fast and thermal neutrons through the filters.

In NAA, the material is bombarded by intense neutron flux, which causes the formation of radioactive isotopes. Radioactive decay path and mechanism of each isotope is well known. Therefore, through these nuclear fingerprints (decay paths) it is rather straightforward to identify the unknown isotope. To demonstrate the NAA, we consider ^{60}Co. Due to its short half-life (5.27 years), ^{60}Co is not found naturally in nature. If ^{59}Co is bombarded with neutrons, ^{60}Co is produced as follows:

$$^{59}_{27}Co + ^{1}_{0}n \rightarrow ^{60}_{27}Co^{*}$$

The decay scheme of ^{60}Co is shown in **Fig. 1**. The unstable ^{60}Co goes through two beta decay emitting 0.31MeV and 1.48 MeV betas with probabilities 0.9988 and 0.0012 respectively. After each beta decay, newly formed nucleus, ^{60}Ni, is in its excited state; emission of gammas (1.1732MeV and 1.3325MeV) takes this nucleus to its ground state. These peaks are shown in **Fig. 2**.

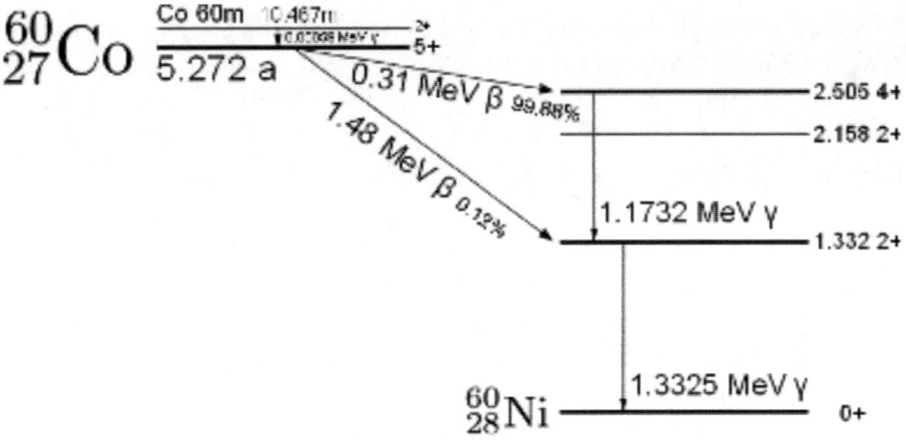

Figure 1 Decay scheme for ^{60}Co [1]

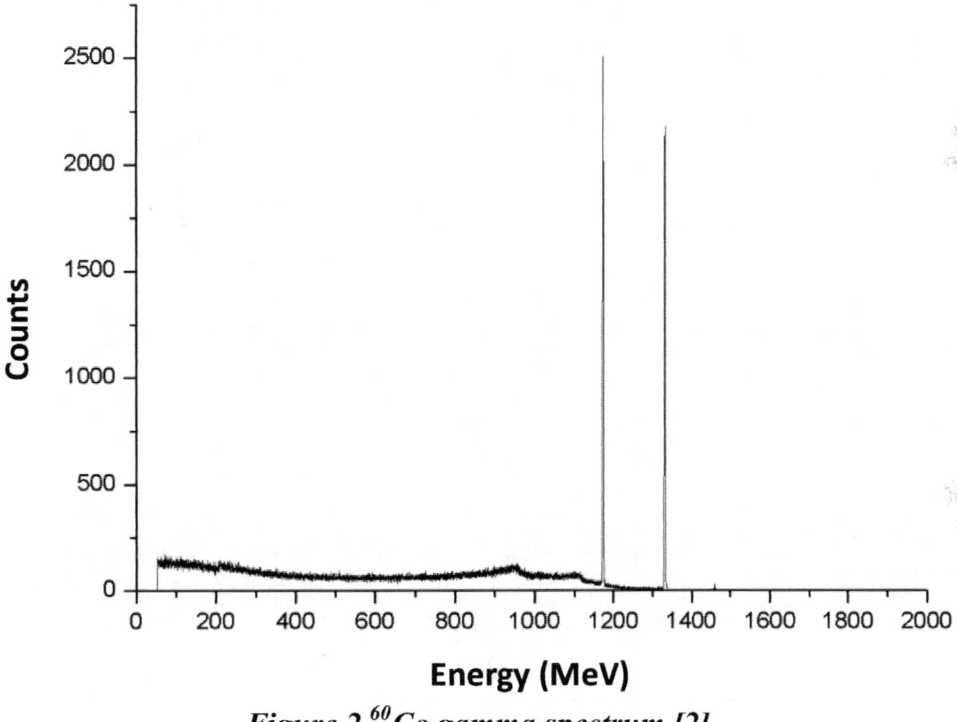

Figure 2 ^{60}Co gamma spectrum [2]

Sample Studies using Neutron Activation Analysis (NAA) at the University of Utah Research Reactor (UUTR)

Utah Research Reactor (UUTR)

Materials and chemicals are examined using NAA to determine elemental composition and concentrations. The Utah Nuclear Engineering Program (UNEP) operates the University of Utah TRIGA Reactor (UUTR). This is a MARK I General Atomics TRIGA reactor licensed by the U.S.

Nuclear Regulatory Commission to operate at 100 kW; the reactor however is usually only operated at 90 kW. The UUTR is a pool type reactor that uses water as both a moderator and coolant. Fuel elements for the UUTR are uranium zirconium hydride (UZrH) rods that are arranged in a hexagonal lattice. At the operating power of 90 kW the neutron flux at the center of the reactor core is approximately 3.6×10^{12} neutrons * cm^{-2} * sec^{-1}.

The UUTR is equipped with four separate irradiation facilities. The central irradiator (CI) is in the very center of the core and can be fitted with a rotary specimen rack. The pneumatic irradiator (PI) is pneumatic system that uses compressed helium gas that moves samples to and from the inner core. The fast neutron irradiation facility (FNIF) and the thermal irradiator (TI) are both located on outside faces of the reactor core and are used for irradiation from neutrons having specific energies.

NAA Samples Preparation

The specimens are always prepared into two small samples of 10 mg and 100 mg ± 0.01 mg. These measured samples are placed and sealed in small sterile plastic pouches or small plastic vials. Plastic is composed primarily of hydrocarbons which have very small neutron absorption cross sections making them unfavorable to neutron capture, thus reducing the probability of that the plastic containers will participate in the neutron irradiation process.

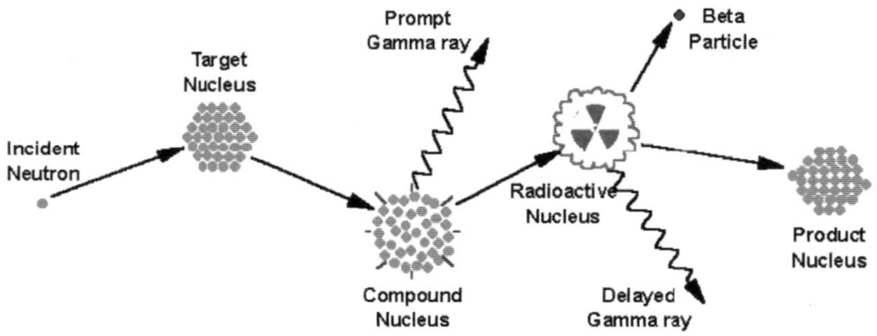

Figure 3 HPGE Diagram representing Neutron Capture and Radioactive Decay [4]

To irradiate the samples the plastic containers are placed in 1"x3" plastic vial which is lowered into one of the UUTR irradiator facilities. For NAA purposes the TI is the better facility. The sample is irradiated for a specific amount of time at a specific neutron flux. The 10 mg samples are irradiated at a flux of 3.6×10^{11} neutrons * cm^{-2} * sec^{-1} for four and a half minutes until they have been exposed to a neutron fluence of about 9.6×10^{13} neutrons/cm^2. The UUTR TI is an aluminum box that is filled with heavy water (D_2O) that acts as a moderator to slow the higher energy neutrons. Fast neutrons from the reactor enter the TI and they interact with the nuclei of deuterium found in heavy water. As they do so they transfer most of their kinetic energy to the deuterium nuclei. Neutrons that are moderated are slowed to the thermal energy range which is far more favorable for neutron absorption. For example the tendency for a neutron to be absorbed by ^{235}U nucleus corresponds directly to the energy of the neutron. The absorption cross section for a ^{235}U nucleus corresponding to a 2 MeV fast neutron, recently born from a nuclear fission reaction, is about 7.18 barns [3]. But the absorption cross section for that same ^{235}U for a 0.025 eV thermal neutron is 680.9 barns. The thermal irradiator allows enough nuclei of the target atoms in the samples to capture the lower energy thermal neutrons, become radioactive and begin to decay and release gamma rays. **Fig. 3** illustrates the path of a nucleus that captures a neutron to become a radioactive isotope. The isotope then releases energy in the form of gamma rays and alpha or beta particles to decay into an entirely different element.

Measurements and Instrumentation

A high purity germanium HPGE gamma spectrometer is used to acquire and count the gamma rays being released from the decaying isotopes. Each of the small 10 mg sample are counted for 10 minutes. Each isotope has a signature decay path that is well documented as discussed in the previous section. The gamma spectra are examined and the corresponding isotopes are identified by matching the sample's gamma peaks, like those shown in **Fig. 2,** to the peaks of known isotopes.

Figure 4 HPGE Gamma Spectrometry System

The larger 100 mg samples are irradiated for much longer time. The neutron fluence for the larger samples is in a range of 50 -160 times greater than for the smaller samples. This is done so that the elements with longer half-lives and smaller neutron absorption cross sections are more efficiently activated. The 10 mg samples are irradiated with the intent to activate and detect radioactive isotopes with relatively short half-lives such as Au, Al, Na, K, Mn, Ti and V. After a longer irradiation time, the 100 mg samples are allowed to decay for several hours, even days, if necessary, so that when the gamma spectra are evaluated, only the isotopes with longer half-lives like Sc, La, Sm, Zr, Zn, Fe, Mo and Co are remaining while the isotopes with short half-lives are almost entirely decayed out. The 100 mg are then counted for 30 minutes each.

By performing NAA using two separate experiment procedures with different masses and different irradiation times broader spectrums of elements are able to be detected and identified. This also allows the radioactivity of the samples to be high enough to detect and count, but low enough to safely handle and manage. Even trace elements can be detected and analyzed in very small samples.

Results and Discussion

Using these procedures the UNEP lab has recently been able to efficiently profile elements found in various products and samples. NAA has been used for several decades as a reliable forensic tool to investigate elemental compositions and concentrations. UNEP has been using the UUTR to perform such research to develop a NAA forensic and material analysis program.

Tobacco

Several tobacco products were activated in the UUTR to study possible additives as shown in Table 1. The products studied were found to have the following elements, Ba, Mg, Mn, Na, Cl, K, Br and Al. Many of these elements are commonly found in plant life like K and Mg. Br is most likely a component of the alkaloids found in nicotine. Al could possibly be contamination from the metal surfaces or the tobacco cutting and shredding devices used in the growing and preparation of the tobacco products. **Figure 5** represents a gamma spectrum acquired from a sample of a Marlboro cigarette. A number of samples from Great Salt Lake area were studied for possible contaminates. It has long been known that industrial and refining pollution has been dumped into the Great Salt Lake over many decades. NAA is a useful tool to find what heavy metals may be present in these wetlands. Neatly a dozen specimens have been irradiated and analyzed. Using the procedures previously described more than a dozen of different elements have been identified, several of which could represent foreign contamination. More detailed studies are required to understand the origin of these contaminants.

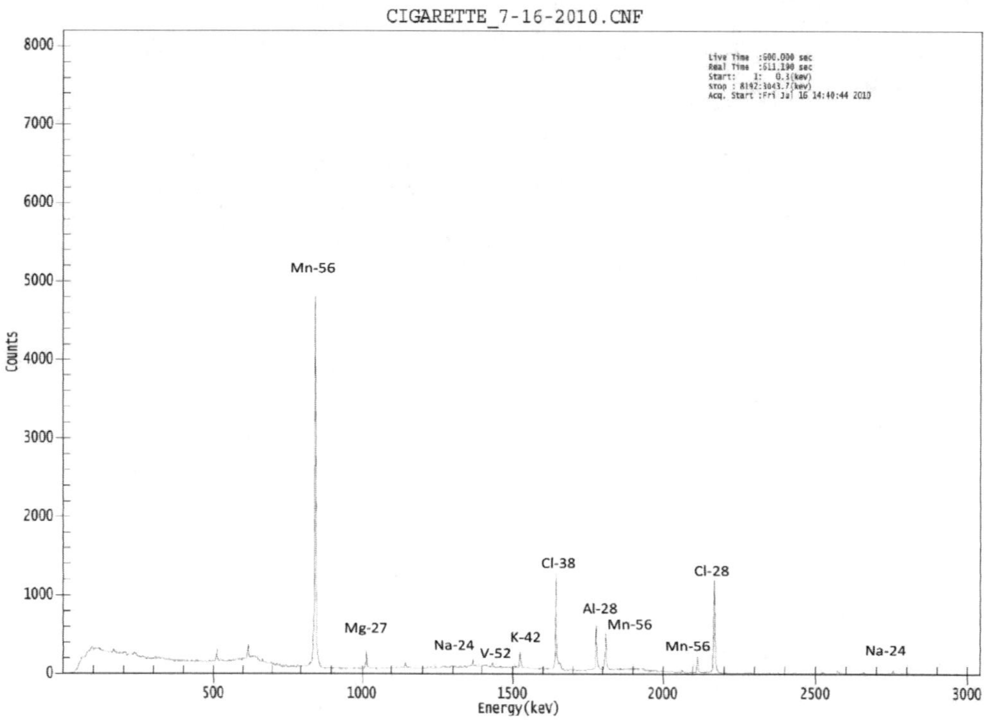

Figure 6 Cigarette Gamma Spectrum

Table 1 Elements Detected in Tobacco Products Using NAA at the UUTR

Product	Elements Detected
American Camel Cigarette	Mn, Mg, K, Br, Cl, Na
American Marlboro Cigarette	Mn, Mg, Cu, K, Cl,
European Camel Cigarette	Ba, Mg, Mn, Ca, Cr, K, Cl,
Cuban Cigar	Mg, Mn, Cu, Ca, Na, Cr, K, Cl, Al
Serbian Cigarette	Ba, Mn, Mg, Ca, K, Cl

Table 2 Great Salt Lake Soil Sample Profile

Sample Name	Elements Detected
S9-1	Sm, Ba, Dy, I, Mg, Mn, Eu, Na, K, Cl, Al, La
S9-2	Sm, Ba, U, Ti, Dy, I, Mg, Mn, Eu, Na, K, Cl, V, Al, La, Sc, As, Br
S9-3	Sm, Ba, U, Ti, Dy, I, Mg, Mn, Eu, Na, K, Cl, V, Al, La, Sc, As, Br
S9-4	Ba, Ti, Dy, Mg, Mn, Eu, Na, K, Cl, V, Al, La, Sc, As, Br
S9-5	Sm, Ba, U, Ti, Dy, I, Mg, Mn, Eu, Na, K, Cl, V, Al, La, Sc, As
S9-6	Ba, Ti, Dy, Mg, Mn, Eu, Na, K, Cl, , Al, La, Sc, As
S9-7	Sm, Ba, U, Ti, Dy, I, Mg, Mn, Eu, Na, K, Cl, V, Al, La, Sc, As
S9-8	Ba, Ti, Dy, Mg, Mn, Eu, Na, K, Cl, V, Al, La, Sc, As
S9-9	Sm, Ba, Dy, Mg, Mn, Eu, Na, K, Cl, Al, La, Sc, As

Chinese Chop Sticks

In addition to the soil samples and tobacco products imported Chinese chopsticks have been analyzed to determine what bleaching agents are used to give the chopstick a lighter appearance. Many bleaching agents can have detrimental biological effects if consumed. It was found that barium sulfate was used to give the chopsticks a lighter appearance. Barium is very abundant in China making it an easy and economical substance for wood bleaching. [For reference, barium sulfate is also used as a radiocontrast in X-ray imaging when diagnosing various medical conditions.]

Conclusion

NAA is an economical and accurate method to analyze materials and chemicals from a large gamut of commercial, industrial, medical and scientific applications and investigations. The samples typically do not need any chemical preparation or modification in order to be tested. Trace elements can be easily detected making this a valuable resource for those looking for elements ranging from rare metals to contaminants in biological tissues. The UUTR, like other research reactors, can be a significant means to carry out NAA for a many entities. NAA is suitable for many tests like necessary periodic product quality control tests as well as customized one-time element investigations.

References

[1] http://en.wikipedia.org/wiki/File:Cobalt-60m-decay.svg

[2] http://upload.wikimedia.org/wikipedia/commons/8/88/60Co_gamma_spectrum_energy.png

[3] http://www.ncnr.nist.gov/resources/n-lengths/list.html

[4] http://www.jeffreycreid.com/Analytical_Methods/NAA_theory.html

[5] Conejos Sanchez, I. (2009). Reverse paintings on glass - A new approach for dating and localization. *Applied Radiation and Isotopes 67(12)* , 2113-2116.

[6] Montero Cabrera, M. (2000). Instrumental neutron activation analysis of rocks from Cayajabos petroleum ore. *Applied Radiation and Isotopes 52(1)* , 143-146.

[7] *Nuclear Science & Technology*. (2008). Retrieved september 12, 2010, from Instituto de Pesquisas Energeticas e Nucleares: Progress Report 2005-2007: http://www.ipen.br/conteudo/upload/201002021820590.4_nuclear_science.pdf

(618a) Reaction Engineering of a Novel Sulfur-Sulfur Thermochemical Water-Splitting Cycle

Alex Yokochi and Nicholas AuYeung, Department of Chemical Engineering, Oregon State University, Corvallis, OR

The Sulfur-Iodine and hybrid (Westinghouse) water-splitting cycles have suffered from the use of water as reaction media in their low temperature reactions. Separation of the various species from an aqueous medium results in significant energy expenditures and materials challenges in practical implementation. For the past few years we have examined an alternative route where the oxidation of sulfur step takes place in an ionic liquid medium. The application of ionic liquid reaction media for the low temperature reactions presents possible solutions to the problems of excess water removal, I_2 solidification, and SO_2 solubility.

The observation that carrying out the Bunsen reaction in ionic liquid media (Eq. 1 below) can be tailored to produce significant amounts of H_2S gas from the reaction mixture has led to the development of a novel Sulfur-Sulfur cycle, shown below.

$$4SO_2 + 4I_2 + 8H_2O \rightarrow 4H_2SO_4 + 8HI \quad \text{Eq. 1}$$

$$H_2SO_4 + 8HI \rightarrow H_2S + 4I_2 + 4H_2O \quad \text{Eq. 2}$$

$$\text{Net Low Temperature Reaction: } 4SO_2 + 4H_2O \rightarrow H_2S + 3H_2SO_4 \quad \text{Eq. 3}$$

$$H_2S + 2H_2O \rightarrow SO_2 + 3H_2 \quad \text{Eq. 4}$$

$$3H_2SO_4 \rightarrow 3SO_2 + 3H_2O + 3/2O_2 \quad \text{Eq. 5}$$

The production of H_2S (Eq. 2) following the Bunsen reaction, a normally undesired side reaction, can be intentionally enhanced under the appropriate conditions (concentration of I_2 and water, temperature and reaction time). The I_2 is regenerated in this reaction without having to transport HI to a decomposition section, and is therefore required only in catalytic amounts. The H_2S spontaneously desorbs from the reaction medium and can then be steam reformed to produce H_2 and SO_2, (Eq. 4) the latter of which is recycled. The thermal decomposition of H_2SO_4 to produce O_2 and SO_2 (Eq. 5) is common to other

Sulfur based thermochemical cycles. The H_2S generation and steam reformation steps (Eq. 3 and 4) have been demonstrated in our lab.[1] An overall thermal efficiency assuming a steam reformation temperature of 1100 K is estimated to be 55%.

This presentation will focus on the parametric dependence of the H_2S production and steam reforming reactions on reactant concentrations, temperature, and residence time, and details of the estimated thermal efficiency calculations will be presented.

---

[1] A. Yokochi, N. AuYeung, "*A Proposed New Sulfur-Sulfur Thermochemical Cycle*" *2009 Meeting of the AIChE.*

See more of this Session: Reactions and Separations for Hydrogen Production
See more of this Group/Topical: Topical 8: Hydrogen Production and Storage

(618b) Simultaneous Solubilities of Oxygen and Sulphur Dioxide in Water: Thermodynamic Data for the Sulphur Family of Thermochemical Cycles

Rachael H. Elder[1], Moises Romero[2], Andrew C. Shaw[2] and Ray W. K. Allen[2], (1)Department of Chemical and Process Engineering, University of Sheffield, Sheffield, United Kingdom, (2)University of Sheffield, Sheffield, United Kingdom

The Sulphur family of thermochemical cycles shows great promise for the massive scale production of hydrogen from water. The decomposition of sulphuric acid is a common stage in the cycles and poses design problems due to the corrosive nature of the solution, the high temperatures involved and the difficult separations required. The work considered here focuses on the oxygen separation stage of the process where, in order for a science based design approach to be taken, thermodynamic data are required for multicomponent phase equilibrium relationships between water, sulphur dioxide and oxygen; H2O-SO2-O2. A bespoke vapour-liquid-equilibrium still is used to measure the simultaneous solubilities of the ternary solution. The results are presented, along with a comparison to a model based on weak electrolyte thermodynamics. The model can be used for flash calculations, necessary for the design of the separation equipment.

See more of this Session: Reactions and Separations for Hydrogen Production
See more of this Group/Topical: Topical 8: Hydrogen Production and Storage

(618c) Storage and Separation of Hydrogen by the Metal Steam Process

Markus Thaler, Viktor Hacker and Matthaeus Siebenhofer, Institute of Chemical Engineering and Environmental Technology, Graz University of Technology, Graz, Austria

Certain metals react with water to build a metal oxide. During the reaction the water molecule is split and hydrogen is released as by-product. Using this behavior metals in their elemental form can be applied as chemical hydrogen storage material. The hydrogen is produced on demand. Only metal and water have to be stored in separate vessels thus hydrogen is not present in its physical form during the storage time.

Also the separation of hydrogen in a gas mixture can be achieved via repeated oxidation (Eq. 1) and reduction (Eq. 2) of metals and metal oxides. Reductive species such as CO can be used for the reduction of the metal oxide to reach a higher yield of hydrogen (Eq.3).

Oxidation step: xMe + yH2O → MexOy + yH2 (Eq. 1)

Reduction step: MexOy + yH2 → xMe + yH2O (Eq. 2) MexOy + yCO → xMe + yCO2 (Eq. 3)

The product gas of the reaction of metals with steam is especially capable for polymer electrolyte fuel cells as the hydrogen is already humidified.

However, some of the metals react at room temperature, others need significantly higher temperatures. To evaluate the materials capable for the storage and separation of hydrogen thermodynamic and experimental validations of the different metals are conducted. Metals as hydrogen storage materials are compared with established energy storage systems and advantages and disadvantages are worked out. By means of thermogravimetric analysis selected metals are reduced with hydrogen and oxidized with steam to confirm the theoretical results. Kinetic parameters are derived from the measurements for system calculation.

See more of this Session: Reactions and Separations for Hydrogen Production
See more of this Group/Topical: Topical 8: Hydrogen Production and Storage

(618d) The Kinetics of H2-D2 Exchange Over PdCu Surfaces

Casey O'Brien[1], James Miller[1], Bryan Morreale[1], Andrew Gellman[1] and Petro Kondratyuk[2], (1) USDOE National Energy Technology Department, Pittsburgh, PA, (2)Dept. of Chemical Engineering, Carnegie Mellon University, Pittsburgh, PA

Very high purity H_2 can be produced from mixed gas streams, such as coal gasification syngas, by the use of dense Pd-based membranes which have the unique ability to dissociate molecular H_2 and allow only atomic H to permeate through its bulk. To improve Pd's mechanical properties and its resistance to minor gas-phase components that can cause deterioration of membrane performance, such as H_2S, Pd is often alloyed with other metals, including Cu.

In this work, we use the H_2-D_2 exchange reaction to probe the H_2 dissociation activity of clean and sulfur-contaminated Pd, Cu and PdCu alloy surfaces. Over fixed beds of clean, diced Pd foils, the rate of H_2-D_2 exchange is limited by the desorption rate of product HD; over clean Cu, the reaction is limited by the relatively large barrier to dissociative adsorption. PdCu alloys display intermediate behaviors.

To provide a basis for understanding membrane performance in environments typical of the separation application, we also present kinetic analyses of H_2-D_2 exchange over Pd and PdCu surfaces in the presence of H_2S. Our results contribute to the design basis for next generation purification devices.

See more of this Session: Reactions and Separations for Hydrogen Production
See more of this Group/Topical: Topical 8: Hydrogen Production and Storage

(618e) Thermo Catalytic Decomposition of Methane Into Hydrogen and Carbon Nanofibers On Ni and Fe/ SiO2 Catalyst

Sushil Saraswat and Kamal Pant, Chemical Engineering, Indian Institute of Technology, Delhi, New Delhi, India

Thermocatalytic decomposition of methane is one of the promising technology for the production of pure hydrogen which is desirable for PEM fuel cell application. Thermo Catalytic methane decomposition to produce COx- free hydrogen and carbon nanofibers was carried out over Ni/SiO2 and Fe/SiO2 catalysts in a fixed bed quartz reactor. Metal components were loaded on SiO2 support by the wet impregnation method. Effect of metals loading on the support was investigated by increasing the metals loading from 20- 70 %. The morphology of the fresh catalysts and the characteristics of the carbon produced were also studied on the basis of the results of BET, Chemisorption, XRD, TGA, SEM and TEM analysis. The catalytic decomposition reaction was carried out at 700 deg C and atmospheric pressure. More than 80% methane conversion was obtained on Ni/SiO2 & Fe/SiO2 catalysts at 700ºC. Scanning electron microscope examinations revealed that the deposited carbon formed was in the form of carbon nanofibers. XRD analysis shows that carbon nanofibers formed are of graphitic nature. Transmission electron microscopy (TEM) studies revealed that filamentous carbon was formed during the reaction.

See more of this Session: Reactions and Separations for Hydrogen Production
See more of this Group/Topical: Topical 8: Hydrogen Production and Storage

(632a) Development of Coal-Carbon Limit in Fluidized Bed Steam Reformer (SBSR) Testing

Dan P. Lambert[1], Alex S. Choi[1], William Eugene Daniel[2], Michael E. Stone[3] and Joshua Mysona[4], (1) Savannah River National Laboratory, Aiken, SC, (2)Engineering Process Development, Savannah River National Laboratory, Aiken, SC, (3)Process Technology Programs, Savannah River National Laboratory, Aiken, SC, (4)School of Chemical & Biomolecular Engineering, Georgia Institute of Technology, Atlanta, GA

Modeling and testing were initiated to determine the maximum carbon concentration in a new waste stream, a product of the Fluidized Bed Steam Reformer (FBSR) at the Savannah Rive Site (SRS). The carbon limit is designed to prevent a flammable carbon monoxide concentration from developing during processing of high-level radioactive waste (HLW).

Historically, carbon in various forms has been present in all of the sludge processed in the SRS Defense Waste Processing Facility (DWPF). Sludge can contain coal and carbonate as the primary carbon species. The new FBSR waste stream is much higher in coal and carbonate than has been processed in the past. The sludge and FBSR product are combined and processed with nitric and formic acid to neutralize the waste, reduce Hg and Mn, destroy nitrite, and modify the slurry rheology, increasing the carbon concentration in the resulting melter feed.

Modeling was used to estimate the carbon limit of the FBSR product stream. A series of experiments with added FBSR product and sludge simulant will be completed in June 2010 in SRS's Aiken County Technical Laboratory. Processing included sludge concentration, a Sludge Receipt and Adjustment Tank (SRAT) cycle and Slurry Mix Evaporator (SME) cycle. This was the first demonstration with the added FBSR stream. A gas chromatograph was used throughout the experiment to compositionally characterize the offgas. The main objective for the experiment was the determination of redox and carbon concentration of the melter feed. Results of this testing will be reported.

See more of this Session: Chemical Engineering Advances in Processing Radioactive Wastes and Nuclear Materials

See more of this Group/Topical: Nuclear Engineering Division

(632b) Electrostatic Impacts On Fouling of Microfiltration Membranes

Richard C. Daniel, Rick Shimskey and Reid A. Peterson, Pacific Northwest National Laboratory, Richland, WA

Over the last two decades, PNNL has been working on separation processes for the Waste Treatment Plant and in particular cross flow filtration. Recent work at the laboratory has focused on determining the mechanisms of reversible and irreversible depth fouling of the filter media for this technology. Fundamental particle properties play an important role in filter/membrane depth fouling and the formation of solids cakes on the filter surface. With respect to depth fouling, penetration of the porous network is governed by particle size and particle filter interactions are governed by standard DLVO (Derjaguin, Landau, Verwey, and Overbeek.) interactions. The surface charge of the filter and particle play a significant role in how strongly the filter fouls (and if that fouling is reversible). When the filter surface and particles have the same charge, fouling is expected to be minimal. When the filter and particles have opposite charges, fouling is expected to be strong. Because the filter medium and the particles comprising the dispersion are typically different materials, charge disparity between the two is likely. However, surface charge is also affected by pH and ionic strength of the slurry supernatant. To advance a fundamental understanding of filter fouling dynamics, the impacts of particle interactions (as governed by zeta potential) were tested and incorporated into existing models for filtration.

See more of this Session: Chemical Engineering Advances in Processing Radioactive Wastes and Nuclear Materials
See more of this Group/Topical: Nuclear Engineering Division

(632c) Experimental Verification of Solidification Stress Theory

Charles W. Solbrig, Matthew C. Morrison and **Kenneth J. Bateman**, Pyroprocessing Technology, Idaho National Laboratory, Idaho Falls, ID

Summary A research program is being conducted to develop a crack-free ceramic waste form (CWF) to be used for long term encasement of fission products and actinides resulting from processing spent nuclear fuel. Cracking usually occurs in the cooldown phase of the glass or ceramic formations. A crack-free formation should have more resistance to leaching than one with many cracks. In the research leading up to producing a CWF, a model was developed that proposes a permanent stress develops when the melt solidifies and that this stress can cause failure as the CWF nears room temperature. This paper reports on how the formation, CWF2, confirms the existence of this stress. The solidification stress is in addition to and of opposite sign of the thermal stress. Its derivation is reported on in Ref. 1. Cracking of the CWF would occur at low temperatures if solidification stress exists but at high temperatures if it doesn't. If solidification stress occurs, then the cooldown rate during solidification should be reduced. If not, it should be reduced when the thermal stresses are highest. Recording cracking sounds confirm the existence of this solidification stress since cracking occurred during the low temperature phase of the cooldown. As a side purpose of this paper, a cooldown rate is proposed that should eliminate cracking in the next experiment, CWF3. CWF2 is a prototype vertical ceramic waste cylinder formed over a period of 10 days by heating a mixture of 75% zeolite, 25% glass frit in an argon atmosphere furnace through melting to 925 C and then cooling through solidification to room temperature. It is approximately 1 m high, 0.5 m in diameter, weighs about 400 kg, and is formed in a stainless steel can 0.5 cm thick. This cylinder developed many cracks on cooldown. At least 15 loud cracks were recorded over a period of 4 days at the end of cooldown when the temperatures were below 400 C. The CWF2 surface and centerline temperatures at mid height were measured which allowed the stress to be calculated. The timing of the cracks was compared to the time the calculated total stress exceeded the tensile stress limit and verified that the cause of the cracking was solidification stress and not thermal stress. Since the CWF is encased in a can in a furnace, the cracks cannot be easily observed but can be detected with sound measurements. Similarly, the stress cannot be measured but only estimated with analysis Temperature Data The cooldown of CWF2 was started but after 4 hours the cooldown and heat transfer from the CWF stopped, resulting in the surface and center temperatures equilibrating over the next hour. This was caused by the coolant fan stopping and staying off for about 8 hours at which time the fan restarted so CWF cooldown started again. The coolant flow is varied to maintain a circulation pump inlet temperature of 100 C or less. The furnace wall continued cooling during this time because the control system kept the heating coils off. As temperatures decreased, the pump stayed on for longer periods of time causing the average coolant flow to increase causing more heat to be removed from the CWF. The largest heat removal occurred during solidification causing a large set-in stress. The temperature difference between the center and the surface during solidification is a measure of the solidification stress which is set-in The CWF3 plan specifies that the temperature difference should decrease during solidification (about 625 C). The data shows the CWF2 temperature difference actually increased. In fact, the largest differences occur during solidification which should have been less than 40 C but the data show a difference near 80 C. This temperature difference resulted in CWF2 cracking. To evaluate the stress, the analytical surface temperature profile was matched to the experimental by varying the heat transfer over time to simulate the pump flow. With this match, centerline temperature, the temperature profile, and stresses developed, both solidification and thermal, could be calculated. Analysis of Stress The thermal stress calculated in the midplane for this cooldown transient. The model used to calculate these stresses are that of Timoshenko and Goodier (Ref. 2). The stresses are shown over 10 evenly spaced radial increments. This stress predicts that the stress will be in tension in the outer radial region of the CWF so that cracking would be expected to start in that region if solidification stress did not exist. The thermal stress at the outer radius exceeds the tensile limit from 32 hours to 48 hours and the stresses slowly dissipate after that. If thermal stress were the cause of cracking in the CWF, it would be expected that cracking would occur during this time period. Sound measurements recorded 15 loud cracks during the course of the cooldown. The first one occurred at 61 hours, much later than the 32 to 48 hours when the thermal stress exceeded the tensile limit. Since all the cracks occurred after the thermal stress exceeded the limit, it cannot be responsible for the damage to the CWF. (Note, only the axial stresses are discussed in this paper. The radial stresses are about the same and the normal stresses are very small. There are no shear stresses.) The solidification stress was calculated using the method presented in Ref. 1. It develops while the CWF is solidifying and occurs because as one layer of the CWF is solidifying, it is attached to an adjacent solidified layer of different length. As the solid then is cooled down to room temperature, all these different lengths are forced to the same length causing stresses. The solidification stress developed is dependent on the temperature profile during solidification but is independent of temperature during the remainder of the cooldown. The maximum solidification stress is about 20000 psi which is well over the tensile stress limit of 12000 psi. The total stress is the sum of the thermal and the solidification stress, and since they are of opposite sign, the thermal stress partially cancels out the solidification stress. Since the thermal stress decreases as the temperature decreases, it cancels out less and less of the solidification stress as the temperature decreases. When the temperature becomes uniform, the thermal stress is zero so the total stress is then equal to the 20000 psi solidification stress at the center of the CWF. This explains why the CWF cracks at low temperature instead of high temperature when the thermal stress is high. That is, the total stress continues to increase as the average temperature decreases and the temperature profile flattens out. Both the solidification stress and the thermal stress add together to obtain the total stress shown in the attached figure. The stress begins at zero at 34 hours after the start of cooldown and increases as the temperature decreases to room temperature. During the initial solidification period, the two stresses almost cancel each other out but the solidification stress is always greater than the thermal stress. Consequently, the inner portion of the cylinder is always in tension. The centerline stress eventually exceeds the tensile limit at 68 hours. The first cracking sound that was recorded occurred at 61 hours. This indicates the stress calculation is close to predicting failure then. The tensile limit shown is only an estimate could be as low as 10000 psi. The damage continued all the way down to room temperature as evidenced with continued cracking, 15 in all. Conclusions A theory has been developed to model a stress which was posited to develop when a ceramic solidifies due to the temperature gradient which exists during the solidification process. An experiment was run which verifies the existence of this stress. Thermal stress alone would have predicted cracking to occur at while temperatures are high but when the solidification stress is added, the total stress calculation predicts cracking of the CWF will occur at low temperatures. Cracking sounds were recorded in this experiment and are used in this paper to show that the existence of this stress is probable since cracking occurred during the low temperature phase of the cooldown. Confirmation of this model allows a cooldown history to be followed which will eliminate cracking. Without including the solidification stress in the calculation, the low cooling rate needed to prevent cracking would be prescribed after solidification and cracking would not be prevented. References 1. Solbrig, C. W., and Bateman, K. J., MODELING SOLIDIFICATION-INDUCED STRESS IN CERAMIC WASTE FORMS CONTAINING NUCLEAR WASTES, to be published in Nuclear Technology 2. Timoshenko, S., and Goodier, J. N. (1970) Theory of Elasticity, Third Edition, McGraw-Hill, New York, NY. Acknowledgement

Work supported by the U.S. Department of Energy, Office of Nuclear Energy (NE) under DOE Idaho Operations Office Contract DE-AC07-05ID14517.

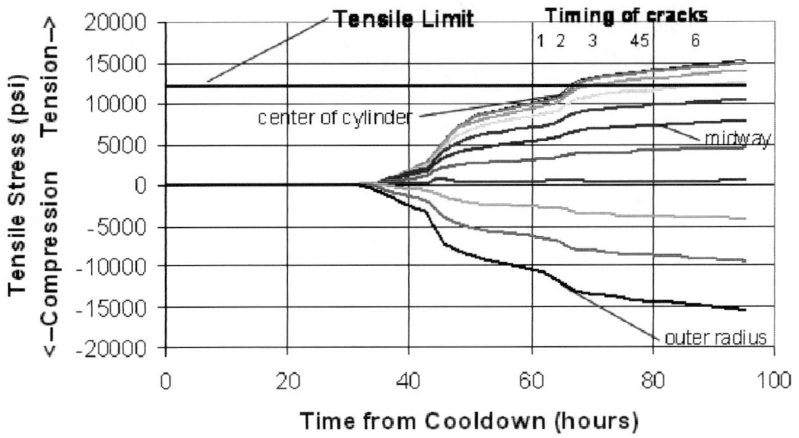

See more of this Session: Chemical Engineering Advances in Processing Radioactive Wastes and Nuclear Materials
See more of this Group/Topical: Nuclear Engineering Division

Reconciliation of Charge-Balance Discrepancies in Hanford High-Level Waste Supernatant Data

(632d)

Daniel A. Greer, Mason A. Anderson, and Jacob G. Reynolds

Washington River Protection Solutions, LLC

P.O. Box 850, Richland, WA, 99352

Abstract

Physical models of Hanford nuclear waste behavior require charge-balanced chemical input data. Unfortunately, real laboratory data is not charge-balanced because of random analytical uncertainty, missing data, and other errors or biases. Therefore, analytical data straight from the laboratory must be adjusted to force charge-balance prior to use in physical models. The way this data is adjusted, however, could impact the outcome of the physical model. This study evaluates a method to charge-balance Hanford waste supernatant data, and is applied to data from Hanford tank AW-101. This tank was selected because it has been sampled six times, yet has remained static, so all of the sample events are measuring the equivalent waste composition. The total charge was within the analytical uncertainty for three of the six datasets, and when averaged across all six datasets. The three datasets with total charge greater than the analytical uncertainty either had missing analytes or apparent gross errors. Gross errors were identified by comparing the charge contribution of each analyte from individual datasets to the average charge contribution across datasets. The results show that gross errors or missing analytes can be identified by individual charge-contributions that are larger than the analytical uncertainty. A method was developed to minimize the adjustments to the data to force charge-balance while ensuring the data stayed within the bounds of the analytical uncertainty.

Introduction

A fundamental requirement of the natural world is that the electrical charges be balanced (sum of positive and negative charges equal zero). Physical models of electrolyte solutions require that the charges balance prior to use. Unfortunately, real analytical data is rarely balanced. This may be solely because of analytical variability, or it may be because of more severe problems such as missing or biased data. Either way, some method of adjusting the data to reconcile the charge imbalance is required before the data can be employed in physical models.

Similar problems exist for mass and energy balance measurements in chemical plants. Much effort has been exerted to develop methods to reconcile mass and energy balance discrepancies in chemical plant data (Romagnoli & Sanchez, 2000). In contrast, there has been minimal development of methods to reconcile charge balance discrepancies. This may be because many systems have only a few electrolytes, which limits the options for making adjustments to the data.

Nuclear waste supernatants at the Hanford site have many electrolytes. Sodium and potassium are the dominant cations in the supernatants, but there are many anions (MIller, 1977). The most prevalent anions are nitrate, nitrite, hydroxide, aluminate, phosphate, sulfate, oxalate, carbonate, chloride and fluoride. In addition to these major anions, there are many other more minor anions (and cations). The present study develops a method to reconcile charge balance discrepancies in Hanford waste supernatant data. The method is applied to data from Hanford tank AW-101.

Data Review

The Hanford Tank 241-AW-101 supernatant data reported in (Disselkamp, 2009) were used in this study. Tank AW-101 has been sampled many times even though it has remained stagnant. In addition, this tank has a salt crust on the surface that inhibits evaporation and CO_2 adsorption. Thus, this tank supernatant composition has remained constant over time and the liquid samples represent the same composition.

Data were obtained for the following liquid sampling events:

- Liquid 1996 Core Sup Seg (Sampling Dates -- March 22, 1996, May 23, 1996)
- Liquid 1998 Grab Composite (Sampling Dates -- April 21, 1998 - April 22, 1998)
- Liquid 2002 Grab (Sampling Dates -- July 12, 2000)
- Liquid Core Composite Data (Sampling Dates -- March 22, 1996, January 20, 2003 - January 28, 2003)
- Liquid Interstitial Liquid (Sampling Dates -- January 16, 2003, January 20, 2003 - January 28, 2003)
- Liquid Supernatant (Sampling Dates -- January 16, 2003, January 20, 2003 - January 28, 2003)

These names of sampling events correspond to the names provided in (Disselkamp, 2009). The specific data provided in the report are the analyte, the method of analysis used, the mean value of the analyte, the degrees of freedom used in the calculation of the confidence limits, the lower 95% confidence limit, the upper 95% confidence limit, and the units of measurement. Note that some analyte mean values are reported as a "less than" detection limit. In this case, the confidence interval data are reported as "n/a".

For each sampling event, the analysis discussed here included calculating the standard deviation of the analyte mean, the charge contribution of the analyte to the total charge, the standard deviation of the charge contribution, the total charge of the sampling event, and the standard deviation of the total charge.

Handling Non-detects

The first challenge in this analysis is determining how the "less than" detection limit values (i.e., non-detects) will be incorporated into the analysis. There are numerous methods for incorporating non-detects, e.g., assigning them the value of the detection limit, assigning them a value of zero, ignoring them (i.e., removing them from the analysis).

In this analysis, there were two issues that needed to be addressed relative to the non-detects. One was to estimate the mean level; the other was to estimate an uncertainty in the measurement (which would then be incorporated into the uncertainty estimate of the charge contribution, discussed later). In order to simply, and somewhat conservatively, provide both of these estimates, it was assumed that the non-detect value was the result of a discrete binary distribution which could assume the values of zero and the detection limit, with a fixed probability (p) of obtaining the detection limit (d). Under this assumption, the mean value is $p \times d$; the variance is $p \times (1 - p) \times d^2$ (Mood, Graybill, & Boes, 1974, p. 538). In the interest of being conservative in estimating the variance (i.e., over-estimate the variance), the variance is maximized when $p = 0.5$. Using this fact results in calculating the mean as $0.5 \times d$ and the standard deviation (the square root of the variance) as $0.5 \times d$ also.

Other distributional assumptions could have been made for the purpose of determining the mean and standard deviation to be used for the non-detects, e.g., truncated normal, truncated lognormal, continuous uniform. The continuous uniform is relatively simple to implement also, resulting in the same mean, but a smaller standard deviation. The truncated distributions are more complicated to use; simple approximations suggest that the standard deviation is also smaller than that provided by the selected discrete distribution. Therefore, it seems reasonable to use the selected discrete distribution as a simple, conservative method for estimating the mean and standard deviation to be used for non-detects.

Calculating Charge Contribution and Standard Deviation

The data reported for the mean value of the analyte are either mass-based (ug/mL) or activity-based (uCi/mL) concentrations. In order to calculate the charge contribution for each analyte, the appropriate conversion to Mol/L, as well as the charge of the particular analyte, is needed. Given this information, the charge contribution for a specific analyte is calculated as the mean multiplied by the appropriate conversion factor multiplied by the appropriate charge.

To calculate the standard deviation of the charge contribution, it is assumed that the conversion factor and the actual charge of the analyte are known constants. (This assumption seems reasonable, given that the conversion factor and charge are based on known specific activities and molecular weights of assumed species.) The standard deviation of the charge contribution is then calculated as the standard deviation of the analyte mean multiplied by the conversion factor multiplied by the absolute value of the charge (Mood, Graybill, & Boes, p. 178ff).

For non-detects, the standard deviation of the analyte mean is calculated as described previously. For all others, the standard deviation of the analyte mean is calculated based on the confidence limits provided, along with the t-value determined by the degrees of freedom provided.

Calculating the Total Charge and Standard Deviation

The total charge for the sampling event is calculated simply as the sum of the individual analyte charge contributions.

The standard deviation of the total charge is slightly more complicated. If the analyte measurements were independent, then the variance of the total charge would simply be the sum of the variances of each of the individual charge contributions. However, given that these measurements all start out as concentration measurements from a sample(s) drawn from one tank, it probably is not reasonable to assume that the measurements are independent. In this case, the variance of the total is the sum of the variances and the covariances (Mood, Graybill, & Boes, p. 178ff). Complicating this is the fact that the covariances are unknown. However, using the fact that the correlation between two variables is always less than or equal to one in magnitude, the covariances for each pair can be bounded from above by the product of the standard deviations. Using this relationship, the variance of the total charge will be less than the square of the sum of the individual standard deviations. Or, more simply, the standard deviation of the total charge will be less than the sum of the individual charge contribution standard deviations.

Note that, due to the approximation used, as well the fact that some of the individual contribution standard deviations are based on a distribution assumption while others are based on actual sample data, the degrees of freedom that would typically be used in the determination of critical values for confidence intervals or hypothesis testing is not well-defined in this situation. For this reason, a rule-of-thumb of plus-or-minus three standard deviations will be used to create an interval to identify whether the charge appears to be large relative to the analytic uncertainty. If the charge is within the interval, it will be considered to be balanced with respect to the analytic uncertainty. As will be seen later, this interval will also be used for adjusting the analyte values to achieve an exact charge balance.

Analysis of Tank AW-101 – Charge balance

Applying the before-mentioned methodology to the six sampling events for Tank AW-101 produced Table 1. The sum of the cation charge is about 11 Mol/L in these samples.

Table 1. Tank AW-101 Sampling Events Evaluated in This Study.

Sampling Event	Number of Cations	Number of Anions	Total Charge (mol/L)	St.Dev. of Charge
AW-101 1996 Core Sup Seg	*3*	*12*	*10.5988*	*0.1099*
AW-101 1998 Grab Composite	7	28	-0.5940	0.7390
AW-101 2002 Grab	8	30	-0.1151	0.4295
AW-101 2003 Core Composite	4	22	5.5458	0.4808
AW-101 2003 Interstitial	3	22	-1.5557	0.4585
AW-101 2003 Liquid Supernatant	3	22	-1.4466	0.6559
AW-101 Tank Average			-0.8439	0.4346

Note that the total charge for sampling events 1998 Grab Composite, 2002 Grab, 2003 Liquid Supernatant, and Tank Average all appear to be small relative to the standard deviation of the charge, suggesting that the charge is essentially zero (i.e., balanced) within the bounds of the analytic uncertainty.

Further inspection of the 1996 Core Sup Seg sampling event indicates that a number of key analytes were missing from the analysis. In particular, Chloride, Hydroxide, Nitrate, and Nitrite were not included in the analysis. Adjusting for these missing analytes by incorporating the tank average charge contribution for each analyte would result in a charge of -0.20 mol/L, which would appear to be within the bounds of the analytic uncertainty for this sampling event.

Further inspection of the 2003 Core Composite sampling event indicates that Hydroxide was not measured in that sample. Incorporating the tank average charge contribution for Hydroxide of -5.80 mol/L would result in a charge of -0.25, which would appear to be within the bounds of the analytic uncertainty for this sampling event.

Further inspection of the 2003 Interstitial sampling event suggests the possibility of several unusual measurements. In particular, Nitrate appears to be somewhat higher in concentration relative to the other sampling events. Coupled with the negative charge, this results in a more negative charge relative to the other sampling events. Sodium appears to be substantially lower in concentration, which coupled with the positive charge, results in a more negative charge. Aluminum appears to be somewhat lower in concentration relative to the other sampling events. Coupled with the negative charge, this results in a more positive charge relative to the other sampling events, somewhat offsetting the impact of the Nitrate and Sodium. Adjusting for these three analytes would result in a charge of -0.76 mol/L, which would appear to be within the bounds of the analytic uncertainty for this sampling event.

Method to Adjust Charge

After inspection of each of the sampling events, it is desired to determine whether it is possible to "adjust" the individual analyte values to make the total charge for the sampling event exactly zero, while maintaining each analyte within reasonable analytic uncertainty bounds. There are numerous approaches to this optimization, which would produce differing results. To further define the problem, it is desired to make the smallest changes in the analyte values that will exactly balance the charge. Mathematically, the optimization is defined as minimizing the sum of the squared differences between the adjusted and measured charge contribution value for the analytes divided by the standard deviation of the charge contribution, subject to the constraint that the total charge equals zero. Note that since the primary objective is dealing with charge of the sample, the optimization is defined in terms of the charge contribution of the analyte, but that the measured value for each sample is the analyte concentration, which is then converted to a charge contribution.

There are multiple approaches that could be used for solving this optimization problem. One approach is to use a numerical procedure (such as., Microsoft Excel Solver add-in (Microsoft Corporation, 2010)). The advantage of the numerical procedure is that it is generally straightforward and simple to use, and can be used in a broad range of problems where other methods might not be applicable. A disadvantage is that it may not produce a global optimum solution, or may not converge to a solution at all.

Analytical methods, using calculus, can provide a closed-form solution, which can be shown to be a global optimum, in certain situations. The disadvantage is that the method may be more complicated to apply, and may not be applicable in a particular situation. Fortunately, an analytical method can be applied in the present application. Using the calculus method of Lagrange multipliers (Mood, Graybill, & Boes, p. 501ff), an analytical method is derived below.

Let:

x_i represent the measured concentration of analyte i;
$x_i^{\sim}$ represent the adjusted concentration of analyte i;
f_i represent the charge conversion factor for analyte i;
s_i represent the standard deviation of charge contribution of analyte i.

Then the objective of minimizing the sum of the squared differences between the adjusted and measured charge contribution value for the analytes divided by the standard deviation of the charge contribution, subject to the constraint that the total charge equals zero, can be written mathematically as follows.

Minimize:

$$L = \sum_i \frac{f_i^2 \times (x_i^{\sim} - x_i)^2}{s_i}$$

subject to:

$$\sum_i f_i \times x_i^{\sim} = 0$$

The method of Lagrange multipliers is used to solve this optimization problem, by transforming this constrained optimization into the following equivalent, unconstrained optimization.

Minimize:

$$L^{\sim} = \sum_i \frac{f_i^2 \times (x_i^{\sim} - x_i)^2}{s_i} - \lambda \times \sum_i f_i \times x_i^{\sim}$$

This optimization is then solved by taking partial derivatives of $L^{\sim}$ with respect to each of the $x_i^{\sim}$'s and λ, and equating them to zero. The solution to this set of equations is then a stationary point; evaluation of the second derivatives will indicate whether the solution is a minimum, maximum or saddle point.

Taking the derivative of $L^{\sim}$ with respect to $x_i^{\sim}$, and equating to zero produces the following:

$$\frac{\partial L^{\sim}}{\partial x_i^{\sim}} = \frac{2 \times f_i^2 \times (x_i^{\sim} - x_i)}{s_i} + \lambda \times f_i = 0$$

which simplifies to:

$$x_i^{\sim} = x_i - \lambda \times \frac{s_i}{2 \times f_i}$$

Taking the derivative of $L^{\sim}$ with respect to λ, and equating to zero produces the following:

$$\frac{\partial L^{\sim}}{\partial \lambda} = \sum_i f_i \times x_i^{\sim} = 0$$

which, when combined with the previous, simplifies to:

$$\lambda = \frac{2 \times \sum_i f_i \times x_i}{\sum_i s_i}$$

Combining this with the previous produces:

$$x_i^{\sim} = x_i - \frac{s_i}{f_i} \times \frac{\sum_i f_i \times x_i}{\sum_i s_i}$$

Defining

$$u_i = \frac{3 \times s_i}{f_i}$$

where u_i represents the analytic error interval half-width for the concentration of analyte i, this then simplifies to:

$$x_i^{\sim} = x_i - u_i \times \frac{C}{M}$$

where

$$C = \sum_i f_i \times x_i$$

is the total charge of the original sample data, and

$$M = \sum_i u_i \times f_i$$

represents the maximum change in the charge possible if all analytes are changed in the most extreme direction, within their respective analytic error intervals.

Application to Tank AW-101 Data

For sampling events with a total charge which was found to be within the analytic error bounds, this adjustment method was applied directly to the analyte values. For those sampling events which were not within the analytic error bounds, the corrections discussed in the earlier section were first performed, followed by the adjustment method. Note that, using this method, it is easily seen whether a sampling event can be exactly balanced, simply by comparing M (the maximum amount the charge could be changed) to the total charge from the sampling event. If the total charge is larger in magnitude than M, then it is not possible to balance the charge while keeping the analytes within their respective analytic error bounds.

Based on the methodology briefly discussed above, every analyte measurement (with a non-zero charge) will be adjusted to exactly balance the charge for the sampling event. For the purposes of discussion, only those analytes that contribute a substantial amount to the change in the charge will be included. Somewhat arbitrarily, this is defined as those analytes which contribute more than 10% of the change in the total charge.

For the 1998 Grab Composite sampling event, the charge can be exactly balanced by adjusting the analyte measurements. Increasing the Sodium value (from 257,000 to 261,509) and decreasing the Nitrate (from 170,000 to 159,516) and Nitrite (from 120,000 to 113,910) values accounts for approximately 84% of the increase in the charge necessary. Incorporating the additional restriction that all non-detect analyte values are zero, the charge can be exactly balanced by increasing the Sodium value (from 257,000 to 261,396) and decreasing the Nitrate (from 170,000 to 159,779) and Nitrite (from 120,000 to 114,063). These changes account for approximately 85% of the increase in charge necessary.

For the 2002 Grab sampling event, the charge can be exactly balanced by adjusting the analyte measurements. Increasing the Sodium (from 259,000 to 260,031) and decreasing Hydroxide (from 102,000 to 101,554) and Nitrate (from 118,000 to 117,051) accounts for approximately 75% of the increase in charge necessary. Incorporating the additional restriction that all non-detect analyte values are zero, the charge can be exactly balanced by decreasing the Sodium (from 259,000 to 258,548) and increasing Hydroxide (from 102,000 to 102,195) and Nitrate (from 118,000 to 118,416). These changes account for approximately 90% of the decrease in charge necessary. Note that for the original sample data, the total charge is negative. When all non-detect analytes values are set to zero, the total charge becomes positive, hence the change in direction of the adjustments. These changes are within the analytical uncertainty of the analytical data.

For the 2003 Liquid Supernatant sampling event, the charge can be exactly balanced by adjusting the analyte measurements. Increasing the Sodium (from 233,000 to 247,775) and decreasing Hydroxide (from 99,500 to 95,868), Nitrate (from 171,000 to 157,593), and Nitrite (from 105,000 to 97,968) accounts for approximately 85% of the increase in charge necessary. Incorporating the additional restriction that all non-detect analyte values are zero, the charge can be exactly balanced by increasing the Sodium (from 233,000 to 247,743) and decreasing Hydroxide (from 99,500 to 95,876), Nitrate (from 171,000 to 157,621), and Nitrite (from 105,000 to 97,983). These changes account for approximately 85% of the increase in charge necessary.

For the 1996 Core Sup Seg sampling event, it is not possible to balance the charge of the original sample data, while keeping the analyte measurements within their analytic bounds. As discussed previously, several analytes were not included in the sample data. Including these four analytes (Chloride, Hydroxide, Nitrate, and Nitrite) at the tank mean levels, and allowing them to be adjusted as well, it is possible to exactly balance the charge. Increasing the Sodium (from 252,000 to 253,300) and decreasing the Hydroxide (from 98,625 to 97,572), Nitrate (from 158,400 to 156,191) and Nitrite (from 104,280 to 103,044) accounts for approximately 90% of the increase in charge necessary. Incorporating the additional restriction that all non-detect analyte values are zero, the charge can be exactly balanced by increasing the Sodium (from 252,000 to 253,233) and decreasing the Hydroxide (from 98,625 to 97,626), Nitrate (from 158,400 to 156,305) and Nitrite (from 104,280 to 103,108). These changes account for approximately 92% of the increase in charge necessary. In this particular example, we knew that data were missing from this dataset because these analytes were available in other datasets. Had we not known that an important ion was missing, we still would have identified that something was wrong because the charge imbalance was greater than the total analytical uncertainty.

For the 2003 Core Composite sampling event, it is not possible to balance the charge of the original sample data, while keeping the analyte measurements within their analytic bounds. As discussed previously, one analyte was not included in the sample data. Including Hydroxide at the tank mean level, and allowing Hydroxide to be adjusted as well, does allow the sampling event to balance. Increasing the Sodium (from 249,000 to 252,982) and decreasing the Hydroxide (from 98,625 to 97,984) accounts for approximately 83% of the increase in charge necessary. Incorporating the additional restriction that all non-detect analyte values are zero, the charge can be exactly balanced by increasing the Sodium (from 249,000 to 252,632) and decreasing the

Hydroxide (from 98,625 to 98,041). These changes account for approximately 85% of the decrease in charge necessary.

For the 2003 Interstitial sampling event, it is not possible to balance the charge of the original sample data, while keeping the analyte measurements within their analytic bounds. As discussed previously, several analytes (Aluminum, Nitrate, and Sodium) appear to be unusual compared to the other sample values. Using the tank mean value for these analytes, and allowing them to be adjusted as well, does allow the sampling event to balance. Decreasing the Hydroxide (from 97,100 to 88,622) and increasing the Sodium (from 246,833 to 248,967) accounts for approximately 78% of the increase in the charge necessary. Incorporating the additional restriction that all non-detect analyte values are zero, the charge can be exactly balanced by decreasing the Hydroxide (from 97,100 to 88,661) and increasing the Sodium (from 246,833 to 248,957). These changes account for approximately 79% of the increase in the charge necessary.

It is noted that there is a consistent group of analytes which are responsible for the majority of the change in charge necessary to exactly balance the sample charge: Hydroxide, Nitrate, Nitrite, and Sodium. It should also be noted that the adjustments made to the analyte values to exactly balance the total charge are relatively small, as desired.

Conclusion

This study evaluated the charge balance discrepancy in supernatant samples from Hanford nuclear waste tank AW-101. The results indicate that the analytical data were charge balanced within the analytical uncertainty for three sampling events, and when the data were averaged across sampling events. A method to force charge balance for these samples was developed. For three other sampling events, the analytical data had large charge imbalances because of missing data or gross errors. These discrepancies could be identified by the fact that the data sets were not charge balanced within the analytical uncertainty.

References

Disselkamp, R. S. (2009). *2009 AUTO-TCR for Tank 241-AW-101, RPP-RPT-42909, Rev. 0.* Richland, WA: Washington River Protection Solutions, LLC.

Microsoft Corporation. (2010). *Define and solve a problem by using Solver*. Retrieved September 08, 2010, from http://office.microsoft.com/en-us/excel-help/define-and-solve-a-problem-by-using-solver-HP010072691.aspx

MIller, A. G. (1977). Laser raman spectrometric determination of oxy anions in nuclear waste materials. *Analytical Chemistry volume 49* , 2044-2048.

Mood, A. M., Graybill, F. A., & Boes, D. C. (1974). *Introduction to the Theory of Statistics.* New York: McGraw-Hill Book Company.

Romagnoli, J. A., & Sanchez, M. C. (2000). *Data Processing and Reconciliation for Chemical Process Operations.* San Diego: Academic Press.

(632e) Process Capacity of the Continuous Sludge Leaching Process

Reid Peterson, Pacific Northwest National Laboratory, Richland, WA

Boehmite has proved to be a difficult aluminum phase to dissolve out of high-level waste (HLW), requiring longer residence times and higher temperatures for more complete dissolution than the gibbsite phase. To provide a simple, compact, effective, and proven method to remove aluminum from HLW sludges, HLW feed can be blended with caustic in a reaction vessel while continuously removing permeate from the reactor. This approach allows for a concurrent addition of fresh caustic and HLW feed while simultaneously removing concentrated reacted slurry. Separation in this manner will keep solids in the reactor for longer periods of time, allowing for higher conversions than could be achieved in a normal batch reactor. Advantages for this continuous sludge leaching process are the reduction of the number of high level waste canisters by one third to one half at the Hanford Tank Waste Treatment and Immobilization Plant, resulting in billions of dollars in life-cycle cost savings. An assessment of the feed variability and the capacity of the proposed system to treat waste from S and SX tank farms was performed. Based on this assessment, a determination was made regarding the sizing of the facility required to treat this waste in a timely fashion.

See more of this Session: Chemical Engineering Advances in Processing Radioactive Wastes and Nuclear Materials
See more of this Group/Topical: Nuclear Engineering Division

CATALYZED AMMONIUM ION FORMATION FROM THE REACTION OF FORMIC ACID WITH NITRATE ION DURING RADIOACTIVE WASTE PROCESSING AT THE SAVANNAH RIVER SITE – paper no. 632f

David C. Koopman, Savannah River National Laboratory, Aiken, SC
Michael E. Stone, Savannah River National Laboratory, Aiken, SC

ABSTRACT

The Defense Waste Processing Facility (DWPF) at the Savannah River Site converts legacy radioactive waste from the Cold War into a stable borosilicate glass. Waste processing produces a significant quantity of contaminated liquid water that is recycled to the evaporators in the tank farm. Until recently only negligible quantities of ammonium ion were detected in the recycle stream. Processing high noble metal and mercury concentrations in the past two years has shown an increase in ammonium ion concentrations to near the maximum levels expected during the DWPF design phase. The cause of the increase was linked to a combination of extended processing times, increased levels of acid addition, and higher concentrations of mercury and noble metals. Both ammonium ion and ammonia vapor were formed during processing. The pH of the aqueous phase in the process determined the preferred form.

INTRODUCTION

Legacy radioactive waste generated at the Savannah River Site (SRS) during production of enriched uranium and plutonium during the Cold War is currently being converted by the DWPF into a stable borosilicate glass waste form for long term storage. The majority of the legacy waste is stored as a mixture of hydroxide and hydrous oxide insoluble solids in large storage tanks at SRS. These 3.5-4.5 thousand cubic meter (900,000-1,200,000 gallon) carbon steel storage tanks also contain 5-7M sodium solutions rich in hydroxide, nitrate, and nitrite anions. The sodium molarity is reduced to about one during preparation of a new feed batch for DWPF.

The preparation of the sixth DWPF sludge batch (SB5) was the first to use large tank aluminum dissolution to reduce the aluminum content of the insoluble solids in addition to reducing the sodium content. This change coincided with the transition from the high iron Purex wastes of the first five sludge batches (SB1A, 1B, 2, 3, and 4) to high aluminum HM wastes. Removal of aluminum led to significant enrichment of other species in the remaining insoluble solids. The major impact was the increased concentrations of Hg, Rh, and Ru. These species are catalytically active during DWPF processing using formic acid. Laboratory-scale preparation activities for SB5 noted lower than expected values of nitrate ion in the processed melter feed, but the cause was not identified. Unusually high concentrations of ammonium ion were subsequently found in the recycle condensate stream leaving DWPF shortly after the start of SB5 processing.

Pilot plant testing of the original coupled flow sheet for DWPF (simultaneous processing of sludge and salt wastes) occurred with formation of ammonium nitrate crystals in the off-gas system. The salt waste processing was believed to be the main source of the ammonium ion, but evidence for catalytic formation of ammonium ion from nitrate during sludge processing had been reported by King et al. in 1997.[1] The DWPF design was modified to include ammonia scrubbers at three key points in the off-gas system to prevent formation of ammonium nitrate deposits. It was subsequently determined that salt waste processing should be done elsewhere at SRS, and it appeared that the DWPF ammonia scrubbers might be superfluous. Recycle condensate sampling during the first ten years of operation indicated minimal ammonium ion content in the composite condensate stream from the DWPF.

EXPERIMENTAL PROGRAM

A lab-scale DWPF simulation program commenced in late 2009 to facilitate preparations for sludge batch 6 (SB6). The main waste component of SB6 was also HM waste. An off-gas ammonia scrubber (packed column) was added to the lab-scale process simulation equipment based on DWPF experience with SB5. Dilute nitric acid was used as the scrubbing liquid. Increased sampling of the Sludge Receipt and Adjustment Tank (SRAT) vessel was included in the test plan to locate the timing of ammonium ion formation and to quantify the moles of ammonium ion in the vessel. Additional sampling also targeted various SRAT cycle condensates and Slurry Mix Evaporator (SME) cycle condensates for ammonium ion. SRAT and SME cycle condensates are eventually sent to the DWPF recycle condensate stream that returns an aqueous solution to the SRS tank farm for processing through the tank farm evaporators. The other main contributor to the recycle condensate stream comes from the melter off-gas system.

The results of the test program show that ammonium ion is formed during the SRAT cycle under certain conditions. Formation starts later than catalytic hydrogen generation which was previously the last major reaction identified in the SRAT chemistry timeline.[2] The amount of ammonium ion produced was found to depend on the amount of excess formic acid added during SRAT processing, as well as on the initial mercury concentration of the radioactive waste feed. Most ammonium ion at the end of SRAT cycle was either in the SRAT slurry or in the circulating acid reservoir for the ammonia scrubber. Significant SRAT slurry ammonium was driven out of the slurry during SME cycle concentration periods and was absorbed into off-gas condensates before the off-gas reached the ammonia scrubber. Active ammonium ion formation continued during the SME cycle in parallel with the stripping of ammonia vapor into the off-gas system.

The proposed overall reaction for ammonium ion formation in the DWPF is given by:

$$5HCO_2H + NO_3^- \dashrightarrow NH_4OH + HCO_2^- + 4CO_2 + 2H_2O \qquad [1]$$

This equation is analogous to equation (2) in reference 1, but it has been written to show the ammonia present as ammonium hydroxide in solution, since the majority of the ammonia was initially found in the aqueous phase of process waste slurry as the ammonium ion (by cation chromatography).

Analytical data from the DWPF SRAT-SME process simulations were used to prepare material balance calculations for different species. These calculations show increased loss of formic acid/formate ion, unexpected loss of nitrate ion, and the formation of additional CO_2 gas commensurate with increased formate loss.

A series of reactions has been proposed based on further studies of ammonium ion formation. The reaction sequence presumes that the reacting nitrogen bearing species (nitrate, nitrite, NO, ammonia) adsorbs/desorbs from a catalytic substrate. Each reaction could be further subdivided into elementary reaction steps. The three main reactions in the sequence add to the overall reaction above. The main sequence includes formation of some nitrite ion as an intermediate.

$$HCO_2H + NO_3^-(s) \rightarrow NO_2^-(s) + CO_2 + H_2O \quad [2]$$

$$3HCO_2H + 2NO_2^-(s) \rightarrow 2NO(s) + 2HCO_2^- + CO_2 + 2H_2O \quad [3]$$

$$5HCO_2H + 2NO(s) \rightarrow 2NH_3(s) + 5CO_2 + 2H_2O \quad [4]$$

The (s) symbolizes a species adsorbed on the catalytic substrate surface. The nitrogen bearing species are free to move into and out of the aqueous phase by reactions such as

$NO_2^-(s) \leftrightarrow NO_2^-(aq)$ (similar for nitrate) [5]

$NO(s) \leftrightarrow NO(gas)$ and [6]

$$NH_3(s) + H_2O \leftrightarrow NH_4^+(aq) + OH^-(aq) \leftrightarrow NH_3(gas) + H_2O \quad [7]$$

The multi-step reaction sequence proposed for ammonia formation involves many species or changes that can be monitored with the available lab and analytical equipment. The following phenomenon could be detected:

- Nitrate should be lost (material balance calculation)
- Atypical NO_x and/or N_2O could be seen in the off-gas (gas chromatography)
- Extra formic acid should be consumed (material balance calculation)
- Extra CO_2 should appear in the off-gas (from formic lost) (gas chromatography)
- Ammonium ion might be found in SRAT slurry samples (cation chromatograph)
- Ammonia vapor might be found in the off-gas (ammonia detector tubes)
- Ammonium ion might be found in the ammonia scrubber solution or other down stream cold traps (cation chromatograph)
- The Nafion moisture dryers in the off-gas system might plug (membrane material adsorbs and swells when exposed to ammonia)
- The pH might be influenced (nitric acid to ammonium hydroxide)

The conversion of nitrite ion to nitrate ion during nitrite ion destruction has generally led to a gain in the number of moles of nitrate and a positive conversion efficiency for nitrite to nitrate. The expected maximum conversion efficiency (moles nitrate made/mole nitrite destroyed) is about 33% from the reaction:

$$3HNO_2 \rightarrow HNO_3 + 2NO + H_2O$$

Measured efficiencies of 10-25% are not unusual, since nitrite is converted to gaseous products alone by two other parallel reactions. Starting with sludge batch 5 (SB5), however, the calculated conversion efficiencies became negative indicating nitrate loss (in spite of the known formation reactions). Subsequently, chilled condensate samples from the guard condenser on the off-gas GC were analyzed and found to contain ammonium ion. This indicated that ammonium ion had been formed and released into the off-gas system during processing.

Testing was initiated to look for other potential markers of ammonium formation, to determine the extent of ammonium ion formation, and to identify where the ammonium/ammonia was at the end of processing. The changes included sampling of the SRAT slurry during processing to test for ammonium ion, sampling of the off-gas for ammonia, the installation of an ammonia scrubber in the off-gas system to collect the majority of the ammonia given off by the SRAT slurry into a nitric acid solution, and additional scrutiny on the NO_x species in the off-gas data.

Off-gas GC data was immediately available from earlier tests. These data were examined, and signs of low level N_2O formation were found after all of the initial nitrite in the feed had been destroyed. These had not been noticed initially because they were only about 2% of the typical concentration levels seen during nitrite destruction. SRAT product slurry samples were also available from earlier tests, and some of these showed ammonium ion (the available cation chromatography instrument has a large interference from sodium that interferes with the detection of low concentrations of ammonium ion). These early data indicated that the process was converting between 0-25% of the nitrate available into ammonia.

At the time that ammonium ion formation is detected in the DWPF SRAT simulations, all of the initial nitrite ion content in the waste feed has been destroyed based. Samples pulled following formic acid addition showed no detectable nitrite or ammonium ion by ion chromatography. This eliminated one of the early hypotheses for ammonium ion formation, which was that the ammonium was being formed from the nitrite ion in the starting waste via reaction [3] above without the need for reaction [2]. Subsequent sample data were conclusive that the period where the ammonium ion was present did not overlap the period when the initial nitrite ion was destroyed, and in fact the two periods were generally separated by at least several hours.

Destruction of nitrite ion by formic acid produced both NO and N_2O. Production of these gases also stopped shortly after the end of formic acid addition. These patterns in the data have been present for over a dozen years of process simulations. The only prior noteworthy exception occurs when the process is switched from dewatering (concentration) to reflux (constant volume) several hours after formic acid addition. The reflux reservoir can contain some nitrite and/or nitrate ion, formed from condensation and absorption of NO_2, and the onset of reflux is often accompanied by a short duration (10-15 minutes) release of some additional N_2O as the reservoir nitrite is reintroduced into the waste slurry.

It was observed in the GC data from certain SRAT simulations that a subsequent episode of N_2O formation occurred after sampling had indicated that the initial nitrite ion had been destroyed. (NO formation is harder to monitor with the GC, since most of it reacts with oxygen to form NO_2 and/or N_2O_4. The latter two gases are not detected by the GC and the small loss of oxygen associated with their formation is also difficult to quantify.)

Figure 1 shows some processed gas chromatography data (an internal He standard gas flow rate and composition is used to convert other measured gas compositions into flow rates that are then rescaled from lab-scale to full-scale for the DWPF customer). The first hour after the end of formic acid addition has the tail end of N_2O produced from nitrite ion destruction.

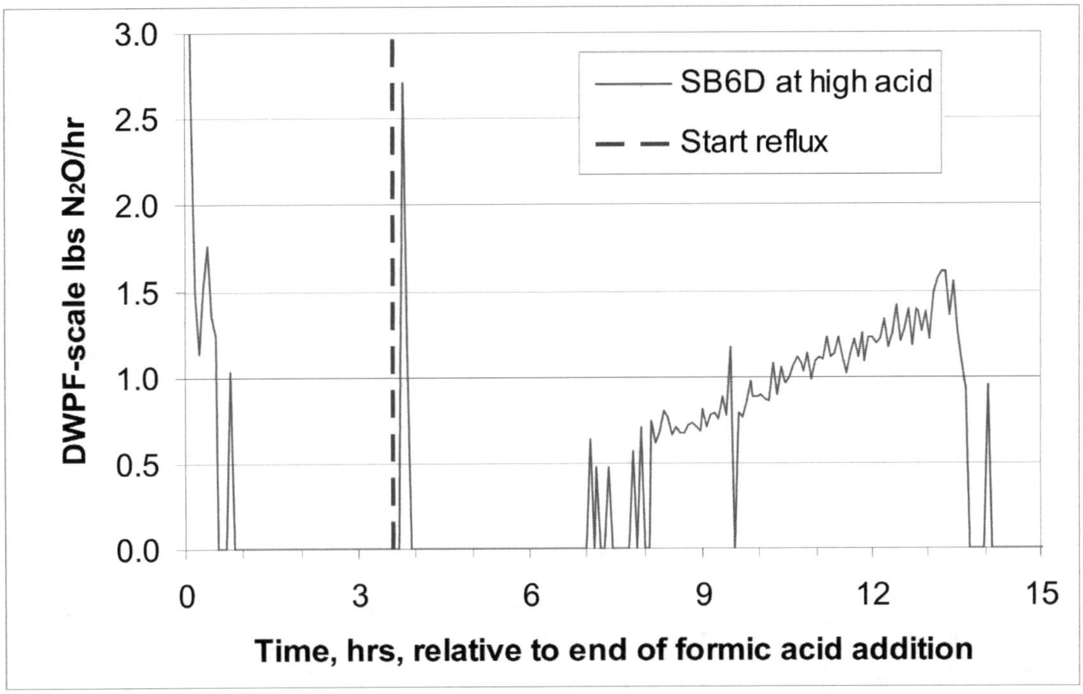

Fig. 1. Unexpected N_2O between 7 and 14 hours after formic acid

Dewatering switched to reflux as shown at about 3.6 hours after formic acid, and this was followed by a short spike of renewed N_2O production, presumably from nitrite accumulated in the reflux reservoir during dewatering. Three hours later, additional N_2O close to the detection limits of the GC were detected, and the concentration increased for seven hours then rapidly fell off. Refluxing at about 101 °C continued until nearly 25 hours after the end of formic acid addition with no further evidence for significant N_2O formation. Although two exiting streams were not sampled for ammonium/ammonia (SRAT condensate and SRAT off-gas), ammonium ion equivalent to 25% of the starting sludge nitrate plus added nitric acid nitrate was found assuming reaction [1] gave the overall stoichiometry (one mole nitrate lost per mole ammonium ion formed). The GC data indicated, however, that some nitrate ion probably failed to make it completely through the multi-step reaction [2]-[4].

Updated compositional analyses of actual SB6 samples mandated a change to the test feed being used to develop the understanding of ammonium ion shortly after the above test was completed. Though such shifts are not unusual during sludge batch preparations, they do tend to complicate experimental design factors such as the timing of samples, comparisons of data from different simulations, etc.

During the initial test above, 96% of the ammonium ion found was in the SRAT product slurry and the rest in the chilled (~4 °C) off-gas condenser condensate. This suggested that most of the ammonium ion being formed was not being converted into ammonia vapor. A new series of tests was performed with a new SB6 baseline composition at four different levels of added acid. In these tests it was found that 15-30% of the accountable ammonium ion was in the chilled condenser condensate, i.e. at least 15-30% of the ammonium formed in the SRAT had gone into the off-gas system. The amount of nitrate destroyed increased with increasing acid additions ranging from about 1% at low acid to 15% at high acid (high acid is limited by catalytic hydrogen generation to stay below about 1 vol% hydrogen in the off-gas). The lab-scale chilled condenser was not designed as an ammonia absorber, although it may have been doing quite well in that capacity (oversized and a lot of surface area typically covered in acidic aqueous condensate). Furthermore, the porous media inside Nafion dryers in the off-gas line, which are used to prevent moisture from reaching the gas chromatograph, are subject to plugging in the presence of ammonia.

Modifications were made to the experimental apparatus based on the issues above. A packed column was installed in the off-gas line after the SRAT off-gas condenser. A reservoir of 0.01 M nitric acid was recirculated through the packed column to absorb ammonia vapor from the off-gas before it reached the chilled condenser, Nafion dryer, and the gas chromatograph. The reservoir was sampled in subsequent runs to monitor the uptake of ammonium ion (this analysis was not subject to the sodium interferences that impact the available cation chromatography column on samples of SRAT slurry which is roughly 1 M in sodium). Calibrated gas sample tubes sensitive to ammonia were also obtained for occasional sampling of the off-gas leaving the SRAT condenser (these could give instantaneous values, while the scrubber reservoir essentially monitored the total ammonium accumulation; the scrubber ammonium content at time t should equal the time integrated total of the time-dependent ammonia flow up to time t assuming 100% recovery in the scrubber).

A third composition shift was made to cover a sludge batch blending scenario. Several additional process simulations were run varying different factors of interest to DWPF (total acid, addition of salt streams, rheological properties, and departures from design cycle time). New concentrations of the catalytically active species in the system accompanied the composition shift which again confounded direct comparisons with the earlier work.

Figure 2 below shows the most comprehensive data set for ammonium ion formation and release of ammonia vapor into the off-gas system. Off-gas ammonia is in ppm in the flowing air purge. Ammonium ion in the SRAT slurry and in the nitric acid reservoir for the ammonia scrubber have been converted into total grams of ammonium using concentration and total mass. The recorded vol % of N_2O in the off-gas stream is given on the same scale (0-1.0). The SRAT slurry pH is shown (multiplied by 100 to put 0-10 readings on a 0-1000 scale). The aqueous SRAT slurry is boiling at atmospheric pressure.

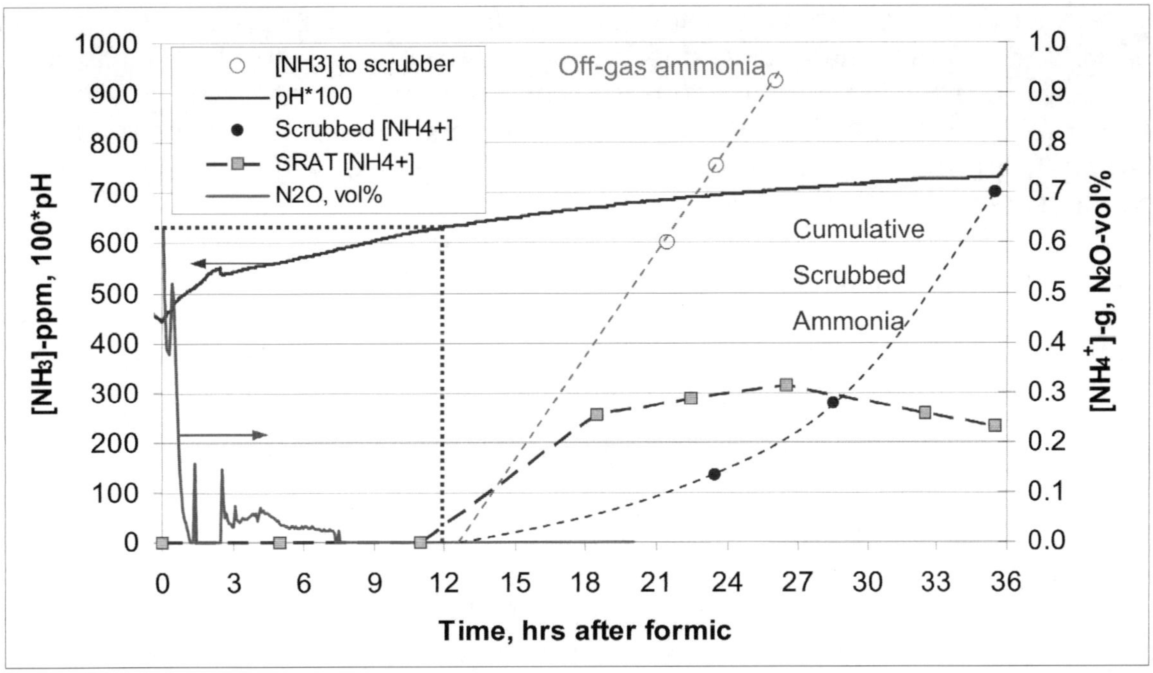

Figure 2. Composite plot showing interrelated data for the ammonium ion formation reaction

The data suggest that nitrate was taken onto the substrate and converted primarily to nitrite ion rather than ammonium ion (reaction eqn. 2) from about 2-7 hours after acid addition. The nitrite ion was released into solution and converted into N_2O by residual excess acid (pH roughly 5.5 to 6). As pH continued to rise, N_2O production ceased, and nothing was detected, though nitrate may have been converting to NO (reaction eqn. 3), and the NO converted into NO_2 (which is not detected by the current GC).

It appears in the case of the test in Figure 2 that ammonium ion formation started about 11-13 hours after acid addition (pH about 6.5; reaction eqn. 4). The concentration of ammonium ion in the SRAT slurry increased until about 26 hours after acid as the pH rose to 7. Some ammonium ion was converted to ammonia and exited into the off-gas system. The rate of off-gas loss increased as pH and dissolved ammonium ion concentrations both rose. The amount of ammonia collected by the nitric acid scrubber responded as expected for a (approximate) ramp function input. The data suggest that equilibrium between species on the substrate and in solution may be very pH dependent.

The production of CO_2 during ammonium ion formation was also quantified. Cumulative carbon dioxide production exceeded four times the ammonium ion/ammonia vapor production during reflux of the SRAT supporting the reaction sequence given in eqns. 2-4. The general timing of the production also seems to align with the ammonium data, although parallel processes destroying formic acid appear to also be occurring (catalytic decomposition of formic acid to CO_2 and either H_2 or H_2O). Testing without the ammonia scrubbers in the off-gas system ultimately led to plugging of the off-gas Nafion dryer elements per the vendor warning.

Early DWPF sludge batches had lower concentrations of mercury and noble metals. The low mercury concentrations allowed the SRAT cycle to last only about 13-15 hours (to accomplish process chemistry; mercury was removed below the maximum product specification in less time). SRAT products often remained acidic due to short refluxing times and low noble metal catalysis of formic acid destruction. Ammonium ion was seldom seen and only at very low concentrations. Current sludge batches have high mercury and noble metal concentrations. Melter feeds frequently are at pH values above 9. Ammonium ion is being routinely observed in the recycle condensate from DWPF to the tank farm evaporators.

The overall chemistry for ammonium ion formation appears to be understood with the exception of whether or not the reaction is being catalyzed. It has been observed that mercury becomes more resistant to steam stripping as the acid stoichiometry increases, and also that ammonium ion formation increases as the acid stoichiometry increases. The correlation between excess acid and increasing ammonium ion formation is clear, but without the catalytic attack on formic acid that raises the pH, it is unlikely that significant ammonium ion would be seen. The range of mercury and noble metal concentrations tested to date have not given a clear indication of whether or not catalysis is contributing to ammonium ion formation.

A shiny gray clay-like deposit was found on the agitator impeller in one of the higher acid runs. The deposit was analyzed and found to most likely be a mercury amalgam with low concentrations of copper, rhodium, palladium, ruthenium, and a trace of calcium. The deposit contained a quarter of the mercury in the initial SRAT feed slurry. The amalgam is a likely catalyst candidate for ammonium ion formation.

REFERENCES

1. Noble Metal Catalyzed Ammonium Generation from Formic Acid Reduction of Nitrate in Simulated Nuclear Waste Media, R. B. King, N. K. Bhattacharyya, H. D. Smith, and K. D. Wiemers, Environmental Science and Technology, Vol. 31, No. 4, (1997) 984-992.

2. Koopman, D., B. Pickenheim, D. Lambert, J. Newell, and M. Stone, "Characterization of Individual Chemical Reactions Consuming Acid during Nuclear Waste Processing at the Savannah River Site", Proceedings of the 2009 Annual AIChE Meeting, New York, NY: AIChE. 2009.

(638a) Degradation Studies of Solid Oxide Electrolysis Cells

Greg G. Tao[1], Bruce Butler[1], Niladri Dasgupta[1], Anil V. Virkar[1] and Manohar S. Sohal[2], (1)Materials & Systems Research, Inc., Salt Lake City, UT, (2)Idaho National Laboratory, Idaho Falls, ID

Hydrogen produced by solid oxide electrolysis cells (SOEC) from steam represents one of the currently available economical means of mass-producing hydrogen, which requires the least purification process comparing to steam methane reformation and coal gasification. SOEC modules with long life and minimal degradation will be critical to reducing the cost of hydrogen (CoH), including but not limited to capital costs and operating costs. A key factor in developing robust SOEC technology is to understand degradation mechanisms and to implement viable mitigation strategies.

Collaborating with Idaho National Laboratory, Materials and Systems Research Inc. has been investigating possible chemical and structural degradation mechanisms in SOECs. Cell materials with a range of compositions have been developed to improve long-term performance of hydrogen production. Negative electrode-supported SOEC cells and short stacks have been fabricated and evaluated under various operating conditions. In this talk, performance in both a cell and stack levels will be presented and discussed.

See more of this Session: Developments in Electrolytic Routes to Hydrogen I
See more of this Group/Topical: Topical 8: Hydrogen Production and Storage

(638b) Renewable Energy Storage by CO2 Recycling Using High Temperature Electrolysis

Joseph J. Hartvigsen[1], Lyman Frost[2], S. Elangovan[2], Carl M. Stoots[3], James E. O'Brien[3] and Stephen Herring[4], (1)Ceramatec, Inc., Salt Lake City, UT, (2)Ceramatec, Inc, Salt Lake City, UT, (3)Thermal Fluids and Heat Transfer Department, Idaho National Laboratory, Idaho Falls, ID, (4)Nuclear Science & Engineering, Idaho National Laboratory, Idaho Falls, ID

We have significantly greater renewable energy resources than can effectively be utilized by the electric grid due to intermittency and remoteness of the resources from large load centers. A further barrier to greater exploitation of renewable energy is that in some locales it may be competing with very low cost generation resources. For example, in many regions off-peak nuclear power is available at very low cost. Of much more urgent concern than electricity generation is the cost and assured availability of transportation fuels. Developing a means of storing intermittent renewable energy and off-peak nuclear energy as liquid transportation fuels offers improved economic prospects and a solution to the energy security of our transportation fuels while simultaneously reducing net CO_2 emissions.

Many countries are now considering taxing CO_2 emissions, which will increase the cost of products that are associated with those emissions. The most common method currently considered for dealing with CO_2 emissions is the capture of the gas, pressurization, and then sequestration in either rock formations or saline aquifers. This is relatively costly in both capital investment and operation of the equipment. Also, there is the possibility that this CO_2 will escape at some point in the future subjecting the CO_2 generators to a perpetual and unquantifiable liability.

Ceramatec and INL have been investigating an alternative approach that converts CO_2 into valuable transportation fuel. Using the solid oxide fuel cell materials set in conjunction with a non-fossil electric energy sources it is possible to generate synthesis gas (CO and H_2) and oxygen from CO_2 and H_2O. This high temperature co-electrolysis (HTCE) process can use both thermal and electric energy inputs to electrolyze the CO_2 and H_2O at 100% thermodynamic efficiency. This is possible because the high temperature co-electrolysis reactions are endothermic, and the heat generated by resistance in the electrochemical device is chemically recuperated in the process. We have demonstrated conversion of this CO_2 derived syngas to liquid fuels via the Fischer Tropsch process.

See more of this Session: Developments in Electrolytic Routes to Hydrogen I
See more of this Group/Topical: Topical 8: Hydrogen Production and Storage

(638c) Electrolysis and Regenerative Operation of the NASA Symmetrical SOFC

Thomas Cable[1], John A. Setlock[2] and Serene C. Farmer[2], (1)NASA Glenn Research Center, Cleveland, OH, (2)Ceramics Branch, NASA Glenn Research Center, Cleveland, OH

Both solid oxide fuel cells and Regenerative SOFCs (RSOFCs) are of interest to NASA for Aeronautics and Space applications. Recent studies performed for future, more electric commercial aircraft have evaluated SOFCs for APUs and have considered RSOFC systems. RSOFCs, based on H_2/O_2, would operate in fuel cell mode while taxiing to the gate and then would perform water electrolysis during flight. The pure O_2 generated in the electrolysis mode could be used for emergency needs. Space applications for RSOFCs are also envisioned, using H_2O for Lunar missions and CO_2 for Mars missions. Interest in H_2 production by DOE has lead to collaboration with DOE/INL with a focus on electrolysis of the hot steam effluent at nuclear power plants.

NASA Glenn Research Center (GRC) has been evaluating a new solid oxide fuel cell concept consisting of a porous YSZ support structure on both sides of a thin YSZ electrolyte, which is later filled with active electrode materials, making it a bi-electrode supported cell (BSC). BSC cells with infiltrated electrodes were evaluated in the regenerative mode of operation, in order to measure the electrochemical voltage efficiency (EVE); which is the ratio of power generated in fuel cell mode (P_{out}) to the required power for electrolysis (P_{in}), measured at constant current density. The highest voltage efficiency measured for a 5 cm OD cell was 90 % at a current density of 200 mA/cm^2 at 850°C. Details of the cell fabrication, testing and electrochemical performance will be presented.

See more of this Session: Developments in Electrolytic Routes to Hydrogen I
See more of this Group/Topical: Topical 8: Hydrogen Production and Storage

(638d) Development of a Hydrogen Home Fueling System

Greg G. Tao, Bruce Butler and Anil V. Virkar, Materials & Systems Research, Inc., Salt Lake City, UT

The use of clean-burning hydrogen for transportation can provide a critical transition in the world's economy, protects public health and environment, while reduces the impact of petroleum price fluctuations and the dependence on non-domestic energy sourcing. In addition, hydrogen production can be decentralized, sidestepping complications in distribution and storage.

Presently, hydrogen is produced via coal gasification, natural gas steam reforming, or water electrolysis for central or forecourt production. Meanwhile, due to the lack of large-scale hydrogen storage, distribution networks, delivery and dispensing infrastructures, remarkable costs from the production plant-gate to the end-consumers (fuel cell vehicle drivers) will be added to the overall cost of hydrogen. Alternatively, the development and adoption of hydrogen home fueling systems can reduce or eliminate the need for the hydrogen storage and distribution. The success of this alternative, however, will depend largely upon the development of economical hydrogen home production technologies, to meet the immediate and long term needs for the quick market adoption of hydrogen-based vehicles.

Sponsored by the U.S. Department of Energy under the Small Business Innovative Research / Small Business Technology Transfer program, Materials and Systems Research Inc. (MSRI) is investigating the development of a hydrogen home fueling system that incorporates the concept of combined hydrogen, heat, and power (CH2P) tri-generation directly using distributed natural gas. Unlike the conventional hydrogen production approaches, this CH2P system is built upon advanced solid oxide fuel-assisted electrolysis cell (SOFECs) and solid oxide fuel cell (SOFC) hybrid technology.

In this talk, MSRI's SOFEC-SOFC technology development will be presented, along with corresponding performance characteristics. A system analysis of a hydrogen home fueling system equipped with this advanced hybrid technology will also be presented, followed by discussions of process economics and merits of the technology adoption.

See more of this Session: Developments in Electrolytic Routes to Hydrogen I
See more of this Group/Topical: Topical 8: Hydrogen Production and Storage

(638e) High Temperature Solid-Oxide Electrolyzers: The Role of Materials in Long-Term Performance Stability

S. Elangovan[1], Joseph Hartvigsen[2], Dennis Larsen[2] and Insoo Bay[2], (1)Ceramatec, Inc, Salt Lake City, UT, (2)Electrochemical, Ceramatec, Inc, Salt Lake City, UT

The materials and fabrication technology that are used for high temperature solid oxide fuel cells (SOFC) are, in general, directly applicable to high temperature solid oxide electrolysis cells (SOEC). Ceramatec, in collaboration with Idaho National Laboratory under the DOE Nuclear Hydrogen Initiative has been evaluating SOEC cells and stacks using SOFC materials set developed at Ceramatec. While the cells and stacks showed good reversibility between the two modes, and the performance scaled as expected from short stacks to tall stacks, a significant difference in long-term stability was observed. The SOEC mode of operation suffered much higher performance degradation.

Long-term performance stability of SOEC is found to be largely materials related. Some potential causes for degradation are common to SOFC device operation while others are unique to SOEC mode of operation. The dominant degradation mechanism that appear unique to SOEC operation was related to interface delamination of the oxygen evolution electrode. Oxidation of metallic components in high temperature and high steam conditions in SOEC mode is particularly severe near the seal areas. Other causes that may be common to both modes are: deposition of species at electrochemical sites to cause electrode poisoning and phase instability of current distribution layer. Under a project funded by the Office of Naval Research, systematic evaluation of various materials was undertaken to address the stability issues. Some of the changes evaluated include the use of Sc-stabilized zirconia electrolyte, new electrode materials set, protective coatings on interconnects and alternative seal options. These materials changes are found to improve stack stability significantly.

See more of this Session: Developments in Electrolytic Routes to Hydrogen I
See more of this Group/Topical: Topical 8: Hydrogen Production and Storage

Implementation of Time-Dependency Into a Spent Fuel Reprocessing Plant Model

John F. Krebs, Argonne National Laboratory, Argonne, IL
Candido Pereira, Argonne National Laboratory, Argonne, IL
Kurt Frey, Argonne National Laboratory, Argonne, IL
Kent E. Wardle, Argonne National Laboratory, Argonne, IL

Abstract

A spent fuel reprocessing plant is comprised of a series of base unit operations including: chopping, fuel dissolution, solvent extraction, concentration, conversion, and solidification. Argonne is currently developing a time-dependent reprocessing plant model capable of tracking transient process changes to evaluate their relevancy to material safeguards. Recent efforts have focused on the fuel dissolution and solvent extraction unit operations. The time-dependent solvent extraction model is based on the AMUSE (Argonne Model for Universal Solvent Extraction) code. The steady-state version of AMUSE has been used for over a decade to design flowsheets for advanced spent fuel solvent extraction processes while more recently being used for identifying off-normal process conditions that potentially may be of a concern from a safeguards standpoint. An overview of solvent extraction codes relevant to the PUREX solvent extraction process will be presented and the election to modify AMUSE rather than other existing solvent extraction codes will be addressed. Development of the time-dependent differential equations for both mass and energy balances that are being implemented into the time-dependent AMUSE code will be briefly discussed. In addition, progress on the development of a dynamic dissolver unit operation model will also be presented.

Introduction

Recently there has been interest in the developing a model of a spent nuclear fuel reprocessing facility. Uses of such a model include: evaluating safeguards by design concepts, examining material control and accountancy (MC&A), design optimization, and as a training platform for operations personnel. A spent nuclear fuel reprocessing plant is comprised of a number of unit operations. Six base unit operations will be modeled in detail: chopping, fuel dissolution, solvent extraction, concentration, conversion, and solidification. Argonne has been developing a time-dependent reprocessing plant model capable of tracking transient process changes. The approach taken has been to develop models for the individual unit operations. This should result in greater flexibility for interfacing the models for the individual unit operations with those of the surrounding plant operations. The model development plan includes an initial verification and validation (V&V) study. The V&V study is critical to model release as it demarcates the strengths and weaknesses of the developed models and lays the groundwork for improving the models. Of the unit operations noted above recent efforts have focused on the fuel dissolution and solvent extraction unit operations.

A time-dependent solvent extraction model, focusing on countercurrent centrifugal contactors, is being derived from the AMUSE (Argonne Model for Universal Solvent Extraction) code. The steady-state version of AMUSE has been used for over a decade to design flowsheets for advanced spent fuel solvent extraction processes. More recently it has been used to identify off-normal process conditions that potentially may be of a concern from a safeguards standpoint. To ensure that conversion of AMUSE to a time-dependent form was the best path forward, a review of solvent extraction codes relevant to the PUREX solvent extraction process was conducted. The review confirmed the initial approach of converting AMUSE.

For the dissolver unit operation, a general continuous stirred tank reactor (CSTR) model was chosen. However, the critical components for the dissolver model are the equations used to describe the dissolution chemistry. A literature review of the dissolution chemistry of UO_2, PuO_2, and spent fuel was conducted. The progress made on the dissolver model will be briefly addressed.

Solvent Extraction Models & Codes

Models of solvent extraction (SX) processes are typically designed for a particular contactor type such as mixer-settlers, pulse columns, or centrifugal contactors (Figure 1). Nearly all models are also based on the countercurrent operation of a contactor. There have been several reviews of the literature covering solvent extraction models and codes. A very early review of computer programs designed for performing solvent extraction calculations was published by Mills.[1] Mills' review covered Olander's code, the DEMISEC code by the Institute for Atomenergi - Kjeller, Norway, the technique developed by Di Liddo and Walsh, examples of codes as given in Chapter 9 of *Computation of Multistage Separation Processes*, and a code from the Hanford Laboratory. Benedict and Pigford noted that SEPHIS and SOLVEX are examples of computer codes that employ the techniques developed in Chapter 4 of the reference.[2] Interestingly, the authors fail to include a significant code, PUBG, developed by Beyerlein and Geldard at Clemson University that was reported in the literature in 1980. A more recent review of currently available solvent extraction modeling codes was written by Tratner and Haefner.[3] Their survey of dynamic simulation programs for nuclear fuel reprocessing covered only SEPHIS, SOLVEX, and AMUSE. Although not an exhaustive review, the authors identify the lack of any reports of solvent extraction process models using commercially available packages such as Aspen Plus® and Aspen Custom Modeler® (Aspen Technology, Inc., http://www.aspentech.com), Pro/II™ (Invensys Operations Management, http://iom.invensys.com), CHEMCAD® (Chemstations, Inc., http://www.chemstations.com), and gPROMS® (Process Systems Enterprise Inc., http://www.psenterprise.com).

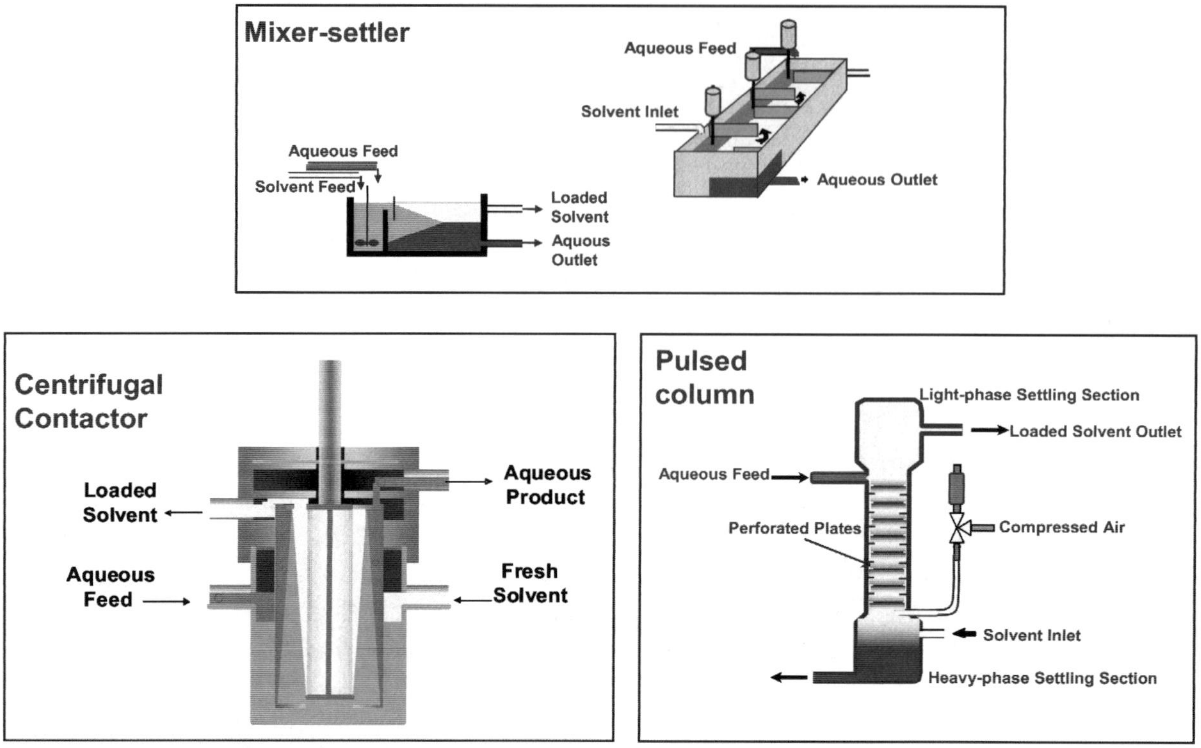

Figure 1. Examples of typical solvent extraction contactors: mixer-settler, centrifugal contactor, and the pulsed column.

Before implementing time-dependent equations in AMUSE, a more thorough review of the literature was conducted to better ascertain the status of solvent extraction models. Of particular interest were the forms of the time-dependent differential equations and the solution techniques employed. Tables 1 and 2 briefly summarize the models that were reviewed. Although several useable models have been used to simulate the PUREX process none appear to have been kept current with the exception of AMUSE. More importantly, none is as flexible as AMUSE as the majority focused on either PUREX or the industrial separation of the rare earths. In some instances copies of the models in the form of computer codes are available in the reports: SEPHIS, ICPSEF, PUBG, CUSEP, and MIXSET. Details of some of the codes developed in other countries are very limited. Of the numerous models/codes three codes are worthy of note: SEPHIS, PUBG, and MIXSET.

Table 1. Solvent extraction models and codes relevant to PUREX.

Code/Model	Year	Contactor[a]	Time Dependency[b]	Ref.
Olander	1961	PC	SS	4
DiLiddo/Walsh	1961	PC	TR	5
$SIMTEX_{UK}$	1963	MS	SS	7
Biery/Boylan	1963	PC	TR	8
Lowe	1968	MS	TR	12
Kikindai	1968	MS	TR	13
SEPHIS	1975, 1979	MS	TR	14
SOLVEX	1975	MS,CC	SS,TR	17
$REPROSY\text{-}P_{Japan}$	1975	MS	SS,TR	18
Bonnefoi	1977	PC	TR	19
Carassiti	1982	MS		22
PUBG	1980-1981	MS	TR	23,24,25
ICPSEF	1982	PC	TR	26
$VISCO_{FRG}$	1983	MS,CC	SS,TR	27,28,29
PULCO	1986	PC		34,35
PUMA	1986	MS	SS	36
PUNE	1987	Contactors	SS	37
CUSEP	1986-1989	PC	TR	38
SASSE	1987, 1994	Mass balance	SS	44,45
GTM	1990	CC,MS,PC	SS	46,47
PULSER	1989-1990	PC		41,42
PULMAT	1990	PC	SS	42
$DYNAC_{Japan}$	1991	PC	TR	48,49
$SAFE_{Japan}$	1992	MS,PC		50,51
$PAREX_{France}$	1992			53
CUSEP-MOD1	1993	PC	SS,TR	43
AMUSE	1997	CC,MS,PC	SS	47
$MIXSET_{Japan}$	1977,1979,1999	MS	SS,TR	31,32,33
EXTRACT/ $SCRUB_{Turkey}$	2003	MS	SS	60
$SIMPSEX_{India}$	1996, 2007			62,63

[a]PC = pulsed columns; MS = mixer-settlers; CC = centrifugal contactors.
[b]SS = steady-state; TR = transient.

Table 2. Solvent extraction models and codes relevant to other separation processes.

Code/Model	Year	Contactor[a]	Process	Time Dependency[b]	Ref.
Hanson	1962	PC	General	SS,TR	6
Sharp/Smutz	1965-1966	MS	rare earths, HNO3, TBP	SS	9,10
Eubanks/Lowe	1968	MS	HNO_3, HCl, Ce, and Cm	SS	11
Sepulveda/Miller	1980	MS	General	TR	20
MacDonald	1981	MS	Zr,Hf	SS	21
Riquelme	1985	MS	Cu,Fe	SS	30
Voit	1996	MS	Ta,Nb	SS	54
RESEX	1996	MS	Rare earths	SS	55
Jiangtao	1999		Rare earths	SS	56
Alonso	1999	MHFM		TR	57
Aminian	2000	MS	Cu	SS	58
Minchem/MUEM	2002		Cu		59
Galvez	2005		General		61

[a]PC = pulsed columns; MS = mixer-settlers; CC = centrifugal contactors; MHFM = microporous hollow fiber modules.
[b]SS = steady-state; TR = transient.

Groenier first reported on the SEPHIS code in 1972.[14] It was designed to model PUREX flowsheets run with mixer-settlers. Mixers were modeled using the assumption of intense, perfect mixing. It was assumed that all mass transfer between the two phases occurred in the mixers and complete separation was obtained in the settlers. Watson and Rainey[15] reported on their modification of the SEPHIS code. The distribution coefficients are calculated using empirical fits to experimental data. In 1979 Mitchell reported on further modifications of the SEPHIS code.[16] Species followed by SEPHIS-MOD4 included U(VI), Pu(IV), Pu(III), HNO_3, and a Pu(IV) reductant. Time-dependent differential equations are used to calculate the concentration profiles during transient and steady-state conditions. SEPHIS-MOD4 uses the Runge-Kutta method to numerically integrate the differential equations describing the mixer. A second, faster integration method using the finite difference form of the mixer differential equations was implemented for calculating steady-state conditions directly. Time-dependent differential equations were developed for the settlers, but were subsequently reduced to finite difference form and solved using iterative methods. Chemical reactions are handled in a separate subroutine, but are not directly included in the mass balance equations. Three different methods are offered to the user to reduce Pu(IV) to Pu(III). Many heat effects, other phase carryover (back-mixing), and phase entrainment were neglected. Bendixsen later adapted SEPHIS-MOD4 for simulating the first-cycle uranium extraction system, conducted on sieve-plate pulse columns, at the Idaho Chemical Processing Plant (ICPP).[26.]

Beyerlein and Geldard developed PUBG for the IAEA.[23,24,25] They used a modified version of the Richardson total nitrate salting strength equation to calculate distribution coefficients of the species. Time-dependent differential equations were used to describe the settlers and were numerically integrated using the fourth-order Runge-Kutta method. The time-dependent, mass balance differential equations were developed for the mixers, but were reduced to couple polynomial equations for steady-state conditions. The coupled polynomial equations were then solved iteratively. A time-dependent differential equation is used to describe heat transfer in the settlers.

MIXSET was developed to simulate solvent extraction processes using mixer-settlers. Known reports for MIXSET are the 1977[31] and 1979[32] versions and the more recent MIXSET-X.[33] The 1977 version, written in FORTRAN IV, handled H^+, U(IV), U(VI), Pu(III), Pu(IV), NH_2OH, and N_2H_4. Distribution coefficients could be provided by the user as either a third-order polynomial or in tables. The 1979 version included two groups of reaction models. The first included reduction of Pu(IV) by U(IV), oxidation of Pu(III) by HNO_3, oxidation of U(IV) by NO_3^- or O_2, and decomposition of HNO_2. The second included reduction of Pu(IV) by hydroxylamine nitrate (HAN), oxidation of Pu(III) by HNO_3, decomposition of HNO_2. The last known version of MIXSET-X handles HNO_3, U(IV-VI), Pu(III-VI-V-VI), Np(IV-V-VI), Am, Cm, Zr, Tc, HNO_2, Sr, Ru, Cs, Ce(III), and Gd(III) species. Overall it includes distribution coefficients of 31 constituents including actinides and fission products (FPs) and 45 chemical reactions. Users can select any distribution coefficients and chemical reactions to be handled and adjust reaction rate constants via the input data. Recycle flow through a buffer tank and TBP degradation calculations are also available. The authors state that MIXSET-X can simulate the solvent extraction processes at the Tokai reprocessing plant in one calculation run. The authors also state that the program structure and mathematical modeling of the system have been entirely changed to obtain calculational stability and program readability.

AMUSE is comprised of two components: SASSE and SASPE. SASSE performs the steady-state mass balance calculations and is based on the work of Leonard.[44,45] Unlike several of the other codes listed in Tables 1 and 2, the current version of AMUSE handles other-phase carryover and contactor efficiency. It also includes models, many based on available literature data, of many of the fission products. AMUSE is best suited for modeling processes conducted on centrifugal contactors or mixer-settlers, but can be used for modeling processes using pulse columns. The distribution coefficients (D-values) of AMUSE are calculated, in SASPE, using thermodynamically based equations as opposed to polynomial equations fitted to sets of experimental data. Because of these features, the decision was made to proceed with implementing time-dependent differential equations into AMUSE rather than modifying another existing code. Initial development of time-dependent AMUSE was performed using the MATLAB® environment. A simple two phase, three stage countercurrent centrifugal contactor flow model was created that initially used fixed D-values. The model calculates the mass balance for each phase at each stage resulting in two time-dependent differential equations per stage. The 4th-order Runge-Kutta method is then used to obtain a solution. The time-dependent mass balance equations were then converted to FORTRAN 90 code and interfaced to the FORTRAN version of SASPE allowing the D-values to be updated during the course of the calculation. The 3-stage model is currently being updated to allow for up to 100 stages to be calculated.

Verification & Validation of Solvent Extraction Codes

The first reported attempt at validation of the SEPHIS code was published by Kyser in 1998.[64] Uranium extraction was over-predicted and the temperature dependence was poorly captured. Kyser concludes that the code can be used to provide only qualitative assessment of various limited scenarios.

In 1990 Chamberlain *et al.* conducted laboratory scale "verification tests" of the TRUEX process that were compared with results of the GTM (Generic TRUEX Model).[65] The purpose of the tests was to verify whether or not the GTM predicts actual extraction behavior. In 1995 Regalbuto *et al.* compared calculated results from GTM against TRUEX data collected at the Chemical Processing Facility, Tokai-works by the Power Reactor and Nuclear Fuel Development Corporation of Japan.[66,67] GTM versions 3.0.1 and 3.1 were used in the study. No statistical evaluation of the comparisons was provided in the reports.

There are no known reports addressing the formal verification and validation of any of the other solvent extraction codes covered in this review. A report that may describe validation of the MIXSET code from Japan has been identified,[68] but at the time of this report the document has not been reviewed.

As there are no good examples of a rigorous validation study of a solvent extraction model/code, examples of models/codes that have under gone a validation study were sought. An excellent example, currently being used as a guideline for the validation of AMUSE, is the seven volume set of documents that describe the validation of fire models that are used by the U.S. Nuclear Regulatory Commission (NRC).[69]

Dissolver Models

There are at least two known reviews of PuO_2 dissolution kinetic studies.[70,71] Table 3 briefly summarizes some of the published work on dissolution kinetic studies on mostly un-irradiated pellets and powders. Table 4 summarizes some of the studies that involved irradiated fuels. The current dissolver model was developed on the Aspen Custom Modeler® platform and is based on the equations of Koga. Model inputs are the initial mass of fuel, liquid volume, temperature, and HNO_3 concentration. The model outputs include the extent of dissolution and the temperature dependent solution density. A basic continuous stirred tank reactor (CSTR) model is currently being developed. The model as a whole is being implemented in MATLAB® to facilitate portability.

Table 3. Dissolution studies involving pellets or powders.

System	Acid	Ref.
ThO_2-UO_2	HNO_3-HF	72
UO_2	HNO_3	73
UO_2, PuO_2, UO_2-PuO_2	HNO_3	74
UO_2	HNO_3	75
PuO_2-UO_2[a]	HNO_3	76,77
ThO_2-UO_2	HNO_3-HF	78
PuO_2	HNO_3	79
UO_2	HNO_3	80,81
UO_2	HNO_3	82,83
UO_2	HNO_3	84,85,86

[a]Samples irradiated to 4.7-99 GWd/mt, 5y cooled.

Table 4. Dissolution rate studies, in HNO_3, involving spent fuel samples.

Reactor	Burnup	Ref.
PWR	31.35 GWd/MTU, 2y	87
LWR	0.2-28 GWd/MTHM	88
Fast	3 GWd/t, 2y	89
Fast	2.2 GWd/t, 3.5y	90
Fast	40.1 GWd/t, 2.7y	91
Fast	100 GWd/t	92
PWR	50 GWd/t, 15y	93
PWR	44.3 GWd/t, 14.3y	94

Plant/Process Simulation

Only a few software programs have been developed or used to model the integrated unit operations within a reprocessing plant, Table 5. The REPROCX code, the TPC Inc. code and the COMPAS code can import fuel compositions from programs like FISPIN or ORIGEN. These programs had the capability of interfacing to solvent extraction codes like SEPHIS, MIXSET, or PULCO. Both the code by TPC Inc. and COMPAS could model the key unit operations of a reprocessing plant. However, the availability of these codes is unknown. As a result this has led effectively to the re-development of these capabilities.

Table 5. Spent fuel reprocessing plant models/simulators.

Code	Year	Process	Platform	Ref.
REPROCX	1986, 1992	PUREX	FORTRAN	95,96,97
TPC Inc.[a]	1994	PUREX	QuickBASIC	98
$COMPAS_{Japan}$	1994-1995	PUREX	FORTRAN	99,100
Cipiti/Ricker	2008	UREX+1a	Simulink®,[b]	101
Smith	2008	UREX+3a	Aspen Plus®,[c]	102

[a]Technical Process Consultants, Inc. (Oak Ridge, TN)
[b]The MathWorks Inc., http://www.mathworks.com
[c]Aspen Technology, Inc., http://www.aspentech.com

Cipiti and Ricker developed a transient flow model of the UREX+1a process.[101] The model is described as being able to track cold chemicals, bulk fluid flow, solids, and individual species. Concentrations in the solvent extraction portion of the plant model were based on static steady-state results calculated by AMUSE. Smith and Dimenna modeled the UREX+3a plant, using Aspen Technology's Aspen Plus® and AMUSE.[102] Both of these models currently employ user defined fixed, fractional values for many of the unit operations and could greatly benefit from more thermodynamically and chemically relevant models.

Summary

Numerous solvent extraction models/codes have been developed over the past 50 years. Many of these models/codes lack the capability of handling critical process conditions such as other-phase carryover (back-mixing) and contactor efficiency. This has led us to use the more current steady-state AMUSE model as a basis for developing a better time-dependent solvent extraction model. In addition, we are currently using available UO_2 and spent fuel dissolution data to develop and refine a dissolver model. Finally, a formal verification and validation effort is implemented as part of the model development work that should facilitate deployment of the models to other users.

Acknowledgements

The submitted manuscript has been created by UChicago Argonne, LLC, Operator of Argonne National Laboratory ("Argonne"). Argonne, a U.S. Department of Energy Office of Science laboratory, is operated under Contract No. DE-AC02-06CH11357. The U.S. Government retains for itself, and others acting on its behalf, a paid-up nonexclusive, irrevocable worldwide license in said article to reproduce, prepare derivative works, distribute copies to the public, and perform publicly and display publicly, by or on behalf of the Government.

Work supported by the U.S. Department of Energy, National Nuclear Security Administration's (NNSA's) Office of Defense Nuclear Nonproliferation, under Contract DE-AC02-06CH11357.

References

1. Mills, A.L. "Review of Computer Programmes for Solvent Extraction Calculations" TRG Report 902; United Kingdom Atomic Energy Authority: 1965; 14pp.
2. Benedict, M.; Pigford, T.H.; Levi, H.W. *Nuclear Chemical Engineering*, 2nd Ed.; McGraw-Hill: New York, 1981; pp197.
3. Tratner, T.J.; Haefner, D.R. "Survey of Dynamic Simulation Programs for Nuclear Fuel Reprocessing" INL/EXT-08-14476; Idaho National Laboratory: 2008; 14pp.
4. Olander, D.R. "Tributyl Phosphate Solvent Extraction Column Design" *Ind. & Eng. Chem.*, **1961**, *53(1)*, 1–5.
5. DiLiddo, B.A.; Walsh, T.J. "Computer Simulation of Pulse Columns" *Ind. & Eng. Chem.*, **1961**, *53 (1)*, 801–804.
6. Hanson, D.N.; Duffin, J.H.; Somerville, G.F. *Computation of Multistage Separation Processes*; Reinhold Publishing Corp.: New York, 1962; Chapter 9.
7. Burton, W.R.; Mills, A.L. "Computer Calculation of Flowsheets for Reprocessing Nuclear Fuels" *Nucl. Eng.*, **1963**, 248-252.
8. Biery, J.C.; Boylan, D.R. "Dynamic Simulation of a Liquid-Liquid Extraction Column" *Ind. & Eng. Chem. Fund.*, **1963**, *2(1)*, 44-50.
9. Sharp, B.M.; Smutz, M. "Stagewise Calculation for the Solvent Extraction System Monazite Rare Earth Nitrates-Nitric Acid-Tributyl Phosphate-Water" *Ind. & Eng. Chem. Proc. Design Devel.*, **1965**, *4(1)*, 49-54.

10. Sebenik, R.F.; Sharp, B.M.; Smutz, M. "Computer solution of solvent-Extraction-Cascade Calculations for the Monazite Rare-Earth Nitrates-Nitric Acid-Tributyl Phosphate-Water System" *Sep. Sci.*, **1966**, *1(4)*, 375-386.
11. Eubanks, I.D.; Lowe, J.T. "Steady-State Solvent Extraction Calculations for Curium Recovery" *Ind. & Eng. Chem. Proc. Design Devel.*, **1968**, *7(2)*, 172-177.
12. Lowe, J.T. "Calculation of the Transient Behavior of Solvent Extraction Processes" *Ind. & Eng. Chem. Proc. Design Devel.*, **1968**, *7(3)*, 362-366.
13. Kikindai, T.; Michel, P.; Rouyer, D.C.; Chazal, G. "Analogical Simulation of Mixer-Settlers" Proceedings of the International Symposium on Analog and Hybrid Computation Applied to Nuclear Energy, Versailles, France, Sept. 16-18, 1968; 214-219.
14. Groenier, W.S. "Calculation of the Transient Behavior of a Dilute PUREX Solvent Extraction Process Having Application to the Reprocessing of LMFBR Fuels" ORNL-4746; Oak Ridge National Laboratory: 1972; 63pp.
15. Watson, S.B.; Rainey, R.H. "Modifications of the SEPHIS Computer Code for Calculating the PUREX Solvent Extraction System" ORNL-TM-5123; Oak Ridge National Laboratory: 1975; 66pp.
16. Mitchell, A.D. "SEPHIS-MOD4: A User's Manual to a Revised Model of the PUREX Solvent Extraction System" ORNL-5471; Oak Ridge National Laboratory: 1979; 79pp.
17. Scotten, W.C. "SOLVEX – A Computer Program for Simulation of Solvent Extraction Processes" DP-1391; Savanna River Laboratory: 1975; 34pp.
18. Tsujino, T.; Kohsaka, A.; Aochi, T. "Calculation Code REPROSY-P for Process Studies on Pu Purification With TBP" JAERI-M 6284; Japan Atomic Energy Research Institute: 1975; 39pp.
19. Bonnefoi, p.; Dargier, C.; Pujol, A.; Rouyer, H.; Zwingelstein, G. "On Line Control of a Liquid-Liquid Extraction Column by Modal Control Techniques" in Digital Computer Applications to Process Control; Van Nauta Lemke, H.R.; Verbruggen, H.B., Eds., The Hague, Netherlands, June 14-17, 1977; 617-622.
20. Sepulveda, J.E.; Miller, J.D. "A Fundamental Approach to the Design and Optimization of Industrial Solvent Extraction Plants" Proceedings of the International Solvent Extraction Conference, Liege, Belgium, Sept. 6-12, 1980; paper 80-233.
21. MacDonald, D.J. "Computer Simulation of Zirconium-Hafnium Separation by Countercurrent Extraction" *Sep. Sci. Technol.*, **1981**, *16*, 1355-1371.
22. Carassiti, F.; Liuzzo, G.; Morelli, A. "A Computer Code for Calculation of Solvent-Extraction Separation in a Multicomponent System With Reference to Nuclear Fuel Reprocessing" in Proceedings of the 3rd Austrian-Italian-Yugoslav Chemical Engineering Conference, 276th European Federation of Chemical Engineering, Graz, Austria, Sept. 14-16, 1982; 373-380.
23. Beyerlein, A.L.; Geldard, J.F.; Chung, H.F.; Bennett, J.E. "Deviations from Mass Transfer Equilibrium and Mathematical Modeling of Mixer-Settler Contactors" in National Bureau of Standards Special Publication 582 - Measurement Technology for Safeguards and Materials Control: Proceedings from American Nuclear Society Topical Meeting, Kiawah Island, South Carolina, November 26-30, 1979; T.R. Canada, B.S. Carpenter, Eds.; National Bureau of Standards: Washington D.C., 1980; 702-711.
24. Beyerlein, A.L.; Geldard, J.F. "Improved Mixer-Settler Models and Applications to Nuclear Material Accounting Systems" Dept. of Chemistry and Geology, Clemson University: December 1981; 114pp.

25. Geldard, J.F.; Beyerlein, A.L. "A User's Manual for PUBG. A New Chemical Model of the PUREX Solvent Extraction Process" Dept. of Chemistry and Geology, Clemson University: December 1981; 87pp.
26. Bendixsen, C.L. "ICPSEF: A User's Manual for the Computer Mathematical Model of the ICPP Purex Solvent Extraction System" ENICO-1124; Exxon Nuclear Idaho Company, Inc.: Idaho Falls, 1982; 76pp.
27. Petrich, G.; Schmieder, H. "Comparison of PUREX-Process Methods to Separate Uranium and Plutonium" International Solvent Extraction Conference1983 (ISEC '83), Denver, CO, Aug. 26 – Sept. 2, 1983; paper 10n; 84-85.
28. Ochsenfeld, W.; Petrich, G.; Schmidt, H.-J.; Stollenwerk, A.H.; Wiese, H. "Neptunium Decontamination in a Uranium Purification Cycle of a Spent Fuel Reprocessing Plant" *Sep. Sci. Technol.*, **1983**, *18*, 1685-1698.
29. Petrich, G. "The PUREX Process Computer Model VISCO and Its Applications" International Solvent Extraction Conference-1986, Munchen, Fed. Rep. of Germany, Sept. 11 – 16, 1986; I-31 – I-39.
30. Riquelme, G.D.; Nunez, P.M.; Triday, J.L. "Mathematical Model for Metal Solvent Extraction Systems, Selectivity Effect" *Chem. Eng. Comm.*, **1985**, *39*, 1-21.
31. Gonda, K.; Fukuda, S. "Calculation Code MIXSET for PUREX Process" PNC-TN841-77-60; Power Reactor and Nuclear Fuel Development Corporation: Tokai Works, Ibaraki, Japan, 1977; 210pp. http://jolissrch-inter.tokai-sc.jaea.go.jp/pdfdata/PNC-TN841-77-60.pdf
32. Gonda, K.; Koichiro, O.; Fukuda, S. "Calculation Code Revised MIXSET for PUREX Process" PNC-TN841-79-26; Power Reactor and Nuclear Fuel Development Corporation: Tokai Works, Ibaraki, Japan, 1979; 250pp. http://jolissrch-inter.tokai-sc.jaea.go.jp/pdfdata/PNC-TN841-79-26.pdf
33. Naito, M.; Suto, T.; Asakawa, K.; Kashiwagi, E. "MIXSET-X: A Computer Code for Simulating the PUREX Solvent Extraction Process" JNC-TN8400-99-005; Japan Nuclear Cycle Development Institute, Tokai Works; Tokai-mura, Ibaraki, Japan, 1999; 273pp. http://jolissrch-inter.tokai-sc.jaea.go.jp/pdfdata/JNC-TN8400-99-005.pdf
34. Gonda, K.; Matsuda, T. "Solvent Extraction calculation Model for PUREX Process in Pulsed Sieve Plate Column" *J. Nucl. Sci. Technol.*, **1986**, *23(10)*, 883-895.
35. Gonda, K.; Matsuda, T. "Calculation Code PULCO for PUREX Process in Pulsed Column" PNCT 841-82-1; Power Reactor and Nuclear Fuel Development Corporation: Tokai Works, Ibaraki, Japan, 1982; 369pp. http://jolissrch-inter.tokai-sc.jaea.go.jp/pdfdata/PNC-TN841-82-19.pdf
36. Geldard, J.F.; Beryerlein, A.L.; Chiu, H.-L. "PUMA – A New Mathematical Model for the Rapid Calculation of Steady-State Concentration Profiles in Mixer-Settler Extraction, Partitioning, and Stripping Contactors Using the PUREX Process" *Nucl. Technol.*, **1986**, *75*, 160-167.
37. Geldard, J.F.; Beryerlein, A.L.; Chiu, H.-L. "Rapid Calculation of Steady-State Concentration Profiles with Nonequilibrium Mass Transfer in Contactors Using the PUREX Process" *Nucl. Technol.*, **1987**, *78*, 151-156.
38. Beyrelein, A.L.; Geldard, J.F. "Nuclear Material Inventory Estimation in Solvent Extraction Contactors" IAEA-R-4086-M; International Atomic energy Agency: 1986; 113pp.
39. Beyrelein, A.L.; Geldard, J.F. "Nuclear Material Inventory Estimation in Solvent Extraction Contactors II" IAEA-R-4441-F; International Atomic energy Agency: 1987; 160pp.
40. Geldard, J.F.; Beyerlein, A.L. "CUSEP – A New Mathematical Model of Pulsed Column Contactors Using the PUREX Process" *Nucl. Technol.*, **1989**, *85*, 172-186.

41. Beyerlein, A.L.; Geldard, J.F. "Nuclear Material Inventory Estimation in Solvent Extraction Contactors III" IAEA-R-4942-F; International Atomic energy Agency: 1989; 189pp.
42. Beyerlein, A.L.; Geldard, J.F. "Rapid Simulation of Concentration Profiles in Pulsed Column Contactors Using the PUREX Process" *Nucl. Technol.*, **1990**, *89*, 318-327.
43. Beyerlein, A.L.; Geldard, J.F. "The Effect of Correlated Flow in Pulsed Column Contactors Using the PUREX Process" *Nucl. Technol.*, **1993**, *102*, 252-258.
44. Leonard, R.A. "Use of Electronic Worksheets for Calculation of Stagewise Solvent Extraction Processes" *Sep. Sci. Technol.*, **1987**, *22(2-3)*, 535-556.
45. Leonard, R.A.; Regalbuto, M.C. "A Spreadsheet Algorithm for Stagewise Solvent Extraction" *Solvent Extraction and Ion Exchange*, **1994**, *12(5)*, 909-930.
46. Vandegrift, G.F.; Chamberlain, D. B.; Conner, C.; Copple, J. M.; Dow, J. A.; Everson, L.; Hutter, J. C.; Leonard, R. A.; Nuñez, L.; Regalbuto, M. C.; Sedlet, J.; Srinivasan, B.; Weber, S.; Wygmans, D. G. "Development and Demonstration of the TRUEX Solvent Extraction Process," Proceedings of WM '93, 19th Annual Waste Management conference, Feb. 28–Mar. 4, 1993, Tucson, AZ; Vol. 2; 1045.
47. Vandegrift, G.F.; Regalbuto, M.C.; Aase, S.; Bakel, A.; Battisti, T.J.; Bowers, D.; Byrnes, J.P.; Clark, M.A.; Emery, J.W.; Falkenberg, J.R.; Gelis, A.V.; Pereira, C.; Hafenrichter, L.; Tsai, Y.; Quigley, K.J.; Vander Pol, M.H. "Designing and Demonstration of the UREX+ Process Using Spent Nuclear Fuel" in ATATLANTE 2004 - Advances for Future Nuclear Fuel Cycles International Conference, Proceedings; Nimes, France June 21-24, 2004; paper 012-01.
48. Nabeshima, M. "Numerical Study of Hydrodynamics in Pulsed Solvent Extraction Columns Using the DYNAC Code" *Nucl. Technol.*, **1991**, *93*, 362-375.
49. Nabeshima, M. "Prediction of Mass Transfer and Heat Evolution of PUREX Pulsed Columns by the DYNAC Computer Model" *Nucl. Technol.*, **1991**, *95*, 33-43.
50. Maeda, M.; Fujine, S.; Yano, H.; Yamagami, S. "Simulation Code SAFE for Analyzing Transient Solvent Extraction Behavior" *Trans. Am. Nucl. Soc.*, **1992**, *66*, 78-79.
51. Maeda, M.; Fujine, S.; Uchiyama, G.; Watanabe, M.; Shiraki, K.; Kashimura, T. "Simulation Code of Extraction Contactor for Analyzing Transient Behavior" Solvent Extraction in the Process Industries: Proceedings of the International Solvent Extraction Conference 1993 (ISEC '93) Vol. 3, London, England, September 9-15, 1993; Logsdail, D.H.; Slater, M.J., Eds.; Elsevier Applied Science: London, 1993; 1517-1523.
52. Maeda, M.; Fujine, S.; Uchiyama, G.; Watanabe, M.; Shiraki, K.; Kashimura, T. "Simulation Code of Extraction Contactor for Analyzing Transient Behavior" Solvent Extraction in the Process Industries: Proceedings of the International Solvent Extraction Conference 1993 (ISEC '93) Vol. 3, London, England, September 9-15, 1993; Logsdail, D.H.; Slater, M.J., Eds.; Elsevier Applied Science: London, 1993; 1517-1523.
53. Dinh, B.; Mauborgne, B.; Baron, P. "Dynamic Simulations of Extraction Operations, Applications in Nuclear Fuel Reprocessing" in European Symposium on Computer Aided Process Engineering-2, Toulouse, France, October 5-7, 1992; Depeyre, D.; Joulia, X.; Koehret, B.; Le Lann, J. M., Eds.; no page numbers available.
54. Voit, D. "Spreadsheet Modeling Tantalum/Niobium Solvent Extraction Circuits" Tantalum, Proceedings of a Symposium held at the 125th TMS Annual Meeting and Exhibition, Anaheim, Calif., Feb. 5-8, 1996 (1996), 29-37. Chen, Edward S., Eds.; Minerals, Metals & Materials Society: Warrendale, Pa, 1996; 29-37.
55. Casarci, M.; Gasparini, G.M.; Sanfilippo, L.; Pozio, A. "R.E.S.EX. A computer Simulation Program for Rare Earth Separation Processes" Proceedings of the International Solvent

Extraction Conference 1996 (ISEC '96) Vol. 2, Melbourne, Australia, March 17-21, 1996; University of Melbourne: Dept. of Chemical Engineering, Parkville, Vic., 1996; 1067-1072.
56. Jiangtao, J; Jianfang, W.; Chunhua, Y.; Chunsheng, L.; Sheng, W.; Biaoguo, L. "Calculation of Extraction equilibrium in Multi-Component System by Newton-Raphson Algorithm" *Journal of the Rare Earths*, **1999**, *17(4)*, 246-249.
57. Alonso, A.I.; Galan, B.; Gonzalez, M.; Ortiz, I. "Experimental and Theoretical Analysis of a Nondispersive Solvent Extraction Pilot Plant for the Removal of Cr(VI) from a Galvanic Process Wastewaters" *Industrial & Engineering Chemistry Research*, **1999**, *38*, 1666-1675.
58. Aminian, H.; Bazin, C.; Hodouin, D.; Jacob, C. "Simulation of a SX-EW Pilot Plant" *Hydrometallurgy*, **2000**, *56*, 13-31.
59. Cramer, K.; Morrison, J.; Moore, T. "Using Avecia's Minchem Copper Solvent Extraction Process Modeling Program for Education" Annual Meeting of the Minerals-Metals-and-Materials-Society, Seattle, Washington, Feb. 17-21, 2002; in Fundamentals of Advanced Materials for Energy Conversion; Taylor, P.R.; Chandra, D.; Bautista, R.G., Eds; The Minerals, Metals & Materials Society: 2002; 715-721.
60. Zabunoglu, O.H.; Akbas, T.; "Flow Sheet Calculations in THOREX Method for Reprocessing Th-based Spent Fuels" *Nuclear Engineering and Design*, **2003**, *219*, 77-86.
61. Galvez, E.D.; Swaney, R.E.; Malatesta, C.M.; Cisternas, L.A. "An ASCEND Library for the Modeling, Simulation and Optimization of Solvent Extraction for Metal Recovery" 15th European Symposium on Computer-Aided Process Engineering (ESCAPE-15), Barcelona, Spain, May 29-June 1, 2005; in *Computer-Aided Chemical Engineering v. 20A-20B*, L. Puigjaner and A. Espuna, Eds.; Elsevier: Amsterdam, 2005; 403-408.
62. Natarajan, R.; Raj, B. "Fast Reactor Fuel Reprocessing Technology in India" *J of Nucl. Sci. Technol.*, **2007**, *44(3)*, 393-397.
63. Kumar, S.; Koganti, S.B. "Dynamic Simulation of High Plutonium Flowsheets Proposed for Reprocessing FBR Fuel" Proceedings of the International Solvent Extraction Conference 1996 (ISEC '96) Vol. 2, Melbourne, Australia, March 17-21, 1996; University of Melbourne: Dept. of Chemical Engineering, Parkville, Vic., 1996; 1073-1078.
64. Kyser, E.O. "Validation of the SEPHIS Program for the Modeling of the HM Process" WSRC-TR-98-00376; Westinghouse Savannah River Company: 1998; 30pp.
65. Chamberlain, D.B.; Leonard, R.A.; Conner, C.J.; Dow, J.A.; Tranovich, M.A.; Van Deventer, E.H.; Wasserman, M.O.; Mrazek, F.C.; Vandegrift, G.F. "Verification Studies" in *Nuclear Technology Programs Semiannual Progress Report, October 1989-March 1990*; ANL-91/42; Argonne National Laboratory: Chemical Technology Division, 1992; 120-135.
66. Regalbuto, M.C.; Aase, S.; Vandegrift, G.F. "Validation of the Generic TRUEX Model Using Data from TRUEX Demonstrations with Actual High-Level Waste" ANL-95/21; Argonne National Laboratory: 1995; 49pp.
67. Vandegrift, G.F.; Regalbuto, M.C. "Validation of the Generic TRUEX Model Using Data from TRUEX Demonstrations with Actual High-Level Waste" Proceedings of the Fifth International Conference on Radioactive Waste Management and Environmental Remediation : ICEM '95, Berlin, Germany, September 3-7, 1995; American Society of Mechanical Engineers: New York, 1995; v. 1; 457-462.
68. Shiyoutarou, R.; Kishimoto, Y.; Shiniti, N.; Takashi, K. "Development of the Centrifugal Extractor - Uranium Test Results" PNC-TN8410-88-051; Power Reactor and Nuclear Fuel Development Corporation: Tokai Works, Ibaraki, Japan, 1988; 105pp. http://jolissrch-inter.tokai-sc.jaea.go.jp/pdfdata/PNC-TN8410-88-051.pdf

69. *Verification and Validation of Selected Fire Models for Nuclear Power Plant Applications, Volume 1: Main Report*, U.S. Nuclear Regulatory Commission, Office of Nuclear Regulatory Research (RES), Rockville, MD, 2007, and Electric Power Research Institute (EPRI), Palo Alto, CA, NUREG-1824 and EPRI 1011999.
70. Cleveland, J. M. *The Chemistry of Plutonium*. Gordon and Breach: New York, 1970; pp. 398-407 and 574-575.
71. Ryan, J. L.; Bray, L. A. "Dissolution of Plutonium Dioxide – A Critical Review." ACS Symposium Series 117; American Chemical Society: Washington D.C., 1980; 499.
72. Bond, W.D. "Dissolution of Sintered Thorium-Uranium Oxide Fuel In Nitric Acid-Fluoride Solutions" ORNL-2519; Oak Ridge National Laboratory: 1958; 23pp.
73. Johnson, H.F. "Analysis of UO_2 Dissolution in Nitric Acid Solution" ORNL-3297; Oak Ridge National Laboratory: 1962; 39pp.
74. Uriarte, A.L.; Rainey, R.H. "Dissolution of High-Density UO_2, PuO_2, and UO_2-PuO_2 Pellets in Inorganic Acids" ORNL-3695; Oak Ridge National Laboratory: 1965; 87pp.
75. Shabbir, M.; Robins, R.G. "The Effect of Crystallographic Orientation on the Dissolution of Uranium Dioxide in Nitric Acid" *J. Nucl. Mater.*, **1968**, *25*, 236-237.
76. Goode, J.H. "Hot-Cell Dissolution of Highly Irradiated 20% PuO_2-80% UO_2 Fast Reactor Fuel Speciments" ORNL-3754; Oak Ridge National Laboratory: 1965; 29pp.
77. Goode, J.H.; Vaughen, V.C.A. "Fuel Reprocessing Tests with Short-Cooled LMFBR-Type Fuel Rods" CONF-720607-16; Oak Ridge National Laboratory: 1972; 20pp.
78. Dyck, R.W.; Taylor, R.; Boase, D.G. "Dissolution of $(Th,U)O_2$ in Nitric Acid-Hydrofluoric Acid Solutions" AECL-5957; Whiteshell Nuclear Research Establishment: Manitoba, Canada, 1977; 50pp.
79. Bray, L.A.; Ryan, J.L.; Wheelwright, E.J. "Dissolution of Plutonium Dioxide Using HCl-HF" PNL-6027; Pacific Northwest Laboratory: 1986; 21pp.
80. Fukasawa, T.; Ozawa, Y. "Relationship Between Dissolution Rate of Uranium Dioxide Pellets in Nitric Acid Solutions and Their Porosity" *J. Radioanal. Nucl. Chem., Letters*, **1986**, *106(6)*, 345-356.
81. Fukasawa, T.; Ozawa, Y.; Kawamura, F. "Generation and Decomposition Behavior of Nitrous Acid During Dissolution of UO_2 Pellets by Nitric Acid" *Nucl. Technol.*, **1991**, *94*, 108-113.
82. Koga, J.; Honma, S.; Kanehira, O.; Matsumoto, S. "Fuel Dissolution Rate and Its Mechanism" in Proceedings of the Third International Conference on Nuclear Fuel Reprocessing and Waste Management – RECOD '91; April 14-18, 1991, Sendai, Japan; 687-691.
83. Homma, S.; Koga, J.; Matsumoto, S. "Dissolution Rate Equation of UO_2 Pellet" *J. Nucl. Sci, Technol.*, **1993**, *30(9)*, 959-961.
84. keda, Y.; Yasuike, Y.; Takashima, Y.; Park, Y.-Y.; Asano, Y.; Tomiyasu, H. "17O NMR Study on Dissolution Reaction of UO_2 in Nitric Acid" *J. Nucl. Sci, Technol.*, **1993**, *30(9)*, 962-964.
85. keda, Y.; Yasuike, Y.; Nishimura, K.; Hasegawa, S.; Takashima, Y. "Kinetic Study on Dissolution of UO_2 Powders in Nitric Acid" *J. Nucl. Mater.*, **1995**, *224*, 266-272.
86. Nishimura, K.; Chikazawa, T.; Hasegawa, S.; Tanaka, H.; Ikeda, Y.; Yasuike, Y.; Takashima, Y. "Effect of Nitrous Acid on Dissolution of UO_2 Powders in Nitric Acid – Optimal Conditions for Dissolving UO_2" *J. Nucl. Sci, Technol.*, **1995**, *32(2)*, 157-159.
87. Campbell, D.O.; Buxton, S.R. "Hot Cell Studies of Light Water Reactor Fuel Reprocessing" CONF-761103-13; Oak Ridge National Laboratory: 1976; 33pp.

88. Johnson, D.R.; J.A. Stone "Light Water Reactor Fuel Reprocessing: Dissolution Studies of Voloxidized and Nonvoloxidized Fuel" DP-1520; Savannah River Laboratory: 1980; 83pp.
89. Campbell, D.O.; Mailen, J.C.; Fellows, R.L. "Dissolution Behavior of FFTF Fuel" CONF-840802-19; Oak Ridge National Laboratory: 1984; 5pp.
90. Fellows, R.L.; Campbell, D.O.; Mailen, J.C. "Dissolution of Low Burnup Fast Flux Test Reactor Fuel" CONF-840802-16; Oak Ridge National Laboratory: 1984; 16pp.
91. Suimei, H.; Hitoshi, O.; Kazutaka, U.; Rikio, H.; Idemitsu, K.; Hiro, Y.; Tsutomu, K.; Kiyotaka, O.; Akira, Y. "FBR Fuels Reprocessing Experiments at CPF (9)" PNC TN845-85-02; Japan Nuclear Cycle Development Institute: Tokai-mura, Ibaraki, Japan, 1985; 219pp.
92. Nemoto, S.; Shibata, A.; Shioura, T.; Okamoto, F.; Tanaka, Y. "Studies on the dissolution of mixed oxide spent fuel from FBR," *Donen-Gihou*, **1995**, *95*, 43-51.
93. keda, Y.; Yasuike, Y.; Nishimura, K.; Hasegawa, S.; Mason, C.; Bush, R.; Takashima, Y. "Dissolution Behavior of Pulverized Irradiated Fuels in Nitric Acid Solutions" *J. Nucl. Sci, Technol.*, **1999**, *36(4)*, 358-363.
94. Mineo, H.; Isogal, H.; Morita, Y.; Uchiyama, G. "An Investigation Into Dissolution Rate of Spent Nuclear Fuel in Aqueous Reprocessing" *J. Nucl. Sci, Technol.*, **2004**, *41(2)*, 126-134.
95. Harrison, J.W.; Mills, A.L.; Wilkins, M. "REPROCX – A Reprocessing Flowsheet Computational System" International Solvent Extraction Conference 1986 (ISEC '86) Preprints Vol. 1, Munchen, Fed. Rep. of Germany, September 11-16, 1986; I-131-I-136.
96. Arm, S. "Fission Product Removal in Liquid-Liquid Contactors" Nuclear Europe Worldscan, 1992, 12(3-4), 51-52.
97. Harrison, J.W.; Mills, A.L.; Hirose, Y. "REPROCX – A Computational Procedure for the Nuclear Fuel Cycle" in *Solvent Extraction in the Process Industries – Proceedings of ISEC '93*, International Solvent Extraction Conference, University of York, York, UK, Sept. 9-15, 1993, D.H. Logsdail and M.J. Slater, Eds.; Elsevier: New York, 1993; 1486-1493.
98. Burch,W. D.; Yarbro, O.O. "Real-Time computer Simulation of a Fast Reactor Reprocessing Plant" in The Fourth International Conference on Nuclear Fuel Reprocessing and Waste Management – RECOD '94, Proceedings – Vol. 3; London, UK, April 24-28, 1994.
99. Homma, S.; Sakamoto, S.; Takanashi, M.; Nammo, A.; Satoh, Y.; Seojima, T.; Koga, J.; Matsumoto, S. "Development of COMPAS, Computer Aided Process Flowsheet Design and Analysis System of Nuclear Fuel Reprocessing" *J. of Nucl. Sci. Technol.*, **1995**, *32(4)*, 357-368.
100. Sakamoto, S.; Homma, S.; Koga, J.; Matsumoto, S. "Flowsheeting System for Design and Analysis of Reprocessing Process" in The Fourth International Conference on Nuclear Fuel Reprocessing and Waste Management – RECOD '94, Proceedings – Vol. 3; London, UK, April 24-28, 1994.
101. Cipiti, B.B.; Ricker, L. "Advancing the State of the Art in Materials Accountancy Through Safeguards Performance Modeling" SAND2008-5100; Sandia National Laboratories: Albuquerque, NM, 2008; 36pp.
102. Smith III, F.G.; Dimenna, R.A. "Modeling the UREX+3a Process Using Aspen Plus Coupled with AMUSE" WSRC-ST-2008-00013; Savannah River National Laboratory: 2008; 3pp.

Improvement of Energy Resolution via Correction on Non-Uniform Light Collection in Large Scintillation Detectors

Haori Yang, Canberra Industries, Meriden, CT

Background

Because of their low cost, high efficiency, moderate energy resolution and robustness, scintillation detectors are quite popular when large active volume is required. Large scintillation detectors are widely deployed in the field. Their applications include homeland security, oil exploration, nuclear physics and etc. In scintillation detectors, measurement of incident photons starts with the interaction of photons with the scintillation materials mostly through three types of interactions: photoelectric absorption, Compton scattering and pair production. As a result of these interactions, energetic electrons are produced, which then deposit their kinetic energies by activating scintillation materials and cause them to scintillate. Excited atoms of scintillation materials return to their low-energy ground level by emission of scintillation photons. The emission spectrum and the decay constant of the emission are material specific parameters that vary dramatically from material to material. Some sort of photon sensors have to be implemented to read these scintillation photons out and convert them to electric signals that can be easily measured and quantified. In scintillation detectors, the energy resolution is affected by many factors. The major players are scintillation efficiency, linearity, light collection efficiency and quantum efficiency of the photon sensors. Good scintillation efficiency is the basis for good energy resolution. Scintillation efficiency is normally defined as the number of scintillation photons generated by unit energy deposition in the scintillation material. Higher scintillation efficiency means lower relative uncertainty on the number of information carriers, thus better energy resolution. The scintillation efficiency is not always a constant. It can vary according to the difference in electron kinetic energy. Non-linearity means that the number of scintillation photons generated dose not only depend on the amount of energy deposition, but also the process of the deposition. Quantum efficiency of the photon sensor also has a big impact on the energy resolution of the whole scintillation detector device. It is defined as the probability for a scintillation photon to generate a photoelectron when it impacts the photocathode. Usually, a photon sensor has an optimal working spectral range. One needs to choose photon sensors wisely to match the emission spectrum of the scintillation material. Last but not least, light collection efficiency contributes to the detector energy resolution as well. Light collection efficiency is the probability for scintillation photons to reach the active area of the photon sensor. In small scintillation detectors, uniform light collection efficiency can be achieved across the whole scintillator by careful surface treatment. In large scintillators, however, it is almost impossible to achieve uniform light collection efficiency. Scintillation photons generated at different locations have different probabilities to be collected by the photon sensor. This can cause severe degradation in energy resolution. In this paper, a method is proposed to alleviate this undesired effect. First, a Monte Carlo model of the scintillation detector is created based on the knowledge of the geometric dimensions and material composition of the detector components. In the model, the radiation source can also be constructed using the tools provided by the Monte Carlo software package. For each incident photon, the Monte Carlo simulation provides information on the amount of energy deposition and interaction position for each step during the absorption process. Secondly, the scintillation crystal is carefully modeled in DETECT2000. In order to do this, one needs to provide material composition as well as the surface parameters. Using DETECT2000, the distribution of light collection efficiency can be calculated within the scintillator. Monte Carlo simulation was then performed for a specific counting geometry, at interested energies. Combining the knowledge of light collection efficiency and interaction history, response functions are formed for the scintillation detector at interested energies, for a specific counting geometry. The response function can then be used to deconvolve the measured spectrum to improve energy resolution.

Monte Carlo Model for Scintillation Detectors

Monte Carlo method is widely used to study radiation transportation. It relies on repeatedly random sampling, validated physics models and reliable data libraries to simulate the interaction of radiation with various materials. Monte Carlo method is only practical when sufficient computing power is available. Along with the development of electric computers, Monte Carlo method quickly finds various applications including space mission, radiotherapy, nuclear criticality and etc.

Currently, two most commonly used Monte Carlo codes in radiation transportation are MCNP/MCNPX and GEANT4. In this paper, the simulation was performed using GEANT4. GEANT4 is a toolkit managed by the international GEANT4 collaboration [1]. It is the first Monte Carlo simulation toolkit to use Object Oriented Programming (C++). It provides a set of tools to cover all areas in radiation transportation, including geometry, particle tracking, detector response, source particle generation and etc. GEANT4 uses a different approach to model real geometric configuration from MCNP. Instead of defining the surfaces that bound the volumetric cells, objects are directly defined as solids in GEANT4. Compared with MCNP/MCNPX, GEANT4 is much more flexible, but it does require some level of C++ programming skill. The GEANT4 source code is freely available, while a license is required to run or modify MCNP/MCNPX.

In order to build a model that can well represent the real geometric configuration, one would need to know the dimensions of the scintillation detector components. Aside from the geometric dimensions, the composition of the materials involved is also needed for the Monte Carlo model. Once the Monte Carlo model of the scintillation detector is created, it can be benchmarked against real measurements under well controlled counting conditions. Quite often, even a model built based on the best knowledge of geometric dimensions and material properties cannot reproduce the measurement results exactly. Based on benchmark results, one can improve the model by changing its parameters, so that it represents the real scintillation detector better. GEANT4 provides a rich set of tools for source definition to model the radiation sources used in the measurements. Using the benchmarked detector model and source model, Monte Carlo simulation can be performed. Similar technique has been implemented in radiation detector industries, for example, Canberra's In-Situ Object Counting System (ISOCS) technique [2]. A Monte Carlo model for a generic stabilized NaI detector is shown in Figure 1.

In GEANT4 simulation, one would need to write his own functions to get the necessary information out. Each incident photon is tracked through its whole lifetime within the scintillator until it is absorbed, or lost. At each step, the interaction position and the amount of energy deposition are recorded. This information is needed for later calculation.

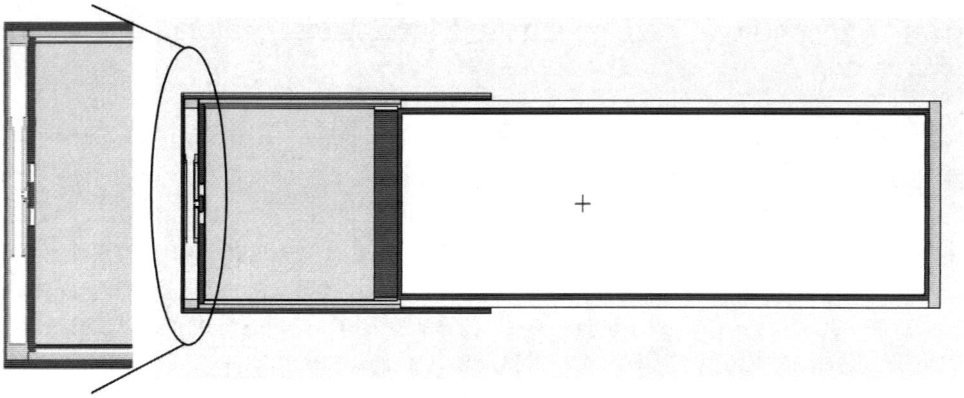

Figure 1. Monte Carlo model for a generic stabilized NaI probe

Calculation of Distribution Map of Light Collection Efficiency

As discussed above, in large scintillation detectors, uniform light collection efficiency is hard to achieve. In this paper, a method to correct for this non-uniformity is proposed. Once a scintillation detector is manufactured, the light collection efficiency distribution inside it is unlikely to change over a short time period. Thus, it can be corrected for if it is well known. Software tools are available to calculate light transportation in optical systems. In this work, DETEC2000 is chosen as the simulation tool. DETECT2000 is a Monte Carlo model of the behavior of optical systems with a special emphasis on scintillation detectors [3]. Individual scintillation photons are generated in user specified regions inside the scintillator. The code then follows each photon in its passage through various components of the model. It tracks various interactions of each photon with surfaces. In DETECT2000, absorption and re-emission by a wave-shifting component are allowed in the simulation. At the end of its life, each photon is recorded as absorption, escape or detection. The scintillation decay time can also be simulated in DETECT2000. To do this, the user needs to provide a probability distribution function so that the code can draw samples from it. Photon sensor response time can be simulated in the same way as well. The wavelength distribution of the scintillation photons and the spectral response of model components are also included in DETECT2000. In addition to the final results of light transportation (absorption, escape, and detection), the user has access to many other simulation results too, such as decay time, delay time, whether or not the photon was wavelength shifted and etc.

The first step in the calculation is to create a model for the simulation. DETECT2000 uses a general syntax for geometry specification. The model is created as a group of components. To define a component, all the surfaces that bound the component need to be defined first. In addition to the geometric dimensions, the optical finish of each surface is also required. The optical surface finish supported by DETECT2000 includes DETECT, POLISH, GROUND, PAINT, METAL and UNIFIED. The details of these optical surface finish models can be found in Reference [2]. Once the bound of a component is appropriately defined, the volume needs to be filled with a material. In DETECT2000, a specific material is defined in terms of its optical properties, including refractive index, bulk absorption mean free path, scattering mean free path and etc. To start a simulation, a scintillation region within the detector needs to be defined. Starting point of scintillation photons are sampled uniformly within this region. The photons can be emitted isotropically, along a specified direction or isotropically within a cone.

Once the model is ready, a map of light collection efficiency within the scintillation detector can be constructed through simulation. A grid is created to cover the whole scintillator. The density of the grid depends on the application and can vary within a scintillator. In regions where light collection efficiency changes rapidly, the grid needs to be dense enough to capture the variation. One simulation is performed at each knot on the grid. In each simulation, sufficient amount of scintillation photons are created. It is assumed that these scintillation photons are emitted isotropically. Each photon is tracked until it is absorbed, detected, or lost. At the end of each simulation, the light collection efficiency can be calculated as the ratio between the number of photons detected and the number of photons emitted. The spectral distribution of the scintillation photons is material specific. DETECT2000 can sample from a user defined probability distribution for this purpose. Once the light collection efficiency at each knot on the grid is determined through simulation, the light collection efficiency at any arbitrary location within the scintillator can be calculated by interpolation. Figure 2 shows a map of the light collection efficiency calculated for the detector shown in Figure 1.

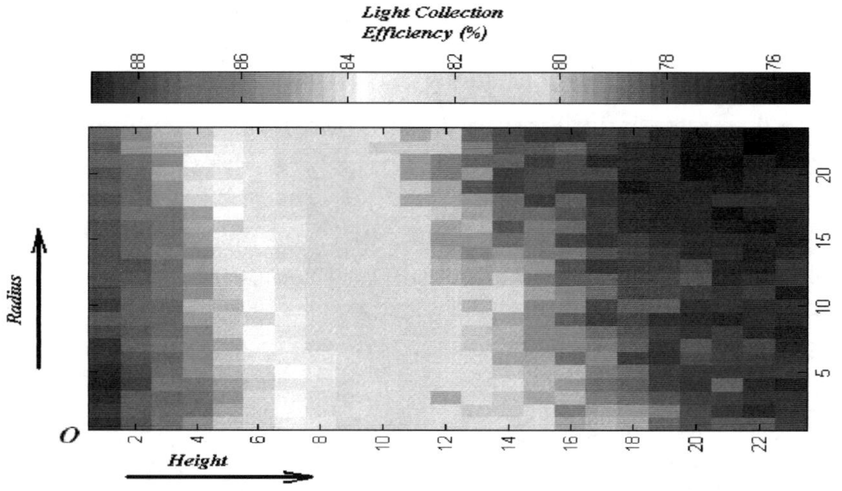

Figure 2. DETECT2000 calculated light collection efficiency distribution inside a NaI crystal

<u>Construction of Detector Response Function</u>

The detector peak response to a mono - energetic incident photon can be closely represented using Gaussian distribution. Detector energy resolution is described by the Full Width at Half Maximum (FWHM) parameter of the Gaussian function. As discussed above, in addition to other factors (statistics fluctuation in creation of signal carriers, electronics noise and so on), non-uniform distribution of light collection efficiency makes contribution to the degradation of energy resolution as well. For a specific counting geometry, a Monte Carlo model can be created including the scintillation detector and the radiation source. Incident photons with interested energies were traced through its lifetime inside the scintillation crystal. Energy deposited at each interaction position was recorded in list mode. For a particular scintillation material, the scintillation efficiency is assumed to be a constant in this study. Knowing the average energy required to produce a scintillation photon inside the scintillator, we can calculate the amount of scintillation photons generated at each interaction position. Based on the light collection efficiency distribution generated using DETECT2000, the 'collected energy' can be calculated for each incident photon. This equivalent 'collected energy' is calculated by summing the product of light collection efficiency and the energy deposition over all the interaction positions. Histogram of this 'collected energy' for a mono-energetic photon source can then be used as the detector response function, which includes the contribution of non-uniform distribution of light collection. An example of this response function is shown in Figure 3.

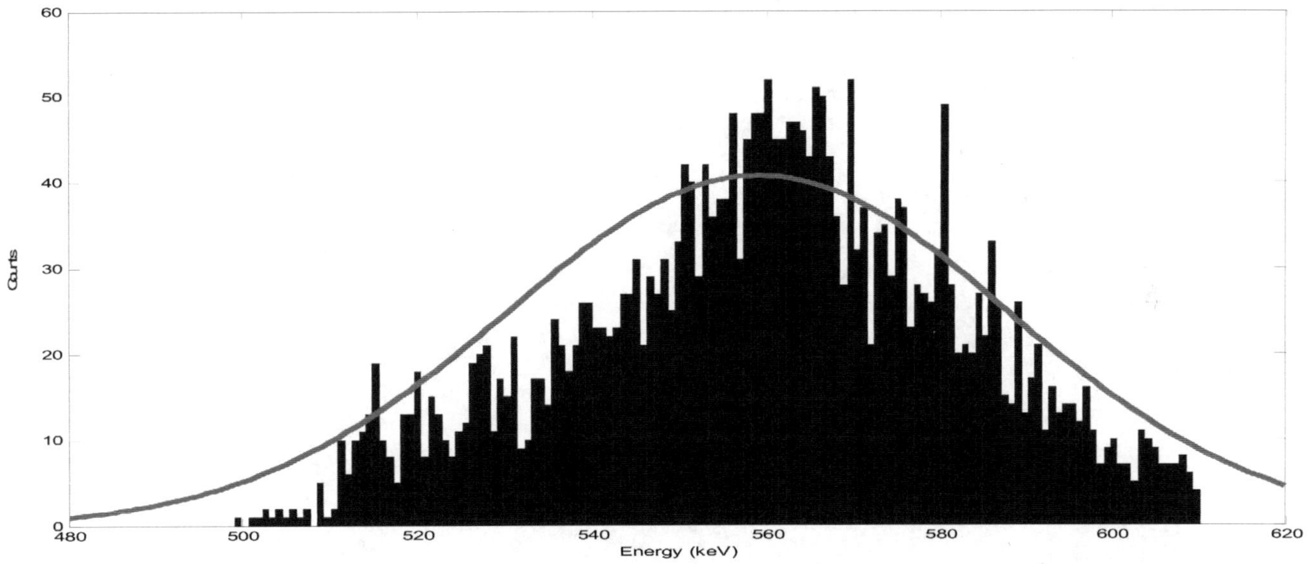

Figure 3. Simulated detector peak response function of a NaI detector to 662 keV Incident Photons and its fitting function

Deconvolution of Measured Spectra

Maximum Likelihood – Expectation Maximization (MLEM) Method was used to deconvolve the measured spectrum. MLEM is a popular method to fit the data to a known statistical model. The parameters of the statistical model can be optimized through iteration to maximum the likelihood between the data and the model predicted value. In this work, the formula used is shown below. This formula has been discussed in details by Shepp and Vardi [4].

$$\hat{f}_j^{k+1} = \frac{\hat{\hat{n}}_j}{s_j} = \frac{\hat{f}_j^k}{s_j} \sum_{i=1}^{I} \frac{g_i t_{ij}}{\sum_{j'=1}^{J} t_{ij'} \hat{f}_{j'}^k}$$

f_j is the incident energy spectrum that we wanted to deconvolve from the measured spectrum g_i . t_{ij} is the probability that an incident photon with energy j being recorded with energy i in the measured spectrum. This probability is directly related to the response function calculated in previous section. Starting from a uniform distribution f^0, the result after 10 iterations is shown in Figure 4 together with the originally measured spectrum. Improvement in energy resolution is obvious. The reason of the slightly shift in the peak position is that the measured spectrum can't be perfectly modeled by a Gaussian function.

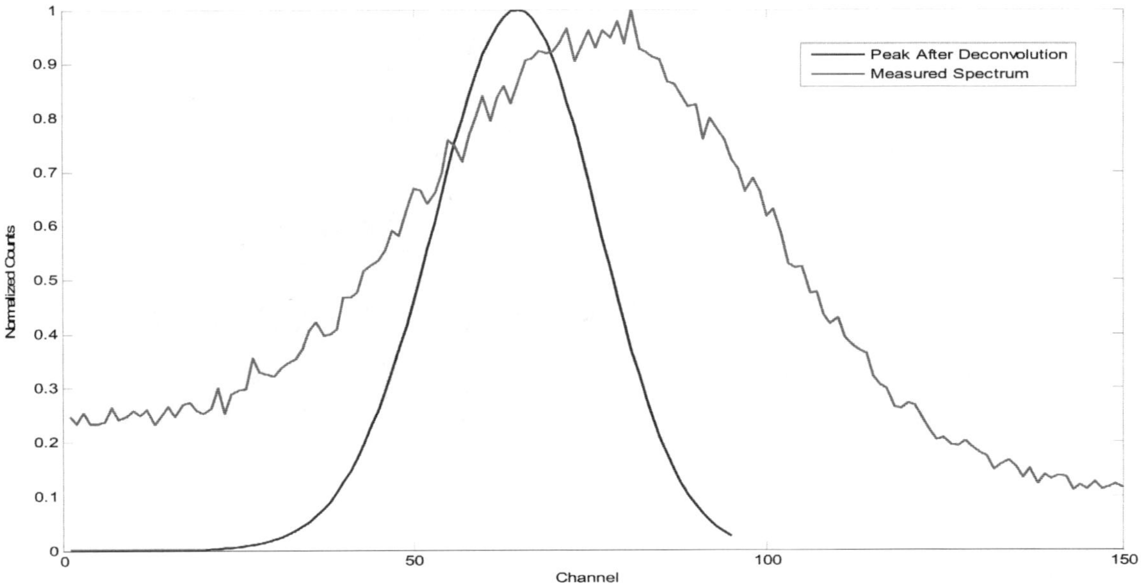

Figure 4. Comparison between measured spectrum and peak after deconvolution

Conclusion

Large scintillation detectors are widely used in field applications, because of its low cost, moderate energy resolution and robustness. In this paper, a method to alleviate the degradation in energy resolution due to non-uniform light collection efficiency is proposed. Using DETECT2000 software, light collection efficiency distribution within a scintillation detector can be calculated. In simulation based on a Monte Carlo model of a specific counting geometry, the interaction position and energy deposition are recorded for each step of individual incident photons through their lifetime within the scintillation detector. A detector response function is then established as the histogram of equivalent 'collected energy'. The energy resolution of a scintillation detector can then be improved by deconvolution between the detector response function and the measured energy spectrum using maximum likelihood method.

References

[1] S. Agostinelli, et al, "G4-a simulation toolkit," *Nuclear Instruments and Methods in Physics Research* A 506 (2003) 250-303
[2] F. L. Bronson, B. M. Young, and V. Atraskevich, "ISOCS mathematical calibration software for germanium gamma spectroscopy of small and large objects," *American Nuclear Society Annual Meeting,* 1997
[3] C. Moisan, F. Cayouet, and G. McDonald, "The object oriented C++ language version of DETECT, a program for modeling optical properties of scintillators," September 2000
[4] L. A. Shepp, and Y. Vardi, "Maximum likelihood reconstruction for emission tomography", September 2000

(676c) Experts in Discrete Element Modeling (EDEM) Validation for Packed Bed Structural Properties

Vaibhav Khane[1], Gary E. Mueller[2] and Muthanna H. Al-Dahhan[1], (1)Chemical Engineering Department, Missouri University of Science & Technology, Rolla, MO, (2)Nuclear Engineering Department, Missouri University of Science & Technology, Rolla, MO

Experts in Discrete Element Modeling (EDEM) is a discrete element method (DEM) based modeling software capable of providing a wealth of information about granular solid flows occurring in many industrial applications. A computational study involving application of EDEM to pebble bed nuclear reactor application is being performed at Missouri S&T. Prior to full involvement with EDEM based numerical simulations, an effort has been made to validate the packing algorithm used in EDEM; particularly for cylindrical packed bed applications. This is essential; as the local flow and transport properties are closely coupled with structural characteristics of a packed bed which in turn depend on the packing of the particles. It is well known that the distribution of particles in a randomly packed bed reactor shows a certain degree of order close to the container wall and dispersion when moved towards the center. Also, the radial voidage profile exhibits a typical shape of damped oscillation, with higher values of voidage at the wall. These structural properties will be used to compare the EDEM based packing algorithm results against the previously reported experimental and numerical simulations data for range of aspect ratios. The results of this validation study will provide recommendations/suggestions to simulate real packing bed structures using EDEM and henceforth perform additional reliable numerical analysis.

See more of this Session: Advances in Numerical Simulations Bridging Chemical and Nuclear Engineering Phenomena or Processes
See more of this Group/Topical: Nuclear Engineering Division

Visualization as a Bridge Between Chemical and Nuclear Engineering Simulations

Thomas Fogal[*], Shanjie Xiao[‡,†], Xue Yang[‡], Jens Krüger[*,§] and Tatjana Jevremovic[‡,¶]

Abstract

Collaborating across disciplines can be a difficult task for a variety of reasons. One of which is the lack of common methodologies and tools for visualization and analysis. Tools developed in one group can rarely be applied directly to data from another group, due to differences in data formats, analysis methodologies, and learning curves associated with the advanced visualization systems necessary for modern scientific data. Despite these difficulties, multi-physics collaboration is key to developing and understanding the next generation of computational models, which couple the dynamic processes and scales of multiple disciplines. As a step towards integrative models, we propose a visualization and analysis system to serve as a common interconnect between chemical and nuclear engineering. This choice allows multiple organizations to fruitfully exchange data and example visualizations in a reproducible manner, providing a foundation for productive discussion. In this work, we demonstrate the utility of a visualization system as it applies to bridging simulation domains and providing for multidisciplinary research in nuclear and chemical engineering.

*Scientific Computing and Imaging Institute, University of Utah
†Radiation Protection, SLAC National Accelerator Laboratory
‡Utah Nuclear Engineering Program, University of Utah
§Interactive Visualization and Data Analysis Group, DFKI Saarbrücken, Germany
¶Civil and Environmental Engineering, University of Utah

1 Introduction

Many file formats have been developed to provide flexible data exchange between systems and software packages. FITS [13] has achieved some success in unifying data formats for the domain of astronomy. Both HDF5 [14, 2] and netCDF [12, 6] have gained some traction as generic containers for data. DICOM [10] has achieved widespread use in the medical imaging community, however nuances and incompatibilities between scanners that export the format have created significant difficulties for processing such data. Despite these developments, file formats for scientific data have repeatedly come and gone without widespread community acceptance. Rather than simply a new format, we propose a new model for handling data which obviates the need for community-wide consensus.

This tiered method of data access allows a visualization and analysis system to take advantage of novel mediums for communication, such as handheld mobile devices. Encarnação et al. [3] discussed the advantages of mobile platforms in a general sense, and others have demonstrated mobile platforms to be useful in the medical domain [8, 11]. We have utilized a Moser and Weiskopf-style renderer [9] to distribute nuclear reactor simulation results, and identified uses of the system which have been fruitful in communicating fundamental and research results.

2 Data Formats and Exchange

A plethora of data formats has arisen in the simulation community. Almost every group uses a custom format which is incompatible with existing tools and practices; common overlap among file formats is largely restricted to groups utilizing or collaborating across a single model or code base. This makes exchanging data among groups a challenging task.

With a plethora of data formats comes a plethora of tools. Research groups or communities build up collections of visualization, analysis, and other tools in order to understand their data. Collaborating across communities can require not just the exchange of data, but the exchange of additional code to understand that data. It is common for that code to be developed "ad hoc": parts of the code base get tested on the machines the researcher is using when some analysis task comes up. Required visualization tasks get tied to specific APIs that can be difficult to compile for certain machines, or, worse, may not be universally available.

A secondary issue is the efficacy of existing file formats for the purposes of visualization and analysis. Simulation and even acquired data are typically written to disk in the form which is most convenient for the program which generates or scans the data. However, the needs of visualization and analysis software are unlikely to be equivalent to that of simulation software: as one example, for visualization purposes it is best to use smaller pieces than would normally be used, to facilitate out-of-core or GPU processing.

2.1 Concerns

The obvious solution in such a system is to get the wider community to agree on a common file format. However, the scientific computing community has been attempting to achieve such a unilateral choice since its inception – with minor successes – yet has never achieved a global standard. While logic allows the possibility that consensus is just hard to achieve in a large committee, we feel such efforts are doomed from the start primarily for one reason: scientists do not want a single file format.

The reasons for this are many and varied.

- **Concerns about the new format.**
  - *Time Investment.* A new file format typically means a new API for writing out data. Simulation code authors can be reluctant to invest the required time to rework their existing output routines to utilize the new library.
  - *Scalability and performance.* Authors of large scale simulation codes are very concerned with the scalability of their code. Matters are made worse due to the sensitive nature of I/O: it is common for these routines to already be a significant component of overall runtime. If the API for a new file format does not have an extremely positive track record for scalability, authors are conversely reluctant to spend the time evaluating an integration of the new library.
- **Learning new tools.** Users who have worked with a toolchain for significant periods of time grow accustomed to it, and have deep misgivings about switching. A new workflow for visualizing and analyzing data can be seen as distracting from their core science.
- **Domain-specific tools.** The ecosystem encourages development of tools which are designed to evaluate and understand specific types of data in specific ways. Asking users and developers to switch means these tools will no longer work, or at best that they must be rewritten, which may require significant effort.
- **Existing data.** There could be many hundreds or even thousands of runs of the code which have occurred over the years. Users are concerned with the applicability of that data once the new format is in place.
- **Simple code.** As the authors of a code base, developers know their I/O code well and enjoy this advantage. They can make changes, output new or additional data, and tweak for scalability with ease. A new API introduces a level of indirection and uncertainty.

While much effort has been expended on nullifying these concerns, at a community level they still remain prominent. A different approach seems prudent to gain traction on a wider scale.

2.2 Adaptors and Tiered Support

To achieve the goal of unified data access we have found a layered model of adaptor and bridge patterns provides a compelling architecture. The premise is to provide multiple tiers which the collaborator can adhere to, with benefits tied to investment in the system.

- **Tier 0**: We include a number of converters which take common file formats and massage them into a form more amenable to visualization and analysis. This allows many research groups to simply pick up the tool and begin working with it immediately.
- **Tier 1**: This consists of raw data layed out in a single large block. Both our tools as well as others can import such data simply by querying the user for simple characteristics, such as the data dimensions and the number of components in the data (e.g. vector data, scalar data). We've found many authors can modify their simulation code to write out data in this form given tens of minutes.
- **Tier 2**: By writing a native converter. This route involves adding some code to our system which acts as an adaptor to tier 1-style data. By iterating this process with a number of groups, we have identified a number of areas which might provide difficulties and provide methods to handle many common cases. The goal for tier 2 data import is that a programmer not familiar with our system can get this working in an afternoon.

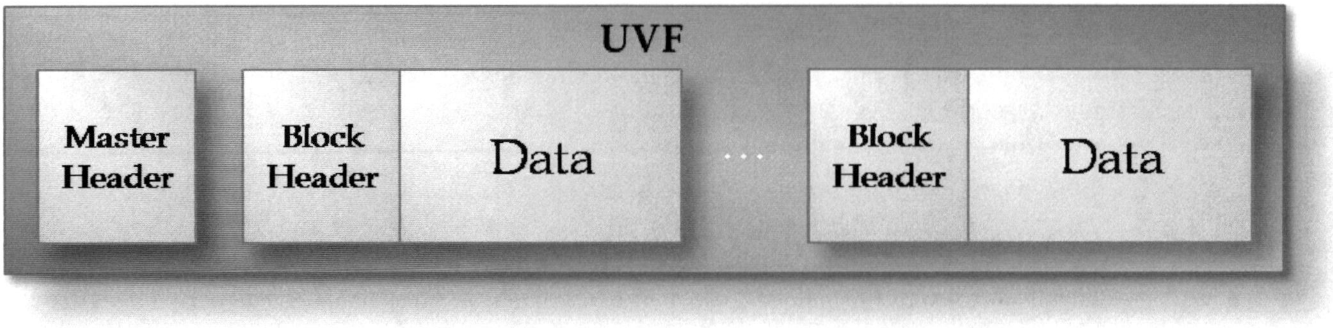

Figure 1: The basic structure of a UVF file.

- **Tier 3**: By writing data out in our native form, Unified Volume Format (UVF). We provide source code to make this process easy, as well as documentation on the file format itself, allowing code authors to implement their own support. Depending on whether the author decides to use our code or not, this process can take some time to implement. The benefit is that visualization and analysis tools can process the data much more quickly.
- **Tier 4**: Implement a bridge from our internal representation to the code format's internal representation. This allows the visualization and analysis tool to read data exactly as it is output by a simulation. The process only makes sense for simulations which utilize a representation amenable to visualization and analysis. Further, as this is the only method which absolutely requires knowledge of our internal representation, it requires a sustained development effort.

Regardless of which path is chosen, existing toolchains can remain largely unchanged, which is a boon to many code authors. Old data is not lost, and in fact new visualization and analysis opportunities are made available for it. Since the layered model imposes little on simulation codes, authors remain free to keep their code simple and understandable, retaining the ability to make quick modifications should they desire. We have found that keeping codes untethered from I/O libraries, which may be viewed as fickle or transient, is a large benefit in the eyes of many.

2.3 Unified Volume Format

Unified Volume Format (UVF) is a general container format which allows an efficient storage mechanism for a wide variety of data types (See Figure 1) [5]. The format is designed to scale from megabytes to petabytes and larger for both structured and unstructured data. UVF supports blocked data access as well as multiresolution representations of data. The former is important for processing data at the largest scale, and the latter is useful in providing real-time feedback on large data.

The goal of UVF is to provide a file format and associated framework which is easy to read, accomodates the needs of current software that must process volumetric data, and provides mechanisms for forward compatibility. Additionally, the format should also be able to store additional data other than volumes, such as accelerations structures, annotations, provenance information, etc.

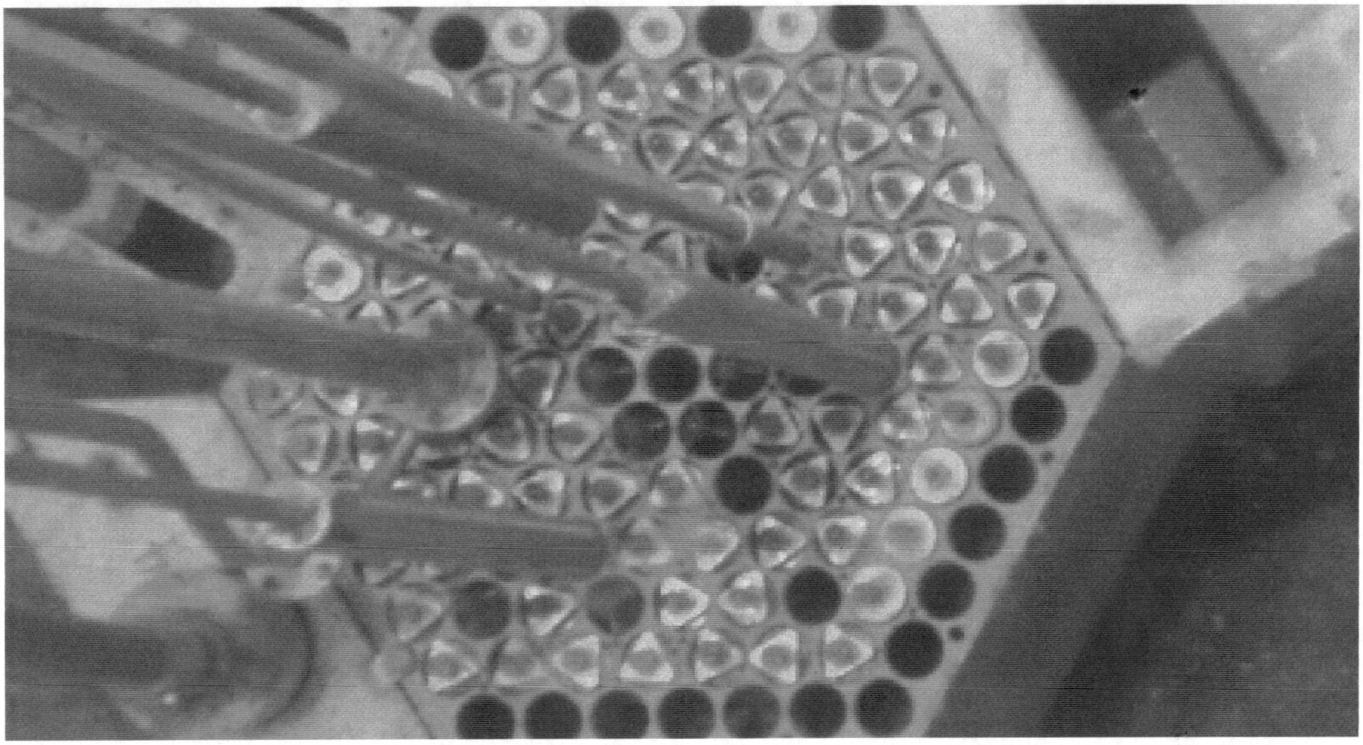

Figure 2: TRIGA reactor at the University of Utah

3 Collaboration

Mobile visualization tools can have broad applicability. For this environment, we have identified two cases where mobile visualization can have a significant impact.

- **Use case 1: Conveying results in a shared environment.** At a conference or other venue where one wants to present preliminary results or expound an issue that occurs in the field, an interactive visualization can be far more intuitive than static images. A presenter could indicate where to obtain some data at the start of their talk, and users could download and visualize the relevant data while the talk proceeds. Later, audience members could engage one another and the speaker by applying analysis routines directly on their mobile devices.
- **Use case 2: Teaching in classroom setting.** Many students learn better if the material engages a variety of mental faculties. Instead of simply presenting equations by writing them on a board and displaying some visual results, students can download and interactively visualize data using mobile devices. By applying analysis routines while learning the material, students understand and retain the knowledge more effectively.

Figure 3 demonstrates a visualization produced on a device running iOS.

4 Application

TRIGA is a class of small pool-type nuclear reactors designed and manufactured by General Atomics for training and educational purposes. The University of Utah owns a 100kW TRIGA research reactor (UUTR), shown in Figure 2, operated by the Utah Nuclear Engineering Program (UNEP). UUTR fuel elements adopt uranium zirconium hydride (UZrH) and

Group Index	Energy Range (eV)	Category
1	$1.35335 \times 10^6 - 1.96403 \times 10^7$	Fast
2	$9118.82 - 1.35335 \times 10^6$	Fast
3	$55.5951 - 9118.82$	Resonance
4	$4.0 - 55.5951$	Resonance
5	$0.625 - 4.0$	Thermal
6	$0.134 - 0.625$	Thermal
7	$10^{-5} - 0.134$	Thermal

Table 1: Energy range of seven group cross section

Library	Collapsed Seven Groups
k-eff	5×10^{-5}
flux criteria	5×10^{-4}
Azimuthal angles	36
Ray seperation	0.02cm
Axial Planes	27
Axial solver	MOC

Table 2: AGENT calculation parameters

are arranged in a triangular-pitch array. Each fuel rod sits at the center of a prismatic cell submerged by water. Three types of fuel with different burn-up, along with graphite, water, heavy water and control rod cells, form the configuration of seven concentric rings.

Numerical analysis of the reactor is an important area in nuclear engineering, because it gives not only the multiplication factor that characterizes the neutron population increasing or decreasing speed, but also the neutron flux spatial distribution that reflects the local neutron population and energy. The Arbitrary GEometry Neutron Transport (AGENT) code [4] is a deterministic neutronics multi-purpose software package developed for detailed analysis of neutronic aspects of various reactors, including arbitrary or less common 2D and 3D geometries. The geometry processing module is built upon R-functions that use logical operations to define complex shapes, and the Boltzmann equation is solved by the method of characteristics.

The UUTR is modeled by AGENT in 3D in the following ways: Seven groups flux weighted isotropic cross sections are collapsed from 172 groups WLUP library [1] by DRAGON code [7]. Table 1 lists the group structure used in DRAGON to collapse the cross sections for AGENT. Then, UUTR is modeled by AGENT using the parameters listed in Table 2.

The seven group scalar flux, fission rate and absorption rate distributions are created from the AGENT simulation. AGENT often simulates complex geometry with irregular shapes, and therefore it is not convenient to output the data for each zone. Thus a uniform 3D fine mesh is used for plotting output data. The scalar flux, reactor rates and coordinates at each fine mesh center are written to a file in which each line represents a set of data at one point. The flux has more of an elliptical shape, because the lower right corner and upper left corner of the core are filled with water and heavy water. The highest thermal flux is obtained in the central irradiation chamber, which is filled with water, while the highest fast flux appears in the second ring surrounding the central irradiation chamber.

Visualizations like the ones in Figure 3 demonstrate the structure of the reactor and variation within multiple energy groups. The inner structures visible in the latter image demonstrate the elliptical distribution of the flux.

Figure 3: Total absorption rate distribution in TRIGA and spatial flux pattern shown on an iPhone.

References

[1] WIMS-D Library Update, 2007. Vienna.

[2] Chilan, C., Yang, M., Cheng, A., and Arber, L. Parallel I/O Performance Study with HDF5, A Scientific Data Package. *TeraGrid 2006: Advancing Scientific Discovery* (June 2006).

[3] Encarnação, J. L., Frühauf, M., and Kirste, T. Mobile visualization: Challenges and solution concepts. In *Proc. CAPE'95* (1995). http://citeseer.ist.psu.edu/159692.html.

[4] Hursin, M., Xiao, S., and Jevremovic, T. Synergism of the method of characteristic, R-functions and diffusion solution for accurate representation of 3D neutron interactions in research reactors using the AGENT code system. *Annals of Nuclear Energy 33*, 13 (September 2006), 1116–1133. http://dx.doi.org/10.1016/j.anucene.2006.07.001.

[5] Krüger, J., Potter, K., MacLeod, R., and Johnson, C. Unified Volume Format: A General System For Efficient Handling Of Large Volumetric Datasets. In *Proceedings of IADIS Computer Graphics and Visualization 2008 (CGV 2008)* (2008).

[6] Li, J., keng Liao, W., Choudhary, A., Ross, R., Thakur, R., Latham, R., Siegel, A., Gallagher, B., and Zingale, M. Parallel netCDF: A High-Performance Scientific I/O Interface. In *In Proceedings of Supercomputing* (Nov 2003).

[7] Marleau, G., Hébert, A., and Roy, R. A User Guide for DRAGON 3.06. Tech. rep., March 2008. ftp://ftp.polymtl.ca/pub/nucl/dragon/dragon-3.06/IGE174R8.pdf.

[8] Meir, A., and Rubinsky, B. Distributed network, wireless and cloud computing enabled 3-D ultrasound; a new medical technology paradigm. *PLoS ONE 4*, 11 (11 2009), e7974. http://www.ncbi.nlm.nih.gov/pmc/articles/PMC2775631/.

[9] Moser, M., and Weiskopf, D. Interactive Volume Rendering on Mobile Devices. In *Workshop on Vision, Modelling, and Visualization VMV '08* (2008), pp. 217–226. https://wwwvis.informatik.uni-stuttgart.de/eng/research/pub/pub2008/vmv08-moser.pdf.

[10] NEMA. Digital imaging and communications in medicine (dicom), 1992-2010. `http://medical.nema.org/`.

[11] Park, S., Kim, W., and Ihm, I. Mobile collaborative medical display system. *Computer Methods and Programs in Biomedicine 89*, 3 (2008), 248 – 260. `http://dx.doi.org/10.1016/j.cmpb.2007.11.012`.

[12] Rew, R., and Davis, G. NetCDF: An Interface for Scientific Data Access. In *Computer Graphics and Applications* (Jul 1990), vol. 10, pp. 76–82. `http://dx.doi.org/10.1109/38.56302`.

[13] Wells, D. C., Greisen, E. W., and Harten, R. H. FITS - A Flexible Image Transport System. In *Astronomy and Astrophysics* (06 1981), vol. 44, p. 363.

[14] Yang, M., McGrath, R. E., and Folk, M. HDF5: A High Performance Data Format for Earth Science. *24th International Conference on Interactive Information Processing Systems (IIPS) for Meteorology, Oceanography, and Hydrology* (Jan 2005).

Comparison of Analytic Moments for the 1-Dimensional Energy Dependent Neutron Diffusion Equation with MCNP5 Moments

Doug Crawford and Terry A. Ring

Idaho National Lab	**Chemical Engineering**
Center for Space Nuclear Research	**University of Utah**
Idaho Falls, ID 83404	**Salt Lake City, UT 84092**

Abstract: The energy dependent neutron flux plotted as a function of energy looks like a Maxwell–Boltzmann distribution[1]. Because the energy dependent neutron flux shows population density function characteristics, the authors of this paper applied the method of moments to the energy dependent neutron diffusion equation (EDNDE) and compared normalized moments 1, 2, 3 and 4 to moments calculated from Monte Carlo N Particle version 5 (MCNP5) for two different bare homogeneous critical spheres. The difference in the normalized moments 1, 2, 3 and 4 varied only by a few percent where diffusion theory is valid and by as much as 20% where diffusion theory is limited i.e. three mean free paths near the boundary. This paper is the beginning of developing the method of moments and in the future the reconstruction of neutron flux with quadrature method of moments or other possibilities as a full spectrum solution for neutron transport.

Introduction

Solving for neutron distributions or power distributions in nuclear reactors is complex and has been studied with various methods, mostly numerical in nature. The main difficulty in solving for neutron distributions lies in solving the neutron transport equation, NTE (the equation can be found on page 114 of Nuclear Reactor Analysis by James Duderstadt and Louis Hamilton)[2]. The complexities and difficulties in trying to solve the NTE arise because the NTE depends on 7 variables: energy of the neutrons (E), angle of neutron interactions (θ and ϕ), spatial (x, y and z) and time (t). Simplifications are made to create a more palatable and solvable equation, but numerical methods are still necessary to solve for neutron fluxes[3] and populations. The Monte Carlo method used in Los Alamos National Lab's MCNP[4] software is widely accepted and respected among nuclear engineers and scientists for determining flux and especially for benchmarking criticality calculations[5]. This paper is focused on comparing analytic moments of the energy dependent neutron diffusion equation (EDNDE) EQN1, and comparing the analytic moments, to moments generated from MCNP5, which is a full neutron transport code. This seems like an apples and oranges comparison, since this is a comparison between NTE and EDNDE moments but MCNP has become the gold standard for neutron flux and criticality calculations so the authors felt this comparison should be made to show that the method of moments does compare to a full NTE solution.

To show proof of concept for the method of moments, two cases of bare homogenous critical spheres were modeled in MCNP and moments were calculated from the MCNP output and compared to analytic moments. The 1st case is a solid ^{235}U sphere and the 2nd is a 94% enriched in ^{235}U GODIVA benchmark[5]. The criticality of the two cases was also compared with analytical moment's calculation of k_{eff}. Noting that if a reactor is to be critical ($k_{eff} = 1$) the neutrons from the previous generation equals the neutrons generated in the next and for one speed theory that the material buckling value (B_m^2) must equal the geometric buckling (B_g^2).[6]

The NTE can be turned into EDNDE with certain assumptions and of course limitations[7]. To find an equation that can be solved analytically and compared to the MCNP cases, the following assumptions/approximations are applied to the NTE: an average angle of scatter for the neutrons, isotropic scattering of the neutrons, Fick's law of diffusion for the neutron current **J** and a homogenous system i.e. the fissile and fertile fuel is uniformly distributed throughout the moderator/reflector. The starting point for the analysis by the method of moments is EQN 1.

$$\frac{1}{\upsilon}\frac{\partial\phi(\vec{r},E,t)}{\partial t} - \nabla\cdot D(\vec{r},E)\nabla\phi(\vec{r},E,t) + \Sigma_T(\vec{r},E)\phi(\vec{r},E,t) = \int_0^\infty dE'\Sigma_S(\vec{r},E'\rightarrow E)\phi(\vec{r},E',t) + S(\vec{r},E,t) \qquad \text{EQN 1}$$

The method of moments (MOM) approach is unique by solving for the moments of a distribution and then reconstructing the distribution from the moments. The neutron flux can be treated as a probability distribution functions (PDF), where the normalized moments provide the mean, variance, skewness and kurtosis[8 9] so once the moments are solved for they can be put into the correct PDF to reproduce the flux. Quadrature method of moments (QMOM) has been shown to be an excellent method to solve this type of problem for particle population balances, aerosols, suspended particles in a fluid within computational fluid dynamics codes. There have been a number of papers written on the success of this method to solve these types of problems for particle population balances[10 11 12 13]. The first step to reconstruct a distribution is to see if the moments are accurate. The MOM derivation and results are presented briefly in this work.

Simplification of EDNDE to find analytical Moments

The starting point for finding an equation to put in a form to get analytical moments is EQN 1. The source term, $S(\mathbf{r}, E, t)$ in equation 1 can be defined this way $\nu(E)\,\Sigma_f(E)\,\phi(r, E, t)$, since the equation to be solved is over the entire fission spectrum, approximated to be from 0 to 10 MeV. The EDNDE looks like the following

$$\begin{aligned}&\frac{1}{\upsilon}\frac{\partial\phi(r,E,t)}{\partial t}-\nabla\cdot D(r,E)\nabla\phi(r,E,t)+\Sigma_T(r,E)\phi(r,E,t)\\&=\int_0^{\infty}dE'\Sigma_S(E'\to E)\phi(r,E',t)+\nu(E)\Sigma_f(r,E)\phi(r,E,t)\end{aligned} \qquad \text{EQN 2}$$

More simplifications to EQN 2 are made by setting the time dependent term, $\frac{1}{\upsilon}\frac{\partial\phi}{\partial t}$ to zero, to find the steady-state solution. Another simplification made to equation 2 is, a homogenous approximation so that a uniform system of fuel and moderator/reflector is the contents of the control volume. This simplification allows for the cross sections to be found by making an atomic weighted average for the system spatially. The homogeneous cross sections are considered to be: uniform, position independent and only a function of energy in this paper. The EDNDE, after the homogenous assumption is applied has the following form in equation 3.

$$\begin{aligned}&-D(E)\nabla\cdot\nabla\phi(\vec{r},E)+\Sigma_T(E)\phi(\vec{r},E)\\&=\int_0^{\infty}dE'\Sigma_S(E'\to E)\phi(\vec{r},E')+\nu(E)\Sigma_f(E)\phi(r,E,t))\end{aligned} \qquad \text{EQN 3}$$

The differential scattering cross section Σ_s (E'->E), is defined such that Σ_s (E'->E) = Σ_s(E)* p (E). By integrating from 0 to ∞, p (E) the probability function is one and Σ_s (E) is the result. The result is equation 4.

$$-D(E)\nabla\cdot\nabla\phi+\Sigma_a(E)\phi(\vec{r},E)+\Sigma_S(E)\phi(\vec{r},E)=\Sigma_S(E)\phi(\vec{r},E)+\nu(E)\Sigma_F\phi(\vec{r},E) \qquad \text{EQN 4}$$

Eliminating the macroscopic scattering term in equation 4 on both sides and the result is equation 5.

$$-D(E)\nabla\cdot\nabla\phi+\Sigma_a(E)\phi(\vec{r},E)=\nu(E)\Sigma_F(E)\phi(\vec{r},E) \qquad \text{EQN 5}$$

An effective approximation to the energy dependent functions D, Σ_a, ν and Σ_f are a set of carefully chosen values for the given system so that the energy dependent functionality of D, Σ_a and Σ_f is retained and creates moments around the energy variable (E). The 1/v or $1/E^{1/2}$ law is a fair approximation to the thermal region of the microscopic cross section data found in the evaluated nuclear data files, ENDF[14] (ENDF data were found at these two web sites http://atom.kaeri.re.kr/ and http://t2.lanl.gov/data/neutron7.html) and over the fast region a constant was approximated. Other researches applying the 1/v or $1/E^{1/2}$ law for neutron cross sections[15] provided the authors of reference 16 with reasonable results for foil activation (a 7% error). Of course the simple approximation of $C_1/E^{1/2}+C_0$ for the microscopic cross sections is somewhat crude but "if we choose the group constants properly, even one-speed diffusion theory could give an accurate description of nuclear reactor behavior" (Duderstadt, James J. and Hamilton, Louis J. page 295 Nuclear Reactor Analysis). To account for the resonance peaks that the general function $C_1/E^{1/2}+C_0$ does not account for, the resonance integral constants labeled C_{RI} from the chart of the nuclides[16] for fission, absorption and scattering are included as a term in equation 8, where they have been lumped in with the 1/E energy term, so that they vanish at higher energies (Lamarsh, John R. pages 31-47) maintaining the general functional shape $C_1/E^{1/2}$ $+C_0$.The functional relationships for D, ν, Σ_S, Σ_a and Σ_F with energy are approximated as follows :

$$\Sigma_a(E)_i \cong \frac{N_i * C_{ai}}{E^{1/2}} + C_{aoi}, \quad \Sigma_F(E)_i \cong \frac{N_i * C_{Fi}}{E^{1/2}} + C_{Foi},$$

$$\Sigma_S(E)_i \cong \frac{N_i * C_{Si}}{E^{1/2}} + C_{Soi} \quad for \quad each \quad isotope(i)$$

$$D(E)_i = \frac{1}{3(\Sigma_{Ti} - \bar{\mu}_{oi}\Sigma_{Si})}$$

$$where \quad \Sigma_T = \Sigma_a + \Sigma_S, \quad \bar{\mu}_o = \frac{2}{3A}$$

$$and \quad A = atomic \quad number$$

$$so \quad D(E) \cong \frac{E^{1/2}}{C_{D1} + C_{D0}E^{1/2}}$$

$$\nu(E) \cong 2.5$$

The factor ν(E) is the average number of neutrons released in fission by an incident neutron of energy E. The function is a linear function where the slope of the line is 0.066 for energies from 0 to 1MeV and 0.15 for energies greater than 1 MeV, given the low slope of the function and for most nuclear purposes only the average number is needed (Duderstadt[2] page 61).

The set of analytical energy dependent neutron moments are found by rearranging equation 5 by putting the functions from figure 1 into equation 4, simplifying and applying the method of moments for the internal ordinate (E). Equation 6 shows the grouping of the energy dependent functions into one function of energy, F(E) listed as EQN 7.

$$\nabla^2\phi(\vec{r},E) + \frac{\nu(E)\Sigma_F(E) - \Sigma_a(E)}{D(E)}\phi(\vec{r},E) - 0 \qquad \text{EQN 6}$$

$$\frac{\nu(E)\Sigma_F(E) - \Sigma_a(E)}{D(E)} = F(E) \qquad \text{EQN 7}$$

Equation 8 can be found by further algebraic simplification, grouping of terms in F(E) and adding in the resonance integrals average values.

$$F(E) = \frac{\nu(E) * \sum_i^n C_{Fi}C_{D1i} - \sum_i^n C_{ai}C_{D1i} + C_{RI}}{E} + \frac{\nu(E) * \sum_i^n C_{F0i}C_{D1i} + C_{Fi}C_{D0i} - \sum_i^n C_{a0i}C_{D1i} + C_{ai}C_{D0i}}{E^{1/2}} \qquad \text{EQN 8}$$

$$+ \nu(E) * \sum_i^n C_{F0i}C_{D0i} - \sum_i^n C_{a0i}C_{D0i} \quad or \quad simplified$$

$$F(E) = \frac{C_{E1}}{E} + \frac{C_{E2}}{E^{1/2}} + C_{E3}$$

Multiplying the simplified form of equation 8 by $\phi(r, E)$ equation 9 is the result.

$$\nabla^2\phi(\vec{r},E) + \frac{C_{E1}}{E}\phi(\vec{r},E) + \frac{C_{E2}}{E^{1/2}}\phi(\vec{r},E) + C_{E3}\phi(\vec{r},E) = 0 \quad \text{EQN9}$$

Apply the definition of a moment (m_k) to equation 9, the result is equation 11.

$$m_k \equiv \int_0^{\infty} E^k \phi(E;\vec{r})dE \qquad \text{EQN 10}$$

$$\int_0^{\infty} E^k \nabla^2\phi(\vec{r},E)dE + \int_0^{\infty} E^k \frac{C_{E1}}{E}\phi(\vec{r},E) + \int_0^{\infty} E^k \frac{C_{E2}}{E^{1/2}}\phi(\vec{r},E)dE + \int_0^{\infty} E^k C_{E3}\phi(\vec{r},E)dE = 0 \quad \text{EQN 11}$$

The Laplace operator depends only on position so it comes through the energy integral and the result is equation 12.

$$\nabla^2 m_k + C_{E1}m_{k-1} + C_{E2}m_{k-1/2} + C_{E3}m_k = 0 \qquad \text{EQN 12}$$

Solution to equation 12 for a 1 dimension case turns into a set of second order ordinary differential equations (ODEs) that can be solved for by the method of undetermined coefficients found in many ordinary differential text books[17]. The moments that provide physical insight for the energy variable are normalized moments: m_0, m_1, m_2, m_3 and m_4. The half moments: $m_{1/2}$, $m_{3/2}$, $m_{5/2}$, and $m_{7/2}$ could be viewed as moments for the speed of the neutrons since speed is proportional to $E^{1/2}$, but the focus is on energy moments, so the moments for energy to the integer power were plotted. The half moments are needed to solve for moments: m_0, m_1, m_2, m_3 and m_4 as can be seen from EQN 12. Equation 12 is solved analytically for a 1 dimensional homogenous sphere with diffusion theory boundary conditions. The boundary conditions from diffusion theory can be found in many nuclear reactor theory textbooks or introduction to nuclear engineering textbooks (see references 1, 2, 6 and 7). The two boundary conditions are: the neutron flux is finite and the second condition is flux is zero at the extrapolated boundary, $\tilde{R}$ where $\tilde{R} = R + r_o$, and $r_o = 2.13 * D(E)_{average}$ where r_o is the extrapolated distance.

Solving for moments, it turns out that C_{E3} equals the geometric buckling constant $\left(\frac{\pi}{\tilde{R}}\right)^2 = Bg^2$ from one speed diffusion theory. Solving EQN 12 for the zeroth moment, the solution is a sinc function ($m_0 = \frac{A_0 \sin(Bg \cdot r)}{r}$). Determining the amplitude; A_0 of the sinc function is found by another relationship which relates the power of the system to the integral of the flux over the volume multiplied by the macroscopic fission cross section Σ_F and a conversion factor E_r, $Power = \int_0^{\infty} E_r \cdot \int \Sigma_f \cdot \phi(r,E)\, dV\, dE$. The power of the reactor is a chosen known value, so after integration of m_0 over the volume, A_0 can be found by dividing the power by the result of the integration. The rest of the amplitudes can be found from the moments themselves by setting each moment from m_1 to m_4 to zero at the extrapolated distance. The amplitudes are normalized by the zeroth moment to give the mean m_1/m_0, variance m_2/m_0, skewness m_3/m_0 and kurtosis m_4/m_0. The normalized plots for the three cases are shown in figures 3-14. Listed below are the raw or un-normalized analytical moments.

$$m_0 = \frac{A_0 \sin(Bg \cdot r)}{r} \qquad \text{EQN13}$$

$$m_{\frac{1}{2}} = \frac{A_{\frac{1}{2}} \sin(Bg \cdot r)}{r} + \frac{A_0 C_{E2}}{2Bg}\cos(Bg \cdot r) \qquad \text{EQN 14}$$

$$m_1 = \frac{A_1 \sin(Bg \cdot r)}{r} + \frac{(A_0 C_{E1} + A_{\frac{1}{2}} C_{E2})}{2Bg}\cos(Bg \cdot r) \qquad \text{EQN 15}$$

$$m_{\frac{3}{2}} = \frac{A_{\frac{3}{2}}\sin(Bg\cdot r)}{r} + \frac{(A_1 C_{E2} + A_{\frac{1}{2}} C_{E1})}{2Bg}\cos(Bg\cdot r) \qquad \text{EQN 16}$$

$$m_2 = \frac{A_2\sin(Bg\cdot r)}{r} + \frac{(A_1 C_{E1} + A_{\frac{3}{2}} C_{E2})}{2Bg}\cos(Bg\cdot r) \qquad \text{EQN 17}$$

$$m_{\frac{5}{2}} = \frac{A_{\frac{5}{2}}\sin(Bg\cdot r)}{r} + \frac{(A_2 C_{E2} + A_{\frac{3}{2}} C_{E1})}{2Bg}\cos(Bg\cdot r) \qquad \text{EQN18}$$

$$m_3 = \frac{A_3\sin(Bg\cdot r)}{r} + \frac{(A_2 C_{E1} + A_{\frac{5}{2}} C_{E2})}{2Bg}\cos(Bg\cdot r) \qquad \text{EQN 19}$$

$$m_{\frac{7}{2}} = \frac{A_{\frac{7}{2}}\sin(Bg\cdot r)}{r} + \frac{(A_3 C_{E2} + A_{\frac{5}{2}} C_{E1})}{2Bg}\cos(Bg\cdot r) \qquad \text{EQN 20}$$

$$m_4 = \frac{A_4\sin(Bg\cdot r)}{r} + \frac{(A_3 C_{E1} + A_{\frac{7}{2}} C_{E2})}{2Bg}\cos(Bg\cdot r) \qquad \text{EQN 21}$$

The plots of the normalized analytic moments compared to the moments calculated from MCNP5 are in the figures 1-8 for the two cases: pure ^{235}U and GODIVA HEU 94% ^{235}U. The pure ^{235}U normalized moments are shown in figures 1-4. Figures 5-8 are the GODIVA moment comparisons.

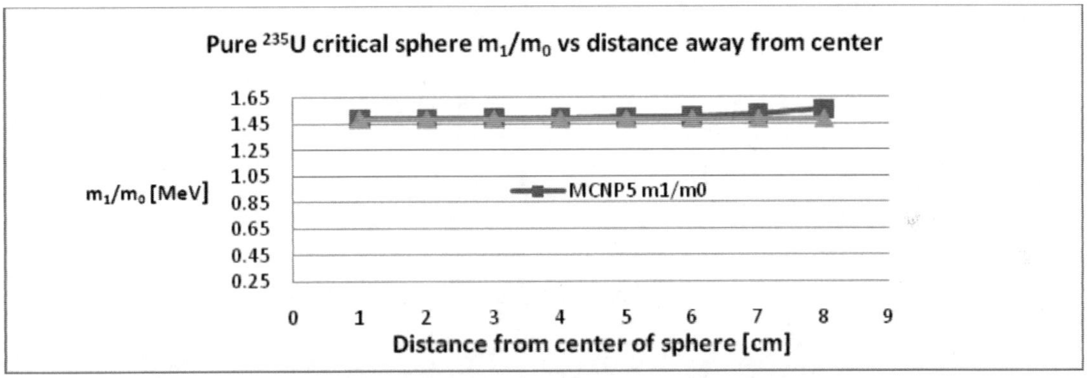

Figure 1 comparison plot of normalized moment 1, Pure ^{235}U case

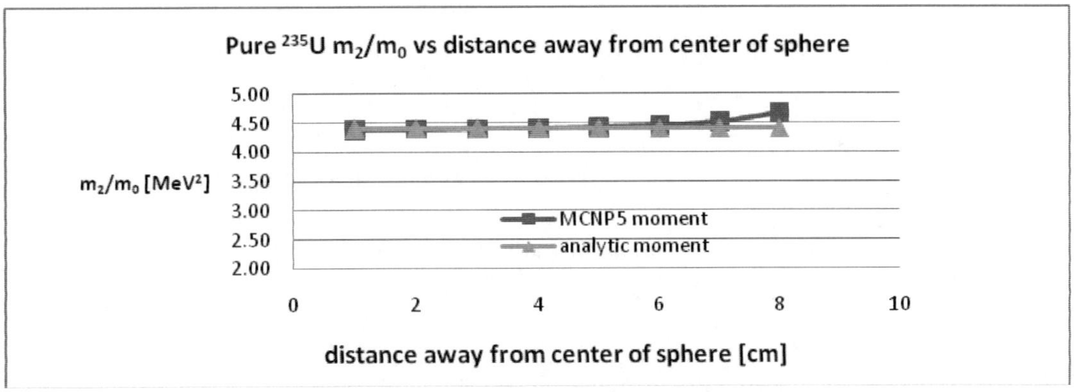

Figure 2 comparison plot of normalized moment 2, Pure ^{235}U case

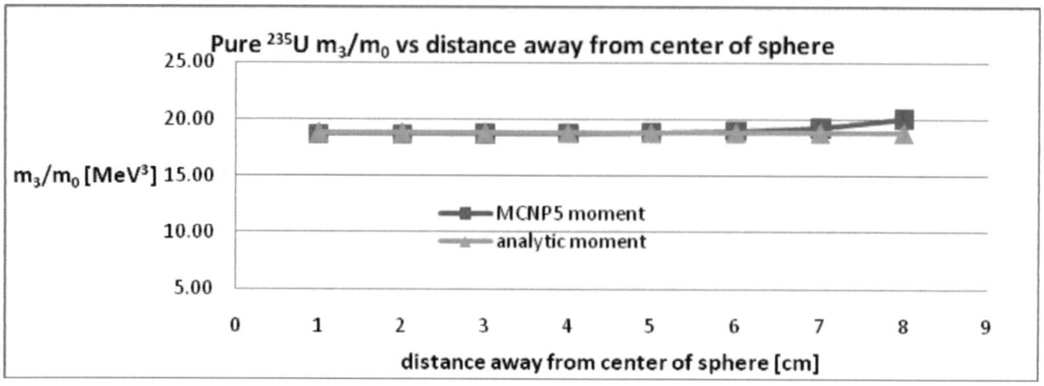

Figure 3 comparison plot of normalized moment 3, Pure ^{235}U case

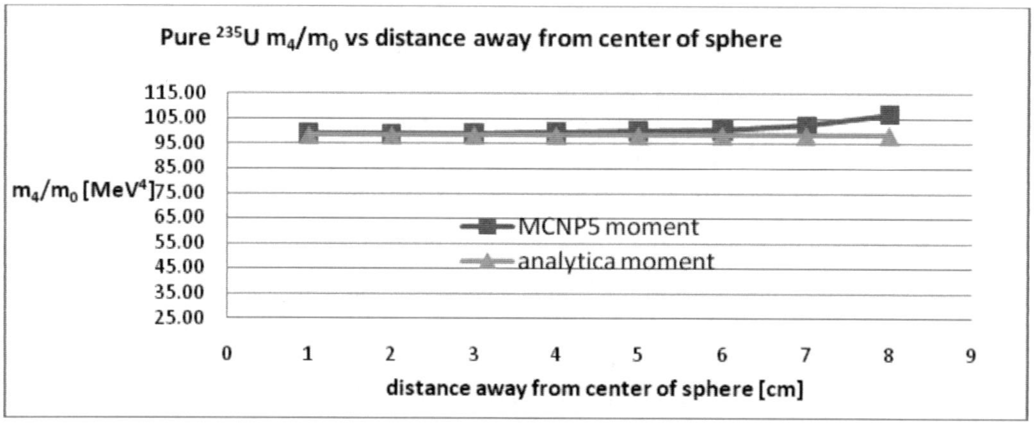

Figure 4 comparison plot of moment 4, Pure ^{235}U case

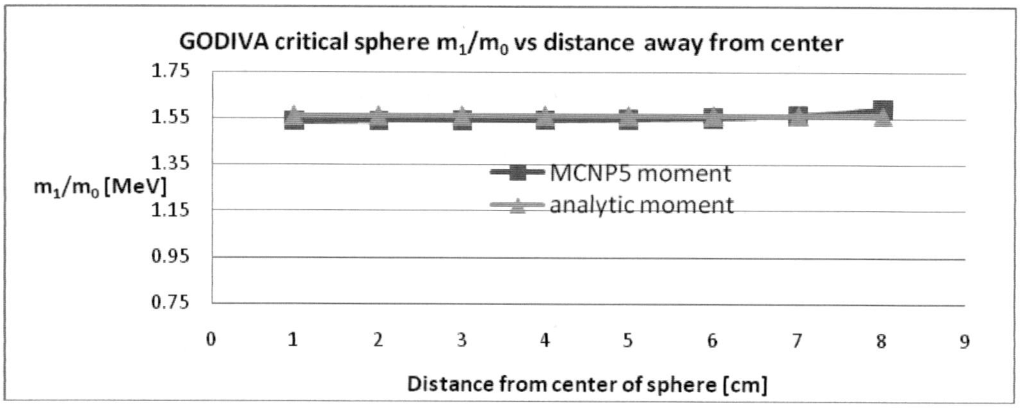

Figure 5 comparison of normalized moment 1, GODIVA case

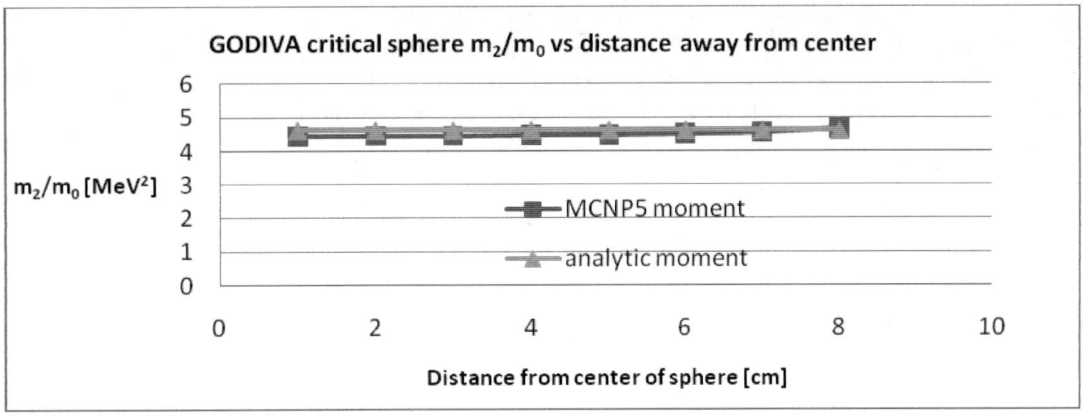

Figure 6 comparison plot of normalized moment 2, GODIVA case

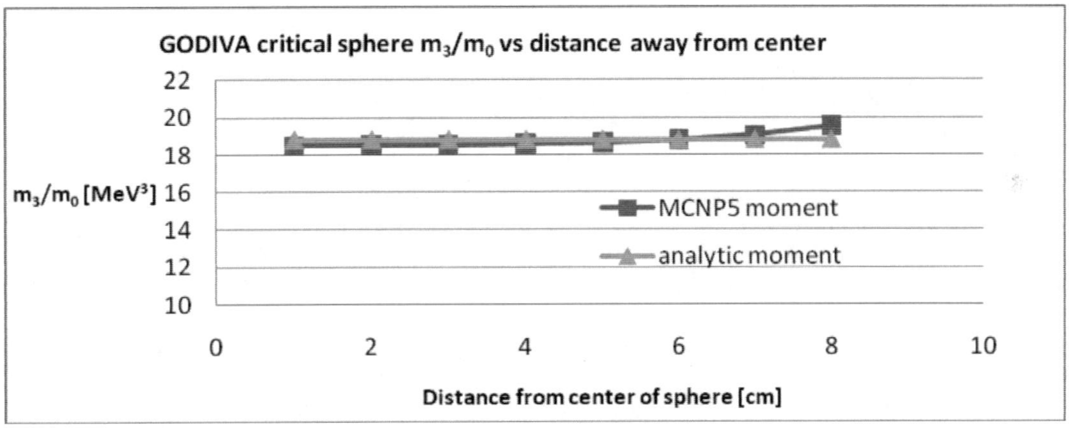

Figure 7 comparison plot of normalized moment 3, GODIVA case

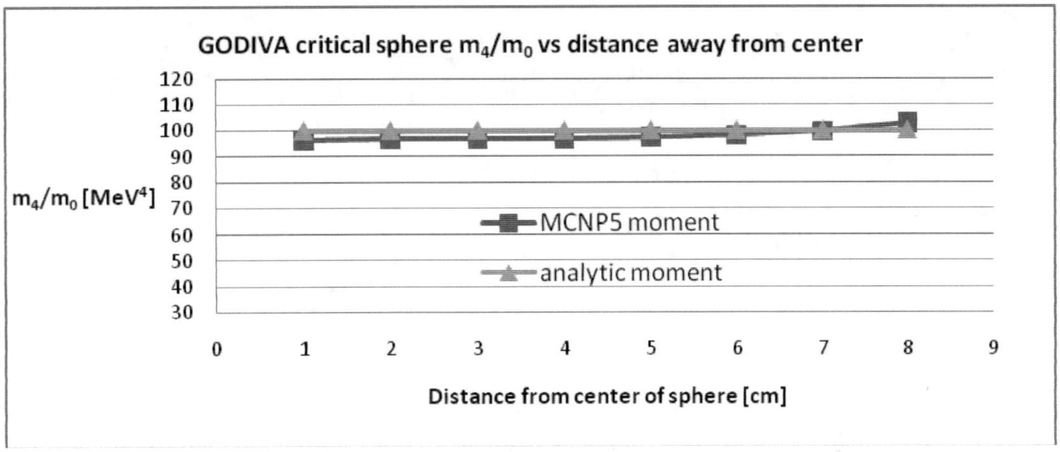

Figure 8 comparison plot of normalized moment 4, GODIVA case

Criticality calculations were also performed from the analytical moments. Criticality or keff can be found from inserting1/ k into EQN 5 and then doing the algebra to isolate k, the result is the following equation 22,

$$k_{eff} = \frac{\nu*(C_F*C_D*m_{k-1}+(C_{F0}+C_F*C_{D0})*m_{k-\frac{1}{2}}+C_{F0}*C_{D0}*m_k)}{C_a*C_D*m_{k-1}+(C_{a0}+C_F*C_{a0})*m_{k-\frac{1}{2}}+C_{a0}*C_{a0}*m_k-\nabla^2 m_k} \qquad \text{EQN 22.}$$

The pure ^{235}U case keff in MCNP5 is 0.99967 +/- 0.00006, analytical moments keff is 0.996 (more work needs to be done for an error analysis). The GODIVA case keff in MCNP5 is 0.99783 +/- 0.00003, analytical moments keff is 1.00 (again more work needs to be done for an error analysis).

Discussions and Conclusion

Method of moments applied to the EDNDE has only a few percent error from a full neutron transport code MCNP5, which is a real surprise since the diffusion approximation was imbedded into the base equation as the starting point for this analysis. Criticality (keff) calculated by the moments for GODIVA and the pure ^{235}U case also differ by a few percent. Even though this was a simple look at neutron diffusion transport with MOM, the comparison of the normalized plots for each moment and keff are encouraging. The comparison plots only differ by a few percent where diffusion theory is valid and 20% where it is not valid. Further research can be done to add in better approximations near the boundary for the method of moments to be more accurate. In addition, further work should be done on neutron systems with reflected boundaries instead of a vacuum boundary, heterogeneous critical fuel assemblies and method of moment comparison with multi-energy group codes like ATILLA.

References

[1] Lamarsh, John R. Introduction to Nuclear Reactor Theory, Addison-Wesley Publishing Company, Inc. Reading, Massachusetts, 1966
[2] Duderstadt, James J.; Hamilton, Louis J. Nuclear Reactor Analysis, John Wiley and Sons, 1976
[3] Lewis, E. E.; Miller, Jr., W. F. Computational Methods of Neutron Transport, American Nuclear Society, Inc. La Grange Park, Il 1993
[4] A General Monte Carlo N-Particle Transport Code - Version 5, Los Alamos National Lab, LA-UR-08-8617
[5] International Handbook of Evaluated Criticality Safety Benchmark Experiments, NEA/NSC DOC(95)03, September 2009 Edition
[6] Lamarsh, John R.; Baratta, Anthony J. Introduction to Nuclear Engineering 3rd Edition, Prentice Hall, Upper Saddle River, New Jersey, 2001.
[7] Weinberg, A.M. ; Wigner, E. P. The Physical Theory of Chain Reactors, University of Chicago Press, 1958.
[8] Kenny, John F. Mathematics of Statistics 2nd Edition, D. Van Nostrand Company, Inc. New York, NY, 1947
[9] Casella, George; Berger, Roger L. Statistical Inference 2nd Edition, Wadsworth Group Duxbury, Pacific Grove CA, 2002
[10] Marchisio, Daniele L., Pikturna, Jesse T., Fox, Rodney O. and Virgil, Dennis R. "Quadrature Method of Moments for Population-Balance Equations," AIChE Journal **49**, 5 (May 2003).
[11] Bin-Wan, Chu; Ring, Terry A. "Verification of SMOM and QMOM Population Balance Modeling in CFD code Using Analytical Solutions for Batch Particulate Processes" China Particuology Volume 4, Issue 5, October 2006, Pages 243-249
[12] McGraw, R. (1997) "Description of aerosol dynamics by the quadrature method of moments". Aerosol Science and Technology 27, 255-265
[13] Marchisio D. L., R. D. Vigil and R. Bo Fox, Implementation of the Quadrature Method of Moments in CFD codes for aggregation-breakage problems, chemical engineering science, v58, n15, August 2003. p. 3337-3351.
[14] http://atom.kaeri.re.kr/ and http://t2.lanl.gov/data/neutron7.html
[15] Morry, S; Williams, N.M.R, "Neutron Flux Perturbations Due to Absorbing Foils", Journal of Applied Physics Volume 5 1972
[16] Nuclides and Isotopes Chart of the Nuclides 16th Edition, revised: Baum, Edward M.; Knox, Harold D.; Miller, Thomas R. Knolls Atomic Power Laboratory, Lockheed Martin, 2002
[17] Edwards, C. Henry; Penney, David E. Differential Equations & Linear Algebra, Prentice Hall, Upper Saddle River, New Jersey, 2001

(676f) Development of a CFD Methodology to Simulate Vortex around Partially Immersed Rotating Shaft: Application to Evaluate Designs of Vortex Breaker

Mahendra Wankhede[1], Prashant Gunjal[2] and **Vivek Ranade**[1], (1)Chemical Engineering, Tridiagonal Solutions Pvt. Ltd., Pune, India, (2)Tridiagonal Solutions Pvt. Ltd., Pune, India

Partially immersed rotating shafts are used in many industrial applications in chemical and nuclear industries. Such rotating shafts may generate deep vortex which may eventually lead to gas entrainment. Gas entrainment needs to be avoided in many applications. Various types of vortex breakers are used in such applications to reduce or eliminate vortex around rotating shafts. In this work, we have developed a new computational methodology to simulate vortex around rotating shafts. The methodology is applied for a specific case and was used to evaluate different designs of vortex breaker. Partially immersed shaft generates inertia dominant flow in the vessel and due to high rotational pump speeds (~300-700 rpm), the flow is turbulent in nature. Centrifugal force due to shaft rotation causes steep radial pressure gradients in the fluid which lead to formation of a vortex around rotating shaft. The shaft rotation may also lead to Taylor vortices near the wall of the shaft. At vortex tip, meridian flow is in vertical downward direction. Pressure gradients are steeper in this region than rest of the vessel. The probability of gas entrainment enhances due to formation of low pressure region. Entrained gas in the form of bubbles gets carried into the vessel. Possibility of gas entrainment can be reduced by damping the swirling flow in the vessel. The design of such dampeners or vortex breakers is not straightforward and may produce adverse effect due to sharp swirl flow gradients. Computational Fluid Dynamics (CFD) can provide detailed flow information which would be useful in several aspects such as, characterization of vortex shape, evaluation of gas entrainment criterion and designing anti-vortex breakers. Conventionally Volume of Fluid (VOF) approach is used for simulating free surface flows like vortex (Ranade, 2002). However, VOF approach requires very fine mesh resolution and small time steps demanding significant resources in terms of CPU time and memory. In this work we present an alternative method to simulate vortex shape based on single phase simulation. The proposed method uses pressure field predicted with the single phase flow model to estimate shape of free surface (which is considered flat initially). The estimated free shape is then used to carry out next iteration of single phase flow simulation. The method converges to the final free surface shape within 5 iterations. The performance of this method is evaluated by comparing the predicted results with those obtained by the VOF approach for simple systems. The developed methodology is then applied to a case of a vortex around a large partially immersed shaft of pump system typically used in nuclear industry. The developed CFD methodology is used to investigate the influence of various parameters on interface depth and shape of vortex in pump vessel. Without considering any internals, maximum vortex depth showed an increasing trend with the shaft rotational speed, the rate of which gradually tapers off (see for example, Figure 1). Efficiency of various internals to reduce the swirling flow and therefore vortex depth are investigated computationally. Sample of results is shown in Figure 2. The results are also used to assess the possibility of gas entrainment and investigate various ways to mitigate such entrainment possibility. Ranade, V.V. (2002), Computational Flow Modeling for Chemical Reactor Engineering, Academic Press, London.

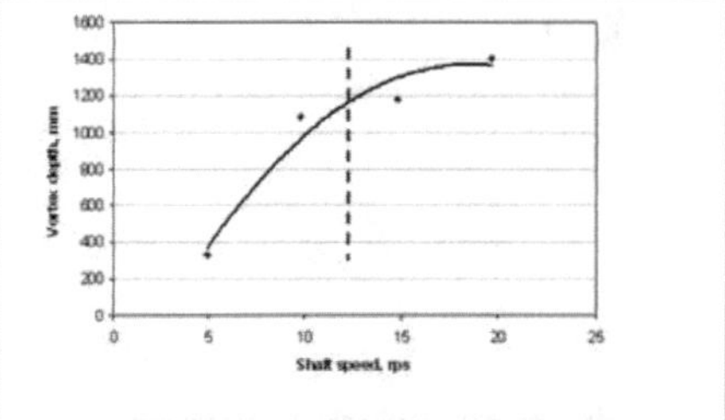

Figure 1: Vortex depth Vs. Shaft rotational speed

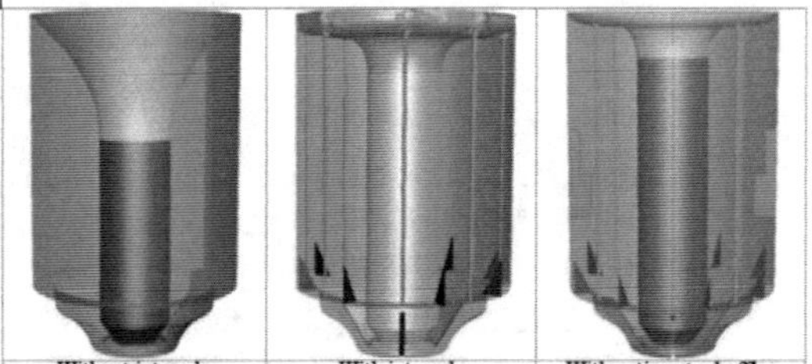

Without internals	With internals	With anti-vortex baffles

Figure 2: Predicted vortex shape in the pump vessel

See more of this Session: Advances in Numerical Simulations Bridging Chemical and Nuclear Engineering Phenomena or Processes
See more of this Group/Topical: Nuclear Engineering Division

AUTHOR INDEX

AUTHOR INDEX

AUTHOR INDEX